Origin and Evolution of The Human Race

Published by:
Research Associates School Times Publications
&

Frontline Distribution International Inc.
Chicago. Jamaica. London. Republic of Trinidad and Tobago.
5206 S. Harper Avenue, Chicago, Illinois, 60615

Origin and Evolution of The Human Race

Published: 2011
Research Associates School Times Publications
Frontline Distribution Int'L Inc.
Published in the Caribbean by: Miguel Lorne Publishers & Frontline

First Published 1910

Library of Congress Control Number: 2011920268
ISBN: 0-948390-31X

Page 86 - First black tribes. Akka is first.

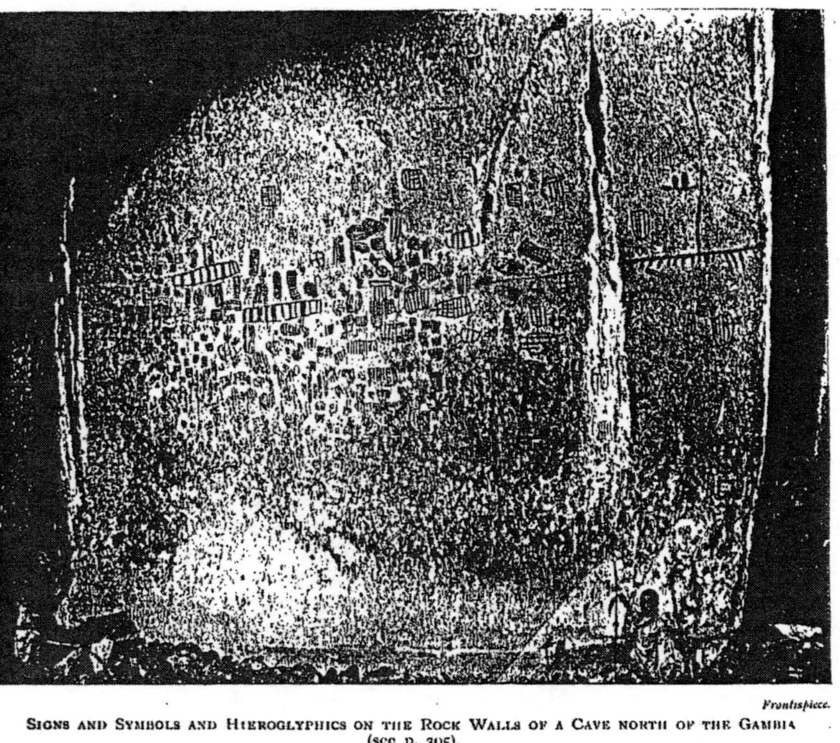

Frontispiece.

Signs and Symbols and Hieroglyphics on the Rock Walls of a Cave north of the Gambia (see p. 305).

I Dedicate

THIS WORK TO ALL PRESENT AND FUTURE ANTHROPOLOGISTS AND ETHNOLOGISTS, TRUSTING THAT IT WILL ASSIST THEM IN THE FUTURE TO ARRIVE AT A MORE DEFINITE AND CRITICAL OPINION ON THE PAST HISTORY OF HOMO THAN HAS BEEN THEIR EXPRESSED CONTENTION HITHERTO

CONTENTS

PART V

LUNAR CULT

LIST OF FULL-PAGE PLATES

ILLUSTRATIONS IN THE TEXT

PREFACE

In publishing this work I must crave the indulgence of my readers on the arrangement of the subject matter: but the book has been written in various parts at various times, and a busy professional life has prevented the better classification of the subject matter which I should have desired.

Nevertheless, I contend that the facts brought forward are true, and will be found irrefutable, that further discoveries will bring this home to those who are sceptical and who at present cannot read the "writings on the wall," and will maintain the correctness of the decipherments I have made, more especially as regards those found in America and Asia.

Anthropologists, Ethnologists, and Geologists will find that their future work will become easier and more intelligible; they will tumble into fewer pitfalls, and will the more easily extricate themselves from those into which they do fall, by adopting my divisions of the Human Race, and by taking into consideration the cults and anatomical conditions found, and the Implements connected with each division as set forth herein.

I am greatly indebted to the Rev. G. L. Harold, B.A., for assisting in correcting proofs, and also to Mrs. K. Watkins for the skill and care she has displayed in executing many of the drawings.

ALBERT CHURCHWARD.

Royal Societies' Club,
63, *St. James' Street*,
London.

PART I

ANTIQUITY AND BIRTHPLACE OF MAN

CHAPTER I

AFRICA THE BIRTHPLACE OF MAN

Know then thyself, presume not God to scan,
The proper study of mankind is man.

POPE.

THE object of this book is to bring before the public such further facts and values regarding the Evolution of Man as have been discovered subsequent to the publication of the "Signs and Symbols of Primordial Man"—first edition published in June 1910, second edition published May 1913—and also to demonstrate that the contention set forth in that publication is, in its entirety, correct: namely, that the Pigmy was the primary Homo, evolved from a Pithecanthropus Erectus, or Anthropoid Ape, in Africa over a million years ago.

From studies I have made during many years, I am fully convinced that the hitherto preconceived ideas of many Scientists regarding the origin of the human race, both as to place and date, are erroneous, and evidence will be brought forward to prove that the human race did not originate in Asia, but in Africa.

This evidence will be objective and subjective—as proofs of my contention against all the learned men of the present day.

The first question which arises is this—

When and where did man make his first appearance on this earth?

Biblical Scholars tell you about 6,000 years ago in Asia.

The Aryanist School, in Asia about 20,000 years ago.

Others, including many Scientists, in Asia—or in some mythical land which has now disappeared. All of them have denied Africa as the "home of man."

```
S A I N Q S V N N P H E O Y P I O C T E C O R A Z
O X L J G K R V E V R W D B P S Y Z O E B E P L F
A C I N I M O D E H A F F K G M Y S X E E G X P W
Q R P F U Y Y N T Q T B W O P J A M A I C A C Z T
X V E H S G E J E X W U L O N D O N J X N P N N R
N C H I G Z L U B N A A O U M X D I I E F Q D B W
X Z R Y U J H E E K D T T S W G B H F K B X T D L
M A H E A J E W C L S H F O I E U F M P A M A C R
P K L B X P Q P Y E T S Z Z G N O S H Y W Q G D F
C A P G N A H C I X E K C R R J I W K X T T W K Q
X E H S X A O Q D M B R T O I R N C P F P X E P K
Y A K W A L E S P B T C G M W I Z Q E W W E D J J
X J R A O V F N V X O J A E J I N B Y X F D O T Y
```

AMSTERDAM	BELGIUM	CORFU	DOMINICA
EGYPT	FUERTEVENTURA	GREECE	HAWAII
ISRAEL	JAMAICA	KENYA	LONDON
MAJORCA	NICE	OSTEND	PARIS
QUEBEC	ROME	SOUTHEND	TORQUAY
UKRAINE	VENEZUELA	WALES	XICHANG
YOSEMITE	ZAGREB		

It is a cardinal tenet of the present work that man originated in Africa, and that our present and past Professors of Anthropology and Ethnology are not correct in their classification of the Homo.

The division into Palæolithic and Neolithic includes many different stages in the evolution of humanity in the former term. These can be separated and easily identified. Then again their classification of the Homo by straight, frizzy and curling hair is entirely misleading, and in my opinion cannot be right, as I shall endeavour to prove. The Pigmy, who has peppercorn hair, has never been taken into consideration at all. As regards the age of man on this Earth, many Anthropologists and Ethnologists, as well as the present School of Geologists, have given a too limited period of time for his first appearance. The fundamental basis of this work is to show the following:—

That it was in Africa the little Pigmy was first evolved from an Anthropoid Ape—in the Nile Valley and around the Lakes at the Head of the Nile (which I will for the sake of brevity style "Old Egypt").

The Pigmy was the first Homo—the little red man of the Earth. From Africa these little men spread all over the world, North, East, South and West, until not only Africa but Europe, Asia, North and South America and Oceania were populated by them.

From the Pigmy, evolution continued progressively in the order shown in the table opposite.

My contention is, that the progress and evolution of the human race can still be studied from the lowest type of the original man, as he advanced up the scale—and that these types are still extant in some parts of the world, where the primary have been driven away into mountains and inaccessible forests, by the Nilotic Negro, and these again into lands where they have been isolated by the Stellar Mythos people, into groups, with little or no intercommunication with others.

I have divided the human race into the following groups because they are easily distinguished and differentiated by their anatomical features—by their cults or beliefs—by their Implements and works of art. By these groupings and divisions humanity, past and present, can be traced all over the world.

PALÆOLITHIC.

Original or Primary Man—age 2,000,000 years.

1. THE NON-TOTEMIC GROUP OR PRE-TOTEMIC PEOPLE.

The Pigmy Group.

These include all the so-called Negrillos or Negritos.

1. They have no Totems or Totemic Ceremonies.

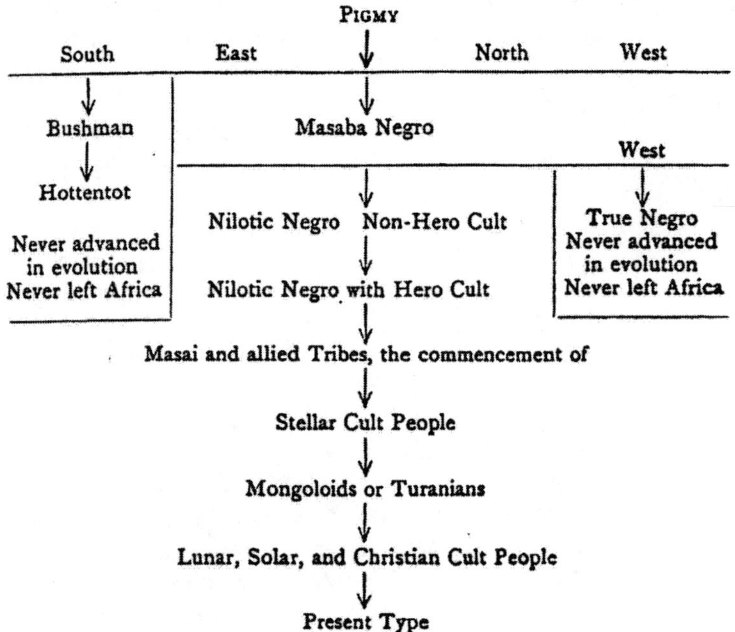

2. They believe in a Great Spirit.

3. They propitiate Elementary Powers and Departed Spirits.

4. Although they have dances, as Sign Language, they have not Totemic Ceremonies.

5. They have no written language, but speak a monosyllabic language, and have a Sign and Gesture Language.
6. They have no Mythology, no Folk-lore Tales.
7. They have no Magic.
8. They have no Initiatory rites.
9. They have no Tribal markings.
10. Their Implements are limited and Primitive.

THE SECOND OR SUB-GROUP, DESCENDANTS OF THE PIGMIES, ARE—

1. The Bushmen, and the Hottentots who descended from the Bushmen.
2. The Masaba Negroes who were evolved from a highly developed Pigmy.

These sub-classes are not found outside Africa—past or present.

All these have no Totems or Totemic Ceremonies.

No Mythology.

No Folk-lore Tales.

No Magic.

They speak a monosyllabic language, have no written language but can draw pictures—Signs and Symbols like the Pigmy.

They believe in a Great Spirit.

They Propitiate Elementary Powers, or Spirits, and the Spirits of their Ancestors.

They have dances—Sign and Gesture Language. Sign Language includes the Gesture Signs to which the Mysteries were and are still danced, or otherwise dramatized in Africa by the Pigmy and Bushmen.

Their Implements are the same as the Pigmy—Primitive.

The Bushman developed from the Pigmy and travelled South and never came North again, and the Hottentot developed from the Bushman and also never came North; these types are only found in South Africa.

The Masaba Negro developed from the Pigmy North, East and West.

None of these are or ever were Anthropophagous.

The Masabas are the connecting Links between the Pigmy and

THE THIRD GROUP—THE TOTEMIC PEOPLE.

All these have Totems and Totemic Ceremonies and are, or were at one time, Anthropophagous.

They are divided into two distinct groups :—

1. THE TRUE NEGROES.

These were developed from the Masaba in the West, and have distinct anatomical features from the Nilotic Negroes—*they never left Africa except as Slaves.*

2. THE NILOTIC NEGROES.

They can be divided into two distinct classes, first, a lower type, and second, a higher type, the latter developed by evolution from the lower, and each can be identified with facility. Let us take :—

1. *The Lower Class.*

    They have no Mythology.
    They have no Hero Cult.
    They have no Tattoo.
    They raise, however, marks on the skin—Cicatrices.

2. *The Higher Class.*

    They have Hero Cult.
    They have Mythology.
    They have Tattoo.
    Both have Magic, Totems and Totemic Ceremonies.
    (Dr. A. C. Haddon, F.R.S., etc., of Cambridge, has used the term Hero *Cult*, which I shall follow, as in my opinion it is a more expressive term than "Hero Worship" for the divination of the Elementary Powers.)

All these *Nilotic Negroes can be found outside Africa*, or at least I should say that their descendants are still to be found in many countries at the present time—descendants of former exodes—they followed the Pigmy all over the world, exterminated him in some places, and drove him into more inaccessible spots in others : but the osteo remains of both are found in many parts of the world in different strata, and types of both are still extant in various countries (see later). All the above may be classed as Palæolithic.

NEOLITHIC.

From the Nilotic Negroes were developed—

THE FOURTH GROUP,

Turanians and Mongoloid—primary, who were those whom I call

The Stellar Mythos People.—They existed (and some are still to be found) for over 300,000 years. They spread over Europe and Asia except the extreme North; North and South America, except the extreme North and South; some Isles of the Pacific and throughout Africa. There are no traces of these people in Australia, New Guinea, New Zealand, the extreme Norths of Europe, Asia, North America, nor Patagonia, in South America. They have left indelible proofs of their great skill and learning, which skill and knowledge may be gauged by the many great cities, and the huge and marvellously finished blocks of stone, the remains of which may be still found in Africa, Central and South America and other places.

Two ages can be separately identified, the first exodus having Sut or Set for their Primary God, and all their first temples are Round and orientated South, and they buried their bodies with their faces to the South, and, (2) the followers of Horus, God of the North, who buried with their faces to the North, and have all their Temples in the form of a double Square, end to end, orientated North. They were all buried in the "thrice-bent position" with characteristic amulets, Implements, etc., placed in their tombs.

After the Stellar, the Lunar and Solar Cult people followed, and various exodes of these left Egypt and followed the Stellar Cult people—but there was much overlapping. The Stellar people's buildings can be identified separately from the Solar people's.

The Stellar were Iconographic, whereas the Solar who followed these were not. The Solar were not bonded, the Stellar were. (They both built with Polygonal shaped stones and with monoliths.) The reason for this was that the Stellar people portrayed their Goddesses and Gods in Zootype forms—all these were pre-human, and in the Solar Cult they had ceased to depict their Goddesses and Gods in Zootype form. They portrayed them in the Human form. It was at the beginning of the Solar Cult that Har-Ur, the Elder Horus, was depicted as a child in the place of the Lamb—the Fish—the Shoot of the Papyrus plant, etc. There was no human figure personalized in the Mythology of Egypt; when the human figure appeared it ceased to be Mythological and became Eschatological.

The Solar Cult people spread throughout Europe, except in the extreme North—through the Southern part of Asia, but not North of Asia—to the South of Africa, and from Yucatan [1] (where they landed from Egypt) down to Peru, in a South-Westerly direction.

None are found, or at least no evidence has been found of these people having gone to the extreme North of Europe, North of Asia, North of America, nor extreme South America, or Australia, or New Zealand, or any Islands of the Pacific.

It is quite possible that they might have gone to the North of Europe, North of Asia and North America, and the whole of the evidence may have been destroyed by ice, etc., at the time of the Glacial Epochs. This cause, however, could not apply to Australia, New Zealand, or Isles of the Pacific. And against any supposition that they did inhabit these extreme Northern regions we have the fact that the present inhabitants are pre-Solar people, who must have been driven South more than once by the intense cold of the Glacial Period, and then driven North again by the Solar Cult people, who did not follow them so far North and exterminate them.

From the Eschatology of the Egyptians, at the downfall of their Empire, Christianity was evolved out of the ashes of this by the Copts, and from these it has spread over various parts of the world, and is supposed to be the highest evolution of all the religious cults at the present time.

Having now divided the humans into their various and separate groups, all of which can be easily identified and

[1] See remarks later on this point.

traced, let us see what evidence we have of man's past antiquity, the proofs of the same, and the opinions of some of our learned Professors.

I am pleased to see, from recent extracts and articles and from his book "Ancient and Modern Types of Men," that Professor Keith, R.C.S., London, has also arrived at the conclusion that man must have existed on this Earth over 1,000,000 years ago; but Professor Sollas, F.R.S., etc. (Professor of Geology, Oxford), states in his recent book, "Ancient Hunters and their Modern Representatives" (a work much lauded by *The Times* and *Athenæum*, which papers so adversely criticized my "Origin and Evolution of Primitive Man," it being a crushing answer critically to Professor Sollas's work), in his Chapter I, that:

"*The world is now as it has always been as regards man and his past history. The Quaternary cannot have exceeded some 300,000 or 400,000 years, during which period there have been four complete oscillations, one of which was of much longer duration than the rest—the Great Ice Age. The recent existence of a Great Ice Age was discovered by Schimper*" (? !).

So Professors Sollas and Schimper ignore the Sun's revolutions around its centre, and the Precession of the Pole Stars, and, if one believed what they have written, there could be no other explanation of this question but that the Sun and Pole Stars had taken a fit into their head and gone for a prolonged holiday, after which they returned to their ordinary roûtine again, in order that we might have one Great Ice Age whose duration was much longer than the others! Can any one for a moment suppose that such a thing could have taken place?—it is simply preposterous. Professor Sollas's "four complete oscillations" were four different glacial epochs out of many.

The old Egyptian Wise Men kept the time, and marked down every stage. They have left records of the same, for at least ten Glacial Epochs, and, although I have studied all the records that I have been able to find, there is no mention that the Sun or Pole Stars ever went for a holiday, and the Egyptians recorded time and the revolution of these for over 250,000 years, during the period of the Stellar Mythos, and they have left the fact recorded.

The arguments, conclusions arrived at, and imaginations of some Geologists, have apparently never taken into account that the sun travels around its centre once in every 25,827 years (roughly 26,000), forming "The One Great Year"; and that during that period the Northern Hemisphere is frozen down to about the 56th degree of latitude for part of the time. There is a great Summer, great Autumn, great Winter, and great Spring in the Sun's year, as in our year of 365 days. It would be useless to enter into any argument on this point, since the above has not been taken into consideration. That the Glacial Period recurs every 25,827 years is a sufficient argument and proof of all that we find, and will explain the differences and the causes of "the bone of contention" amongst Geologists.

Possibly the above Professors have arrived at their opinions because they do not know that a Glacial Epoch takes place every 25,827 years—in fact they cannot, if they believe only in "One Great Ice Age."

But the sun travels around its centre once in this time at the rate of about 40 miles per second. We know from observations for the last 300 years that the Sun has passed through space from one given point to another, for example  300 YEARS. I have reckoned this distance, and the distance that the Sun travels is 33 trillions of miles to perform its circle of one great year.

The pole and equinox are travelling *pari passu*, one in the upper circle of the heavens, the other in the lower circle of the ecliptic, and the shifting of the equinox was correlated more or less exactly to the changing of the Pole Star.

The circuit of precession first outlined by the movement of the seven Pole Stars of the Little Bear (Ursa Minor) was their (Egyptian) circle of the eternal, or seven eternals, that was imaged by the Shennu ring, also by the Serpent of eternity, figured with tail in its mouth and one eye always open at the centre of the coil = the Pole Star, i.e. their circumpolar Paradise.

When Herodotus was in Egypt the Mystery Teachers of the heavens told him that during a certain length of time (11,340 years) "no divinity appeared in a human form, but they do not say the same of the time anterior to this account. During the above period of time, the sun, they told him, had four times deviated from its ordinary course, having twice risen where he uniformly goes down, and twice gone down where he uniformly rises. This, however, had produced no alteration in the climate of Egypt; the fruits of the earth and the phenomena of the Nile had always been the same, nor had any extraordinary or fatal diseases occurred" (Herod., "Euterpe," cxlii).

Now none of our astronomers throughout the world have ever been able to explain this, nor have they taken into consideration the knowledge of the Wise Men of Egypt, except, probably quite unknown to himself, the late Major-General Drayson. He, however, in his works corroborates the facts that were known to these old Wise Men in many points (although I do not think he has mentioned the Mystery Teachers).

He has written and proved certain facts which will coincide with their registered observations, which also proves my contention set forth in this, and my other works, with regard to the glacial period recurring regularly once in every 25,827 years, as computed and recorded by the Egyptians.

Major-General Drayson has asserted in his works:—

1. That the seasons of the year, Spring, Summer, Autumn and Winter, are due to the position of the obliquity of the earth's axis. Let us see how far this can be proved.

I will now quote from a little pamphlet written by Major R. A. Marriott, D.S.O., etc., which he has been kind enough to send to me, in which he agrees with Major-General Drayson's beliefs.

"The seasons, winter and summer, are due to the position of obliquity of the earth's axis. It is reasonable to suppose that if a uniform change is taking place in the seasons, the position of the axis has something to do with it. We know that at some period before history began, occurred the Ice Age or Glacial Period, when the whole of the Northern Hemisphere down to our latitudes was invaded by ice, and that an ice sheet lapped over the east coast as far south as Suffolk, while other blocks borne by land ice were stranded close to London at Highgate. Now, there must be a definite reason for this state of things. It was no casual óccurrence, because there are distinct traces of three or more successive glacial periods stamped on the rocks, and recognized by geologists. Has this former severity of the winter seasons anything to do with the position of the earth's axis? What do astronomers tell us?

"Briefly, astronomers have not hitherto admitted any important change in the obliquity of the axis of the earth, but ascribe the occurrence of the glacial periods to a combination of aphelion conditions with a change in the direction (not obliquity) of the earth's axis. This only occurs after vast periods of time, the last ice maximum having been reached 240,000 years ago, and the last ice period having finally passed away some 80,000 years ago in the Northern Hemisphere. It is important to observe that, according to them, cold conditions in the North Arctic corresponded in time with genial conditions in the Antarctic, and vice versa. This idea has now been falsified by the observations of the South Arctic expedition.

"The above is not at all satisfying to geologists, as the evidences of a much more recent time than 80,000 years ago are plain and convincing. Rocks smoothed and marked by ice and then subjected to weathering by the elements and further disintegration by vegetable life could not, after 80,000 years, look as if they had been laid bare only yesterday; and the signs are everywhere and too numerous for the assertion that the rocks have been preserved fresh under deposits of other material, and have so escaped weathering.

"Geologists have, however, done what they could. Without any unit of time they have had recourse to other methods, and by investigations regarding the retrogression of the Falls of Niagara and of the St. Anthony and corresponding geological evidence, have arrived at results which placed the age of ice not many thousand years back, in no case exceeding 30,000; and indeed the mean of three separate computations gives a round figure of 15,000 years ago as the date of the glaciation. The important corroboration which this furnishes to the subject of this pamphlet will be seen hereafter.

"Geologists and Astronomers are thus at variance. The latter can give the former no definite assistance, and, indeed, have never given any very exact account of the time required for one of the factors that they reckon on, namely, the revolution of the earth's axis, to complete its cycle. The cycle of revolution has been variously stated during the last thirty years, as it has been based on an observed rate of change, which has never been a uniform one.

"Accustomed as we are to the achievements in precision of Astronomers in other directions, this vagueness comes somewhat as a surprise, and when certain statements bearing the stamp of authority are regarded more closely their ambiguity is certainly remarkable, and astronomers cannot complain if they are called in question."

Drayson's Theory.

It is now proposed to give an account of the work of Major-General Drayson, entitled "The Earth's Past History" (Chapman and Hall), published in 1888, in which he traverses the position taken by astronomers regarding the movement of the earth's axis, and states that the earth has a second rotation besides the diurnal one, and that he has computed all the elements of this second rotation. He also adduces proofs of this by his ability to make certain astronomical calculations that are at present beyond the power of astronomers to work out.

Briefly, he accounts with his second rotation for a cycle of 31,682 years, during which the obliquity of the earth's axis varies from a minimum of 23° 25′ 47″ to a maximum of 35° 25′ 47″. The important point of this is that we are now some 15,450 years removed from the height of the Glacial Period, when the contrast between summer and winter in temperate latitudes was inconceivably great, and only 385 years removed from the position of minimum obliquity, when the contrast between the seasons will be least. This, it will be seen, is of immediate interest to us.

Many years ago, when he was a Professor at the Woolwich Academy, a question was asked him by a cadet, which first called his attention to certain conflicting statements regarding the movement of the earth's axis, or, as it is usually styled, the "pole of the heavens." The statements in question are first, that we are told that the earth's axis traces a circle round the pole of the ecliptic *as a centre*, keeping constantly at the same distance from it of 23° 28′ (Sir John Herschel's "Outlines of Astronomy," Art. 316). Then we are told that it does vary its distance from this centre to the amount of 48″ or 46″ per century!

These statements seemed to Drayson so unsatisfactory that he set himself to study and find out what was the real centre round which the pole of the heavens (the earth's axis) revolved, and after some twenty-five years of investigation he arrived at the following most important conclusion:—

Whilst there is a rapid or daily rotation of the earth taking place round an axis which is fixed in the earth, there is also a slow rotation round a second axis which is not fixed in the earth, but appears fixed as regards the heavens. This second rotation causes the two *semi-axes* of the earth to describe cones during one slow rotation, just as the earth's daily rotation causes a line from any locality to the earth's centre to describe a cone during twenty-four hours.

The conical movement of the semi-axes is in accordance with what one would naturally expect if a second rotation has been imparted to the globe; and Drayson gives us the real centre, not the shifting centre, round which the earth's axis revolves, and this is 60° removed from the pole of the ecliptic.

In the words of Drayson: "The pole of the second axis of rotation is distant from the pole of the heavens 29° 25′ 47″, and is situated not very far from a celestial meridian of eighteen hours, or 270° of right ascension."

The consequence of this discovery is of manifold importance.

We have here a very different movement. No longer can the earth's axis be stated practically to maintain the same angle through all time, i.e. through an entire revolution of the equinoxes, but is uniformly changing from a maximum obliquity of 35° 25′ 47″ to a minimum obliquity of 23° 25′ 47″. The former would entail very hot summers in combination with very cold winters, a condition indisputably requisite for the rapid evaporation of moisture for transference to the polar regions as snow.

Effect of Maximum Obliquity.

This maximum obliquity, which took place in 13,544 B.C., was the culminating point of the Glacial Period. All latitudes greater than 54° 34′ 13″ would therefore have been within the Arctic Circle. In other words, all places immediately to the north of Yorkshire would have been under the present climatic conditions of Nova Zembla and northern Greenland, whilst at midsummer similar latitudes would have been subjected to a tropical summer. At midsummer the sun's meridian altitude would be 70° 25′ 47″, that is, 12° greater than at present; but the sun during twenty-four hours would not set, and a midnight sun would therefore be visible from all parts of Scotland.

In midwinter, on the other hand, the sun would not rise above the horizon of latitudes of 55° and upwards during several days, and the whole of Scotland and the north of England would suffer from an arctic climate. The evidence of glaciers and of large masses of ice having been formed, and of boulders having been carried by this ice into distant localities as far south as Middlesex is in accord with this. In North America, owing to a different distribution of land and sea, the evidences of ice action extend even to latitudes of 40°; and in Alpine regions the volume of ice was so enormous that blocks from the Alps were carried even over the high mountains of the Jura to great distances beyond them.

This is all a matter of precise geological history, but without these new astronomical data, the period assigned by astronomy has been too remote to provide an efficient unit wherewith to measure geological time. Such computations as have been hitherto made by geologists have lacked the support of astronomers, and their ardour of research has thus been cooled by the apparent impossibility of arriving at a working agreement. Their efforts in this direction, in the light of the theory of Drayson, however, bear a very different complexion, and the precision of their conclusions is a striking testimony to the truth of geological presumptions.

While the date 13,544 B.C. marks the date when glaciation was at its height, the date 5,624 B.C. marks the time when the Northern Hemisphere began to enter into milder conditions. From that time forward the change was probably rapid. When once the ice began to melt, vast regions in our latitudes must have been in a condition when the summer sun could overcome the well-known resistance of ice to melting, its so-called *latent heat*. The resulting deluge, owing to overflowing rivers and the bursting of ice barriers, might well have spread universal terror, carrying destruction even to southern latitudes, causing the survivors to hand down the dread tradition to future generations of mankind.

That the ancients had some faint knowledge of this great change would appear from the fact that Aristotle mentions that the Chaldæans had a belief of a period termed "a great year (or period)," during the summer of which there was an *ekpyrosis* or conflagration, and during the winter a *kataclysmos* or deluge.

It does not require an expert knowledge of astronomy to adjudicate on facts partaking of the nature of circumstantial evidence such as might be placed confidently before any jury, in expectation of a decision in accord with common sense; but as a preliminary guide to the matter in dispute, it is considered advisable to explain by a simple illustration what the terms "Precession of the Equinoxes," "Obliquity of the Ecliptic," etc., on which so much depends, really are.

Firstly, let the reader imagine the earth to be represented by a sphere of any material, light enough to float just half immersed in water, and let the water surface represent the "plane of the ecliptic." The sphere may be supposed to have a stick passed through its north and south pole to represent the imaginary line of direction of the "pole of the heavens," as it is called, and if this stick is pulled out of the perpendicular to the extent of an angle of 23° odd, the inclined (oblique) position of the sphere to the water surface represents the "obliquity of the ecliptic."

The points where the equator of the sphere cuts the

water surface are called the equinoctial points, and during the earth's annual orbit the sun is on two occasions at the zenith of these two points, causing the equinoxes.

If there was no movement of the pole of the heavens the equinoxes would occur annually exactly at the same times of the year, but there is a small movement, which makes them occur each succeeding year a little in advance of the due time, and this causes the "precession of the equinoxes."

This movement is circular, taking many thousands of years to complete, and can be fairly represented by the stick (the pole of the heavens) being moved round in a circle parallel to the water surface, while keeping its oblique position.

Drayson's contention shows that the pole of the ecliptic is not the centre, but that another point is.[1] An important result is thereby obtained; in that, with Drayson's centre, the obliquity will vary sufficiently to cause an Ice Age, whereas with the hitherto supposed centre of the movement no reason for any change of climate due to astronomical causes can be postulated.

When this supposed conical movement of the axis symmetrically to the pole of the ecliptic was accepted as complete and satisfactory, it was not known that there was any variation in the obliquity of the ecliptic and since the angular distance of the pole of the heavens from the pole of the ecliptic is the exact measure of the obliquity, and as this obliquity was assumed to be invariable, it was therefore asserted that the earth's axis traced a conical movement round the pole of the ecliptic *as a centre.*

More accurate instruments and more careful observers proved that the obliquity was not invariable (though it will be seen that Sir R. Ball, in the work quoted in the pamphlet, attempts to minimize the importance of this), but had decreased during the past 2,000 years at least.[2] It followed, therefore, that the pole of the heavens could not trace a circle round a supposed centre, from which centre it continually decreased its distance.

[1] 6° east is the true centre of the circle travelled by the pole.

[2] And is now steadily decreasing.

In order to meet this contradiction, theorists asserted that the pole of the heavens did always move round the pole of the ecliptic as a centre, but that this centre moved so as to accommodate itself to the theory. When these theorists had altered the position of the pole of the ecliptic, it followed that the plane of the ecliptic, 90° from its pole, must also be altered in a corresponding manner, and, consequently, that the latitude of stars must also be altered to agree with this movement. When, however, ancient catalogues of stars were compared with modern catalogues, it was found that the latitude of stars had not varied in the manner they ought to have done in order to agree with these theories.

Here was another difficulty, but it was overcome by stating that as the stars were not situated where by theory they ought to have been, therefore the stars themselves must have moved, and every star whose position did not agree with the theory was assigned a proper motion of its own.

Another science, however, now made itself known, namely, Geology, and geologists stated that there was evidence on earth of very great changes of climate in the past history of the earth, far more than could be accounted for by any local changes of land and water, or by ocean currents, and astronomers were asked for an explanation—as these changes were seasonal and periodic it was as much an astronomical question as was the construction of the calendar—but astronomers could give no explanation. It was stated that the pole of the ecliptic could not vary its position more than 1° 21′, and therefore that the obliquity could not vary more than that amount. That a very slightly different movement of the earth (as demonstrated by Drayson) from that which had been assumed would account for very great changes of climate, although no movement whatever occurred in the ecliptic, did not seem to occur to any one.

A singular contradiction, however, still continued to be practised in astronomy. Whilst it was asserted that every movement of the earth was known with such minute accuracy that it could be affirmed that no changes of climate due to the obliquity could have occurred for hundreds of thousands of years, yet perpetual observations were still necessary in order that the position of each star relative to the pole and meridian should be assigned for some five or six years in advance.

If the actual movements of the earth were *really* known, the position of each star could be calculated for one or five hundred years in advance, without any reference to observations.

If the daily rotation of the earth were known and its annual revolution round the sun, and the sun's declination for the day, then the sun's altitude for any given latitude can be *calculated* for each hour and minute of the day. The same results as regards the stars can be calculated for hundreds of years when the slightly different movement of the earth's axis, discovered by Drayson, is known and understood.

Method of finding the Obliquity and Pole of the Ecliptic.

In order to find the extent of the Tropics and Arctic Circle, or what is termed the obliquity of the ecliptic, astronomers measure the greatest meridian altitude of the sun at the period of the summer solstice, and its least altitude at the period of the winter solstice. The difference of these two altitudes divided by two is taken as the obliquity of the ecliptic.

Whatever this obliquity is, the angular distance between the pole of the heavens and the pole of the ecliptic must, by a geometrical law, be of the same value; consequently, if the obliquity were 24°, then the angular distance between the pole of the heavens and the pole of the ecliptic would be 24°. If the obliquity were 23° 27′, as now, the angular distance between these poles would be 23° 27′.

From the very earliest date down to the present time observation has proved that the obliquity of the ecliptic has slowly decreased. There is no *visible* point in the heavens to indicate where the pole of the ecliptic is situated, consequently the position of the pole has been assumed from time to time in the following manner:—

Since it was assumed that the pole of the heavens traced

a circle round the pole of the ecliptic *as a centre*, when it was found that the pole of the daily rotation did not vary its angular distance from a star, it was assumed that the pole of the ecliptic must be on the arc of this star and distant from the pole of the heavens as many degrees and minutes as the obliquity was found to be at that date.

As the pole of daily rotation is carried round by the second rotation, other stars would be found whose polar distance did not vary. Consequently the pole of the ecliptic was assumed to move from point to point in the heavens and in the manner shown in the following diagram:—

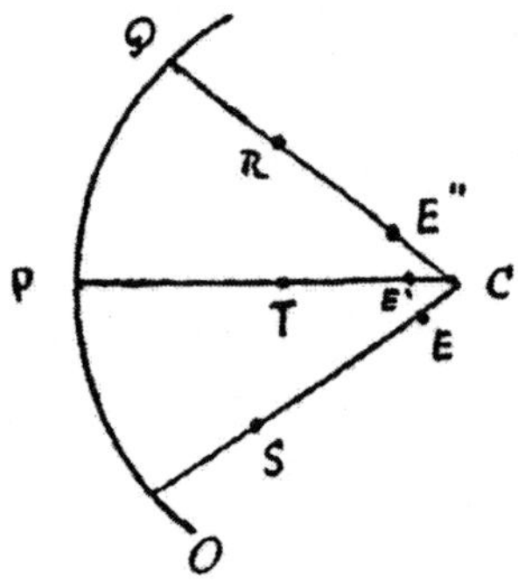

Fig. 1.

Take OPQ as the course traced by the earth's axis in the sphere of the heavens and due to the second rotation, and the pole of the second rotation C 29° 25′ 47″ from the arc OPQ.

When the pole was at O the star S was found not to vary its polar distance, and the pole of the ecliptic was supposed to be on the arc OS, produced, and distance from O whatever the obliquity was found to be at that date. Suppose the obliquity was 24°, as it was about the beginning of the Christian Era, then E, the pole of the ecliptic, would be assumed to be at E on the arc OC, and 24° from O.

As the pole of the heavens was carried round to P by the second rotation, and a star T was found not to vary its polar distance, the pole of the ecliptic would be assumed to be on the arc PTC, and distant from P an angular distance equal to the obliquity at that date.

Taking the obliquity at 23° 40′, as it was found to be

about a thousand years ago, then the pole would be assumed to be at E′, 23° 40′ removed from P.

In like manner the pole of the ecliptic would be assumed to be at E″, 23° 28′ from Q when the pole of the heavens was at Q, and the obliquity 23° 28′.

Hence it has been assumed that the pole of the ecliptic has a movement laterally, and at the same time towards the pole of the heavens, in order to account for the decrease in obliquity or angular distance of these two poles, whereas there is not the slightest evidence to prove that any such movement occurs in the pole of the ecliptic.

The results of this assumption are most serious, many details in astronomy requiring perpetual correction; such, for example, as inserting a few minutes of purely imaginary time between the end of one year and the beginning of the next in order to make facts and theory agree; this was the case in 1833–34, as is mentioned in the last note of Sir J. Herschel's "Outlines of Astronomy."

What has been already said will serve to show firstly, that the assertion that the pole of the ecliptic was the centre of this movement was a mere assumption; secondly, that it was also assumed that the pole of the ecliptic had moved, when it was discovered that it did not remain the centre; and, thirdly, a final assumption was involved, by which it was deemed that the stars had "proper motion."

This confusion, and consequent adjustment and readjustment of figures with the contradictory statements put forward from time to time regarding the period of one entire revolution, indicates that there is something wrong. Drayson shows how simple the cause is, though obscured by the coincidence that just during the centuries when most of our astronomical knowledge has been gained, it happens that the real movement of the pole is very nearly identical with the supposed movement. Had observations been made as closely 5,000 years ago, the real motion would most certainly have been discovered. If the diagram is considered it will be noticed how routine and preconceived ideas are apt to blind one to realities and how near astronomers have been to this important discovery. In assuming the pole of the ecliptic to be at a certain spot at one period and in other positions at other periods, if any one had correlated the positions and produced the arcs beyond the ecliptic poles, they would have been found to meet at one point C, and thus the whole matter of the second rotation would have been revealed!

Drayson goes on to say, "Routine, however, won the day, and the theory that the pole of the heavens always traced a circle round the pole of the ecliptic as a centre remained accepted as a fact. Then it was supposed that by employing more observers and building additional observatories, a solution might be found of these mysteries which were actually self-created."

Improbable though it may seem that so important a fact escaped notice, such appears to have been the case, because Drayson with his wizard-wand of precise knowledge proves that he can solve any question regarding polar distances of stars, the obliquity of the ecliptic at various dates, etc., and gives a confident and definite answer to the riddle of the Ice Age, fully explaining the cause thereof, the dates of waxing and waning and the time of greatest intensity; whereas orthodox astronomy can only direct inquirers to seek for a cause elsewhere, as the movement of the pole of the heavens is said to preclude all possibility of a general change of climate being directly due to it.

Thus we see Drayson's reasons for Glacial Epochs. But the earth moves round the sun, and the sun revolves around another greater sun or centre. The earth takes 365 days (about) to perform its revolution. The sun takes 25,827 years to perform its cycle. (In this, as regards the time of the sun's cycle, I have given that recorded by the Egyptians.)

If my readers who are interested in this subject will get some coloured balls representing the earth, sun, and centre of sun's revolution, and place them on a stand so that they can revolve around each other, this objective demonstration will give them a very good idea of what takes place, and will become an obvious fact to them that the statements I have made are critically correct.

I give on the next page a slight diagrammatic portrayal of what I state.

This is diagrammatic, but will give the reader, I trust,

a good idea of what has been and is occurring, and the vast expanse of the universe that the sun must pass through in

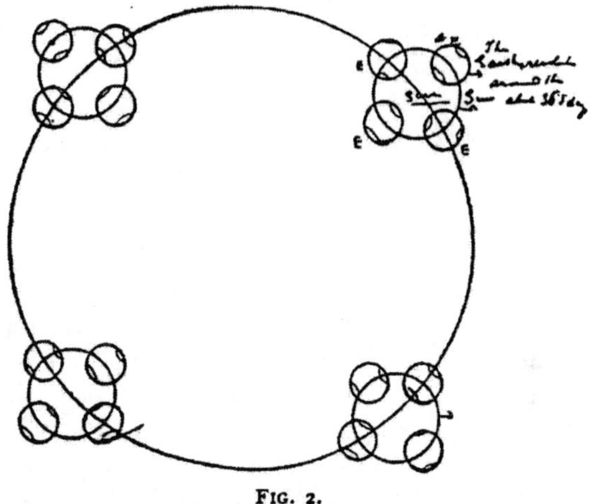

FIG. 2.
Drawn by Author.

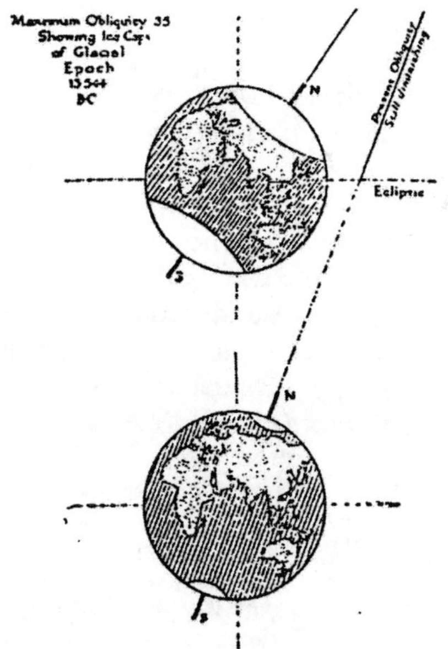

FIG. 3.
By Major R. A. Marriott, D.S.O., late R.M.A., to whom I tender my very sincere thanks for permission to reproduce it.

performing its cycle, and its relation to the other stars, which the old Egyptian priests have noted and recorded, and which our present astronomers do not appear to have taken into consideration. Yet Drayson had worked out a not inaccurate idea of the subject and found the reasons of the exact polar distance of a star for fifty years in advance.[1] The Old Wise Men of Egypt had worked this out and found that one revolution of the Sun corresponded with one revolution in precession or rather re-cession of the Pole Stars (Ursa Minor), and the records of this fact are still extant. Our present Astronomers have never attacked this subject and our present Professors of Geology are, therefore, unacquainted with the same, hence the false doctrines they teach or try to teach.

Again, possibly the above Professors have arrived at their opinions on account of the Great Boulders found in some parts of the North of Europe and because some parts of the North of Europe have sunk to a lower level and risen again; but there is no proof that the whole of the North of Europe was submerged under the sea. The amount of ice and snow which must have fallen in the North, would be sufficient to carry the huge boulders we find in some parts of Europe away from their parent rock, similar to that found, on a smaller scale, in the glaciers which are melting and travelling along, or receding, at the present day. The "*boulder*" *alone* remains *and the evidence of the sea bed is not there.* The forms of the Mountains in Norway and Sweden alone are sufficient proof that they have not been submerged and upheaved, as we find in the Australian continent, and although there is evidence of the rise and sinking of part of this land, and rising again, it has not been submerged as a whole.

In his Chapter II, p. 29, Professor Sollas states: "The dawn of the human race is supposed to belong to a past more remote

[1] Drayson was a man far advanced in the knowledge of real astronomy, as was Gerald Massey in the Mythology. They have not been equalled since the passing of the Old Wise Men of Egypt. Both were ignored and scorned during their lifetime, and branded as "faddist and idealist," but their knowledge will in a future generation be recognized and appreciated, whilst "the present and past professors" will be consigned to the dead.

Drayson's like Massey's works will live when the others are dead and forgotten; they will live the second life, like the souls of the good, whilst others now of great reputation will not. These will be like the souls of the bad, "and die the second life," and their names even will not be remembered in a few generations hence—except as ignorant and foolish men.

than the beginning of *the Great Ice Age; yet of the existence of man antecedent to that epoch, not a vestige of evidence forcible enough to compel universal belief has up to the present time been discovered.*" I do not know what may be the standard of his "evidence forcible enough to compel universal belief," but we have sufficient evidence to prove the fallacy of his statements. Let us bring forward objective, and some subjective, evidence to support our argument.

1. A mile north of Ipswich, *beneath* the Boulder Clay, in the glacial sands (the Boulder Clay was formed by the retreat of the Ice Sheet) the remains of a human skeleton were found. Therefore, this man must have existed here before the last glacial epoch, and from the evidence of the surrounding country, etc., before the last two glacial periods: and this man was a "Stellar Mythos Man."

Also Mr. J. R. Moir found Flint Implements beneath the Red-Crag formation of Suffolk, formed during the Pliocene or Tertiary age—dating probably 600,000 years ago. These Implements prove that the "Nilotic Negroes," if not the Stellar people, existed here then.

But a much more conclusive proof of man's earlier existence was found by the discovery *of Skeletons* at Castenedolo, near Brisca, Lombardy, in Italy, by Professor Rajazzoni, between 1860 and 1880, in undoubted *Pliocene Strata,* and these Professor Sergi describes as of "*modern type*" (we shall refer to this later).

We have the remains of a human found 87 feet beneath a bed, which was formed by running water, in Germany, and the skulls found at Gibraltar, and others well known, such as the Galley Hill man, the strata around which contained flint implements of Stellar Mythos man. The so-called Pithecanthropus of Java (Skull c.c. 855 found in Pliocene Strata) was human; proof of this being a Pigmy—see later.

Also a Palæolithic Human Skull and Mandible in Flint-bearing gravel overlying the Wealden (Hastings bed) at Piltdown, Fletching, Sussex (further evidence of this see later).

That Implements of man have been found in this country in the Pliocene formations is undoubtedly true.

The late Professor Prestwich, a geologist whose great knowledge all recognize, found Implements in the uplands of Kent between the Thames and the Weald (and I have found some good specimens there myself). According to his opinion these deposits were of Pliocene date.

In the Pliocene formation of the Cromer beds of East Anglia, Mr. Clarke has found many also.

In the Norwich crags and under the Red-Crag, Mr. Reid-Moir has discovered some.

Thus in England we see man existed here during the Pliocene formations. How many times he was driven south again by each glacial epoch is a question of how many years ago these were formed; if this was formed about 600,000 years ago he would have advanced and retreated over 23 times.

M. Rutot, of the Royal Natural History Museum in Brussels, states that he has found human instruments in the Miocene and even well into the still older formation, the Oligocene. Personally I must doubt the latter formation containing any human implements "except by accident" for the following reason. The discovery in the oldest Oligocene formation of the Fayoum, Egypt, by Dr. Max Schlosser, two years ago, of the teeth and jaws of "three Higher primates," two of which are allied to the South American Apes, and the other is a *forerunner of the Gibbons*, rather precludes the evidence that man existed at that time. *It does prove, however, that at this early date the South American Apes and the pre-Gibbons were already in existence in Africa.* Here we have an animal which is close to the stock giving rise to man, highly formed at that period. In the Miocene we know that the Great Anthropoids were in existence, and with the appearance of the Great Anthropoids the appearance of the first human is probable. Professor Keith is of the opinion that "From every point of view it is most probable that the human stock became differentiated at the same time as the Great Anthropoids," and I quite agree with him, and if this was so, as I believe evolution shows, man's first appearance would be Miocene and not Oligocene. It really is a question whether the Anthropoids were first developed, or differentiated, from the primates during the Oligocene—I do not know that we have any evidence that they were, personally I think the weight of evidence is against it. Later discoveries may settle the question definitely. The above evidence alone is sufficient proof to show the fallacy of Professor Sollas's beliefs and statements, contained in his little book so highly eulogized by *The Times* and *Athenæum*. These are objective evidences found for any one to see. Now let us take some of the subjective evidence.

We have the fact, proved by Professor Petrie's "diggings" at Abydos, that the Osirian Cult was at its zenith 20,000 years ago, in Egypt, and that it existed in Heliopolis and other Nomes before Abydos was built, and Lunar Cult preceded Solar, and Stellar preceded Lunar. Manetho stated that the Great Pyramids were built by the followers of Horus during the Stellar Mythos, and evidence of this is borne out by the Ritual of Ancient Egypt and the characteristics of the buildings found.

There is evidence that the Egyptian Priests had recorded the time of the Precession of the Pole Stars (Ursa Minor) and, therefore, the revolution of the Sun around its centre for 258,000 years, up to a period of part of the Stellar Cult. And we know, as a fact, that the Precession of the Pole Stars, and the revolution of the Sun round its centre takes about 25,827 years to perform this one cycle, (records left written on the stones of the Great Pyramid and on Papyri), during which time we have one Great Ice Age, or Glacial Epoch. Can Professor Sollas imagine for one moment that all the deposits over these skeletons were formed in less than 26,000 years, or that man had not existed a million years before the Solar Cult? The fact is he has written this book "Ancient Hunters and their Modern Representatives" upon pure theory and guesswork, and has been much eulogized by critics who are as ignorant as himself upon these matters, thus misguiding the general public and spreading false information under the name of a Professor of Oxford and Past President of the Geological Society. I offered the Geological Society my books, "Origin and Evolution of Primitive Man" and "Signs and Symbols of Primordial Man" for their Library, and a short paper in answer to Professor Sollas's above-named work, and how to distinguish and differentiate Palæolithic Implements, criticizing after proofs, but the President of that very learned Society after due consideration refused the lot without thanks. Can there be a greater instance of what narrow-minded men some of the Professors of Oxford and London are at the present day? Each one in his own particular Cabbage-Patch, afraid to lift his head to see if there is any fruit in the next garden, and when attention is drawn to this, mumbles "Weeds—weeds—only weeds"!!!

If the above facts that I have brought forward are not sufficient "to compel universal belief," then the present humans will still continue to wallow in the mire of ignorance, as they have in the past dark and degenerate age that we have passed through since the downfall of the old Egyptian Empire, as far as science is concerned, in this branch of it at least. The Glacial Epoch occurs to the whole of the Northern and Southern Hemispheres. Europe, Asia and North America are frozen and covered with ice down to the latitude of about the South of France. How could the origin of man be in Asia? He could never have worked out his Astro-Mythology in Asia. Nor could he do so, if we placed his original home where Haeckel and his followers have placed it, namely, from a centre which influenced Egypt and Babylonia in the West, and America in the East—a centre in close proximity to Polynesia and Java; it will not bear critical examination after all the proofs to the contrary I bring forward.

They could not take the observation day after day, month after month, and year after year, of the precession of the Heavenly bodies as the old Egyptians did; the climate would not allow this—where you have the whole of the heavens obscured for weeks at a time during the rainy season. And the Egyptians have left records showing how patiently they must have worked for 250,000 years, proving that their observations were continuous and accurate.

Then again, although the Pigmies are in this part of the world, and also the Nilotic Negro, the connecting links of the Bushmen and Masaba Negroes are not here, neither is the true Negro. As a Glacial Epoch occurred, so man who had gone North would be driven back South, and as that epoch passed away he would go North again. Only in this way can any geologist give a definite and final answer as to what he finds in different strata, and in this way alone, ever remembering that Old Egypt was sending out better humans in every way, further developed in evolution, in stature and brain-power and knowledge, as exodus succeeded exodus. Old Egypt was the home and birthplace of the human race, and as each generation grew up so each generation grew in evolution and knowledge into a higher type. It was here man first learnt his language—Monosyllabic, and Sign and Gesture Language—learnt to draw signs and symbols to represent Elementary Powers, for which he had at that time no words or language to express his ideas and thoughts; gradually he grew in stature, and brain cells began to develop; from his being able to think in things he advanced to drawing things, objectively at first, and formed the Hieroglyphic Language—and then an alphabet, and so on progressively until he spoke and wrote—an agglutinated, and finally an inflected language. From his rude Club, or Stone, first used only in the hand, chipped on one side, and then on both roughly, he fashioned these into beautiful polished axes, arrow-heads, knives, etc., and began to work on metals, built temples and gradually advanced in knowledge and wisdom in Astronomy, Eschatology, etc., and formed a Code of Laws—Moral, Religious and Civil—that has never been surpassed. The earliest people who lived in these islands, and North of this latitude, were driven South, or if not died here. This will recur again, and the huge city of London will be ground to atoms by the weight of ice, and nothing living will endure in these islands, only the remains of ancient cities, skeletons, etc., of men and animals for our descendants to discover, as we now find part of the "skeletons of the corpses" of our forefathers and their works, who lived here in the dim, bygone ages. No one can be quite sure if man first existed a million or two million years ago, but from the fact that skeletons and implements of man have been found in certain localities it proves that he must have existed here at least at the time of the geological formation we find him in, "except by accident." Mr. Comyns Beaumont in an article in *The Times*, June 27, 1911, states "from a scientific standpoint a 'Furfooz' man or a 'Cro-Magnon' man is worth nothing." We do not agree with him; our opinion is that Osteo-remains of Ancient man wherever found "prove a great deal and are worth something."

Palæolithic man certainly lived here in these islands and in Europe generally over 1,000,000 years ago. Many of his Implements have been found in the gravel-pits, more than eight feet below the surface at Biggleswade, along with the huge teeth of the "Elephas primigenus," which was a common animal at that time in Bedfordshire, as well as the discoveries at Neanderthal, Spy, La Chapelle-aux-Saints, La Ferrassie, Gibraltar, and the remains discovered in undoubted Pliocene strata at Castenedolo, near Brisca, Ipswich and in Sussex. Progress in the acquisition of knowledge of our past history has made much advance in the last fifty years, characterized by prudent and impressive deliberation; every year we have unearthed more and more traces of what our ancestors were like in past geological ages, he who, in a hard and cold climate, inhabited the caverns of the Quaternary valleys, hunted the cavern bear and the long-haired mammoth and left his bones there, thus enabling us to reconstruct the anatomy of our venerable ancestors, who, although very low down in the scale, far removed from us, anatomically, physically and intellectually, had emerged by evolution from the animal and were "primordial man." He was raised above the mere animal by virtue of his intelligence and his slow development towards better things. He knew how to prepare for his use an outfit of unrivalled arms which enabled him to dominate the rest of the Zoological world. The position in which certain remains have been found shows that he already had "funeral ceremonies" of a kind, unknown to animals, and probably certain rites in his attitude towards the dead.

To understand evolution fully, one must go back to the primitive Pigmy people of small stature and great muscular development; the first little earth-man or red man, i.e. Palæolithic man in his earliest form, who was born probably

2,000,000 years ago. As they spread from Africa farther North, the cold northern climate would cause their muscular tissue to develop greater strength, and they would become hardier than those of hot climates. The whole world was peopled by these "Negroids" first, certainly over a million years ago. As the cold became too intense to live in the Northern latitudes, they were driven back South again. Some, no doubt, went farther South than others. The cold would induce them to live in caves and underground places, and they had to think how they could keep warm and procure food, how to fasten skins together, from the animals they killed, to wear and enable them to retain their natural warmth, and a commencement of mechanical devices to suit, and thus they would have to use their brains more and more which would gradually tend to increase therein in size and cell development. As soon as the extreme cold had passed away they would spread Northwards again, and would be followed by those who had developed both in stature and brain-power in Egypt and the Nile Valley. These would drive out and exterminate the first primitive people, as we find the Ainu did in Japan, and the Australians did in Australia. This probably took a very long time, and all would not be exterminated at once. Some, no doubt, would mingle with the conquerors, and in many cases they would naturally die out. Marriages and intermarriages must be taken into consideration in further developments and evolution, as well as environment. The first Exodus of the lower class Nilotic Negro who followed and came directly after the Pigmy, would kill these men when they came in contact wherever they could find them. But the second Exodus of the Nilotic Negro, or those of a higher type (Hero-Cult People) would not kill the Nilotic Negroes of the lower type, or those of the first Exodus, because all these were "Totemic People." They probably would intermarry but not exterminate each other, as we find at the present day in Africa, Australia and Torres Straits people, and other parts of the world.

Then the Stellar Mythos People, who were the next to leave Old Egypt, a finer race of men who had developed here, would war against the Nilotic Negroes wherever

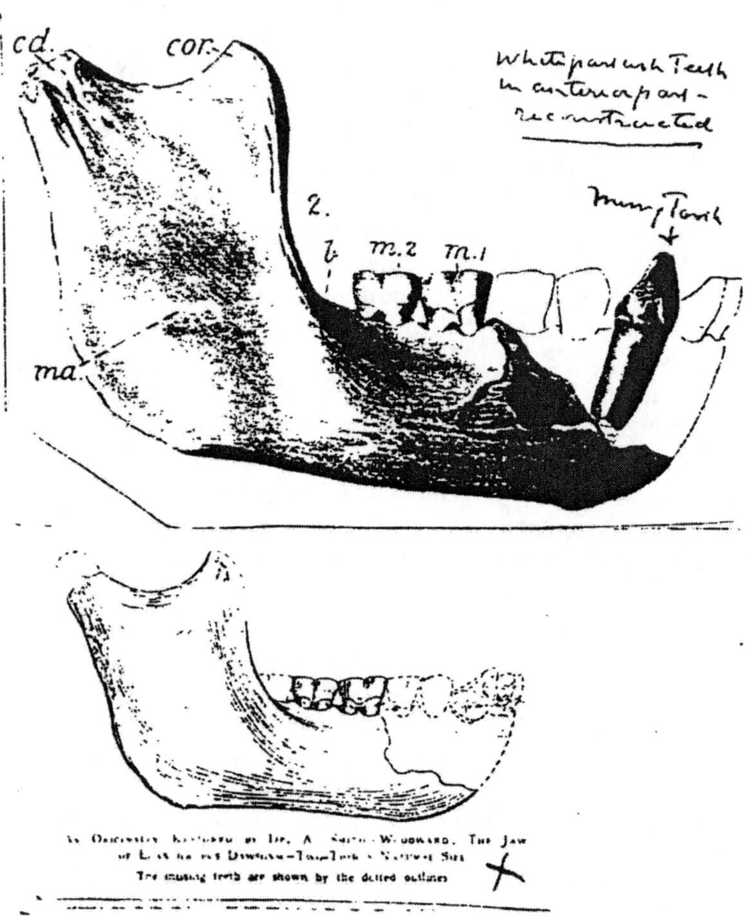

PLATE I.

Piltdown skull and canine tooth.

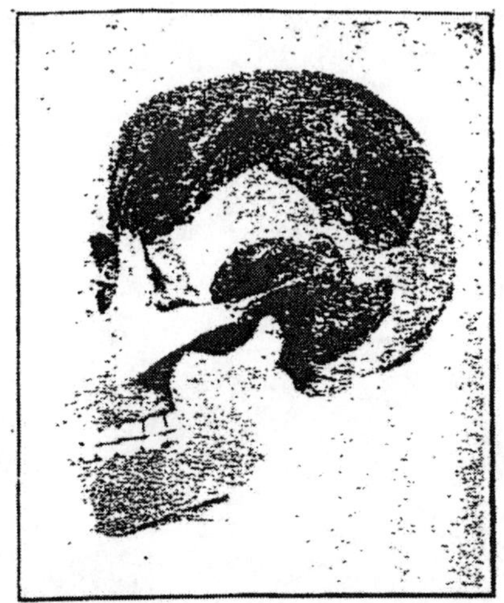

As Restored by Professor Arthur Keith: The Skull of the Piltdown Man—Man-Like Both in Jaw and in Brain-Capacity.

As Originally Restored by Dr. A. Smith-Woodward: The Skull of the Piltdown Man—Ape-Like in Jaw, and of Small Brain-Capacity.

PLATE II.

Piltdown skull reconstructed.

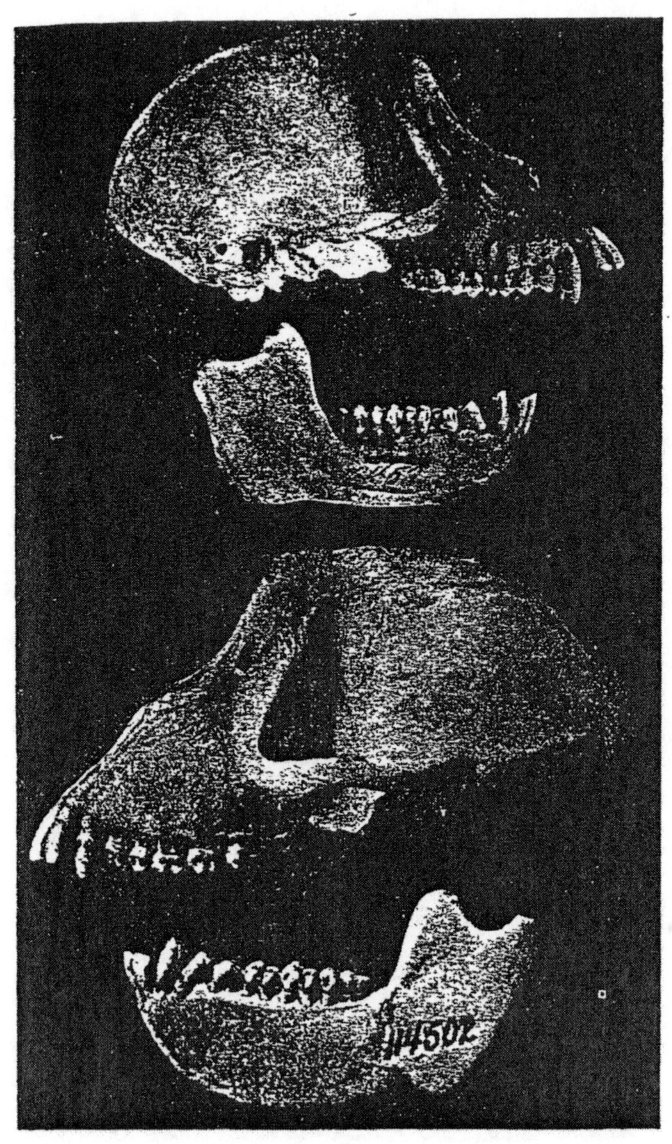

PLATE III.

Jaws of *Symphalangus syndactyles.*

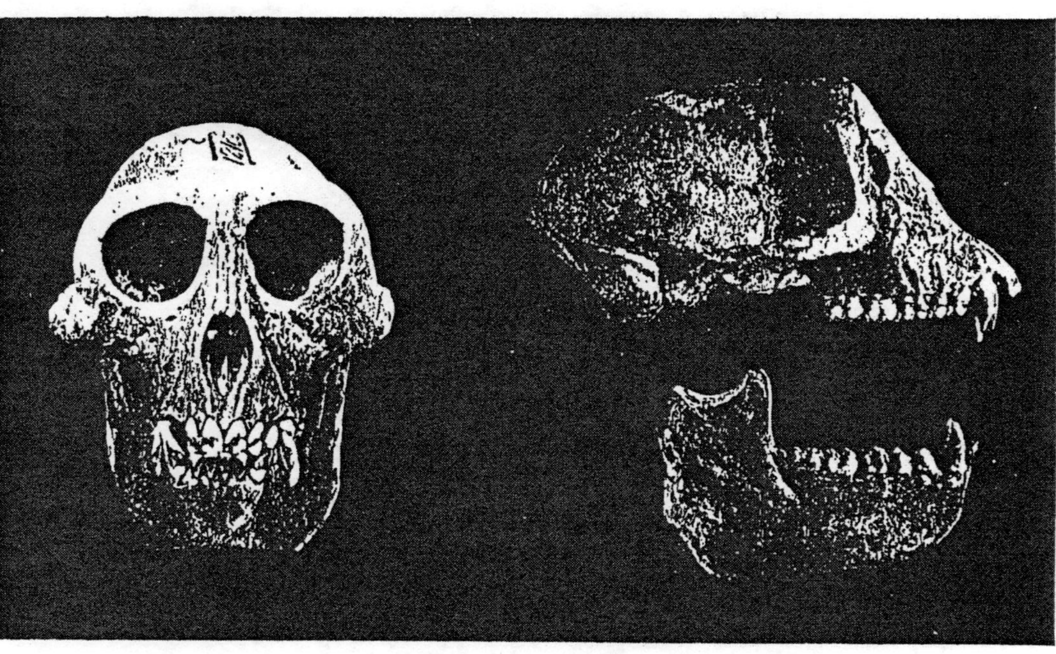

PLATE IV.

Jaws of *Symphalangus syndactyles.*

these latter obstructed or interfered with them in their advancement towards the North.

So eventually all the Pigmies and Nilotic Negroes would be exterminated, in Europe first, then in Asia as far as these Stellar people travelled. To Australia, New Zealand, and some other parts of the world the Stellar people never went, hence you still find the inhabitants of these countries "Nilotic Negroes," descendants of the original exodes from Old Egypt.

My readers will see that for a vast period of time both the Pigmies and Nilotic Negroes must have existed at the same time and contemporarily with the Stellar Cult people in the same continent—the Pigmies in the Hills and most inaccessible places where the Nilotic Negro would not follow them, and the Nilotic Negro giving a wide berth to the Stellar Cult people, who were his superior enemies.

Hence the reason for the different Osteo remains of man found in different countries, and in different strata, which would lead to an erroneous opinion being formed as to the character, ages they lived, and development in evolution attained, if the key to man's origin and evolution was not known.

I have sought in this work to give that key to the world; hitherto it has been lost, and hence the many errors and fallacies that have arisen.

CHAPTER II

THE PILTDOWN SKULL

THERE has been much argument as to what manner of man (or woman) it must have been who owned that part of the jaw and portion of the skull which were found in a gravel deposit near Piltdown Common by Charles Dawson.

Two different reconstructions have been made, and fierce controversy has reigned in the camps of the followers of the reconstructors, namely, Professor A. Smith-Woodward, F.R.S., keeper of the Geological Department of the British Museum, and Professor Arthur Keith, conservator of the Museum of the Royal College of Surgeons.

I give here the two different reconstructions (Plate I), with a newly discovered canine tooth, which Dr. Smith-Woodward believes proves the truth of his restoration. I listened to a very interesting lecture he gave at the Royal Societies Club, but could not agree with his reconstruction, as I stated there. The form of restoration of the mandible as depicted approaches too closely to that of the chimpanzee, and I fail to find any reason why this jaw should be reconstructed as one belonging to the higher apes, i.e. if it is the jaw belonging to the skull found. There are many reasons to suppose that it is, and no reason to suppose that it is not.

It is not possible to agree with his formation of the incisors, canine and premolar teeth, or to the length he has added to the anterior part of the mandible, so that he may place these teeth in position to form his idea of the reconstruction.[1]

[1] Some of the Pigmies and Nilotic Negroes have much larger teeth than the ordinary present type of man (see Pigmies).

I give here his earliest and latest reconstructions (Plate II), and also some types of the jaw from which his does not differ very much (Plates III and IV), which my readers may study for themselves. These are low-type apes, in fact the jaw of the *Symphalangus Syndactyles* in the National Museum of New York is still nearer his latest ideals.

We must now examine the skull to which this jaw is supposed to belong.

In Dr. Smith-Woodward's reconstruction, he gives for the brain the capacity as 1,100 c.c. (cubic centimetres).

Professor Keith in his reconstruction gives the brain capacity as 1,500 c.c. and states as his reason for this, "By some mischance the groove for the median blood channel, which runs along the roof of the skull, was displaced nearly an inch to one side."

In Dr. Smith-Woodward's reconstruction, the bones of the right and left sides are nearly in contact, in Dr. Keith's they are rather widely separated in order that the groove for the venous channel may fall in its natural position, namely, in the middle line of the skull.

Dr. Smith-Woodward appears to have founded his reconstruction on the flattening and thickening of the forward reaching part of the bone, and so has given the human skull the mandible of an ape. But if the bones are placed in Dr. Keith's position, this flattening is not apparent.

Dr. Smith-Woodward *is supported* by Dr. Elliot Smith of Manchester and Sir Ray Lankester, K.C.B., F.R.S.; the latter in finishing up an article on the pros and cons in the *Daily Telegraph*, December 2, 1913, states as follows: "In regard to such questions, there is a regrettable tendency among writers of no credentials to make assertions, and anticipate conclusions which can only be justified by further discoveries." He also backs up Dr. A. Smith-Woodward as follows: "He rightly enough emphasized the conclusion already favoured by many anthropologists, that among the very ancient races of prehistoric man, some must have had well developed brains, and that there were probably several very divergent types of primitive man in early Pleistocene and Phocene times, some more ape-like, some like modern man, and some (the Neanderthal and the Heidelberg man) showing peculiar features and not lying in the direct line of ancestry of any of the races of modern man."

As a writer of "no credentials" to whom I believe he refers, let me point out to Sir Ray Lankester the fallacy of Dr. Smith-Woodward's reconstruction; not by any "assertion" or "anticipated conclusion" has the result been arrived at, but by study and examination of the heads and skulls of humans past and present for the last twenty-five years.

1st. *The highest cubic capacity of the skull of any ape is not more than 600 c.c.* The average c.c. of Pigmies is 900 c.c. The average c.c. of the low class of Nilotic Negroes is 1,100 c.c., and of the Hero-Cult Nilotic Negroes 1,250 c.c. to 1,300 c.c.

This skull, according to Dr. Smith-Woodward, is 1,100 c.c. and considered by Dr. Keith to be 1,500 c.c.

I have personally examined the skull and jaw, and should put the c.c. of the skull at 1,200 c.c., perhaps 1,250, but not more. Dr. Keith has left too much space, and Dr. Smith-Woodward has not left any for the central vessels.

The c.c. alone therefore proves this to be a human skull of the Nilotic Negro class, probably a non-Hero Cult. Types of them are still found in Africa and Australia, and we may find them still in some isolated islands or places not yet explored by the white man. Moreover, one of the temporal bones, including the glenoid fossa, was complete, and Professor Smith-Woodward pointed out how closely this bone and fossa resembled the corresponding parts of modern man. We must remember that the configuration of the glenoid fossa in man was such as to adapt this for articulation with the human jaw, and not with the jaw of a chimpanzee. Therefore, it would be in the temporal bone one would have expected to find some variation in structure from the present day condition.

Then again the Nasal bones which have been found are of the same type and formation as those of the low class Nilotic Negroes; they are thickened at the upper border, proving that this man or woman had a massive and overhanging, or beetling brow, like the Non-Hero Cult Nilotic Negroes. Indeed, this is one of the most distinguishing osteo characteristics of this class of homo (see Australian types later).

Professor Elliot Smith's note in the *Geological Journal* on the brain is interesting, as showing that there were distinct evidences that the area of the brain occupying the position where the modern human brain is developed, "the territory which recent clinical research leads us to associate with the power of spontaneous elaboration of speech, and the ability to recall names," was an obviously expanding area. As the Pigmy, even at the present day, has only about 200 words and the lower class Nilotic Negro only about 300 to 500 words, one would not expect to find this area as greatly expanded as in the Hero-Cult Nilotic Negro with about 900 words—much fewer by many than used by the present type of man.

Professor Elliot Smith, F.R.S., read a paper on "The Brain of Primitive Man, with special reference to this skull," before the Royal Society, in which he made known his beliefs, "that he had no difficulty in reconciling the ape-like jaw with other fragments of the remains. He held that the small brain of this skull, though definitely human in character, represented a more primitive and generalized type than that of the genus homo (!!!). Nevertheless, it could be regarded as a very close approximation to the kind of brain possessed by the earliest representatives of the real man, and as the type from which the brains of the different primitive types of men, Mousterian, Tasmanian and Australian, Bushmen, Negro, etc., no less than those of other modern human races, had been derived, as the result of more or less well-defined specializations in various directions" (!!!). He was confident that the size of the cast could not be more than about 1,200 c.c.

Contrasting the brains of Pithecanthropus and this skull, Professor Smith stated "that they should both be looked upon as the divergent outlying specializations of the original genus of the human family, rather than actually representing the main stem. He believed that this skull belonged to a real genus and represented as close an approximation to the ancestor of man as we could expect to find. Pithecanthropus represented the unprogressive branch which survived into Pleistocene times before it became extinct, whereas this Piltdown Skull was the progressive phylum from which the genus homo was derived"!!!

I should certainly prefer to remain Sir Ray Lankester's "writer of no credentials" than be a Professor and F.R.S. to state such opinions before the F.R.S.'s.

Pithecanthropus had a skull with brain capacity of 850 c.c., a Pigmy in my opinion. Piltdown, according to our learned friend, was 1,200 c.c., and if 1,200 c.c. he was certainly a Nilotic Negro.

Then he mixes Mousterian, Tasmanian, Bushmen, Negro, all together, and calls them all the brains of the different primitive type of man derived from the type of Piltdown skull ! ! ! or in other words, from a human being with a skull capacity of 1,200 c.c. were derived the Pigmy, and others, with skulls of 850 to 1,100 c.c.

I do not think it is necessary to make further comment. One can only feel sorry that our Professors of Anthropology and Ethnology "who have credentials" should still rank in the same state of knowledge as the poor 1,200 c.c.

The Palæoliths found in the same bed are of mixed ages, some undoubtedly Pigmy Implements—"Chipped on one side only," with some chipped on both sides—Nilotic Negro. I believe this to be a low-class Nilotic Negro skull. The first to be found in these Islands, probably not the last.

I must still class this man, or woman, as *Homo Sapiens*, and fail to find any reasons for the name *Eoanthropus* and would suggest *Homo Dawsoni*.

That there are "some different points," anatomically, between this skull and those of Pigmies in the Royal College of Surgeons' Museum, would not alter the fact that it was a Nilotic Negro of the lower type, or yet a Pigmy of the higher type. There are higher and lower types of the Nilotic Negro, as well as of Pigmies, living in Africa at the present day, amongst whom Professor Smith-Woodward will find great variation as far as the jaws are concerned.

The fact that the brain c.c. is at least 1,100, the massive beetling brow he or she must have possessed, and the mixed type of Implements found, is fairly conclusive evidence that it was a low class Nilotic Negro. I quite agree with Professor Smith-Woodward when he states that "he did not think that the difference between the Heidelberg and the Piltdown mandibles necessarily implied differences of geological age," and, as he rightly states, "the swamps and forests of the Weald in the early Pleistocene (or Pliocene as regards that) times may have been a refuge for a backward race."

There can be no doubt that the Pigmy existed in Europe in the Pliocene age, and probably also the Nilotic Negro (Heidelberg man was Nilotic Negro). He also may have existed here up to comparatively a much later time, having been driven into forests where he would not be exterminated, and probably up to "a recent period" in some places. To my mind it is not now a question of a Pigmy, on account of the beetling brows and brain c., but one of the first exodes of the lower type of Nilotic Negro. Taking into consideration all points of the facts that have been found, I should say that this was *the skull of one of the Nilotic Negroes of the lower type without Hero-Cult*, which followed the Pigmy out of Egypt, the forefather of whom must date back at least 1,000,000 years ago.

The discovery of the prehistoric skeleton near the Kentish village of Halling is of great interest, owing to the further proof it shows of my contention as regards the evolution and origin of the human race.

I reproduce here a drawing of the skull and mandible (Plate V, from the *Graphic*, April 19, 1913), which shows this to be a "Modern type of man," and from the strata it was found in and the position of burial, i.e. lying on its back in thrice-bent position with head *to the East*, one of the Solar Cult race. It is of peculiar interest because we have now found in England—

(1) The Sussex man or woman at Piltdown—a Nilotic Negro.

(2) Several Nilotic Negroes (Walton-on-the-Naze).

(3) Harlyn Bay and others—Stellar Mythos.

(4) Halling man—Solar Cult man. Brain capacity 1,500 c.c.

That this is a Solar Cult man there cannot, in my opinion, be any doubt, from the position of burial.

That he is found in older strata than some others of a lower type, Stellar Mythos and Nilotic Negro, would only help to prove my contention. We know that the Solar Cult existed more than 30,000 years ago and probably 100,000 years. That Stellar and Solar Cult people both existed here in these Islands [1] at the same time is also certain; probably all the Nilotic Negroes had not been even exterminated at the early advent of the first Solar man here, and all must have been driven back South several times at the advent of the Glacial epoch, every 25,827 years, and have returned again North after the Glacial period had passed away. Those whose remains we find were probably "caught by the cold," and died before they could return, or from accidental death, except where we find them buried in the thrice-bent position, as *the Stellar people at Harlyn Bay* and this Solar one at Halling.

I give here three reconstructions by Trevor Haddon, R.B.A., under my direction, and I am greatly indebted to Mr. Trevor Haddon for his patience and skill in reproducing all the details, as instructed, in such an elegant and accurate form as he has done in these.

1. A Reconstruction of Primary Man from the cranium here reproduced (Pithecanthus Erectus) found by Professor Du Bois in Java, in 1890, over which there has been much discussion as well as differences in opinions (Plates VI and VII).

Undoubtedly this is part of the remains of a primary man; I quite agree with Professors Du Bois and Keith in their opinions that it is so.

As a further proof of what Professor Keith has already pointed out, I would add that the cranial capacity is critical evidence—viz., the cranial capacity of this is 850 c.c.—although Professor Keith states that this is too low an estimate.

The highest cranial capacity of any Anthropoid is 600 c.c. and the average Pigmy is 900 c.c., therefore this would be the cranial capacity of a low type of Pigmy. Professor Keith and others have never taken into consideration the cranial capacity of the Pigmies, which must be considered, these being the first "humans" formed from the Anthropoid by evolution.

[1] If these were islands at the time, but probably they were joined to the Continent of Europe.

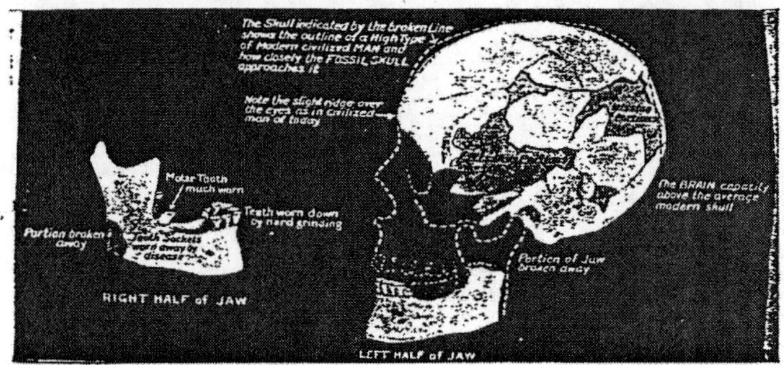

PLATE V.

Skull of modern type of man found at Halling burial-place and in different strata.

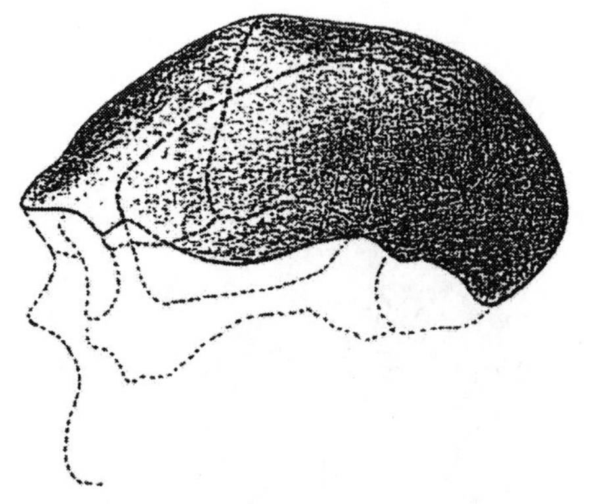

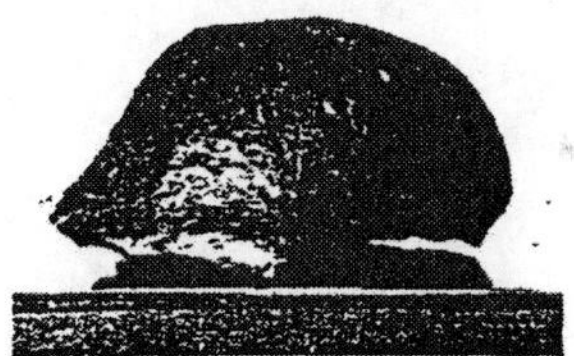

Skull of *Pithecanthropus Erectus*.

Plate VI.

That from which a kind of ape was reconstructed: the skull of the "Pithecanthropus" (found in Java in 1891), which Dr. Dubois declared to be an ancestor of man.

Dr. Dubois, a Dutchman, brought the cranium to Europe in 1891. A year later he discovered two great molar teeth and a femur, which seemed to be human, at a distance of fifteen metres from the scene of the former find. Concluding that the bones might be deemed to belong to the owner of the skull, he reconstructed a kind of ape, which he called "Pithecanthropus." This caused a good deal of comment at the Paris Exhibition of 1900.

PLATE VII.
A Reconstruction of Primary Man—*Pithecanthropus Erectus.*

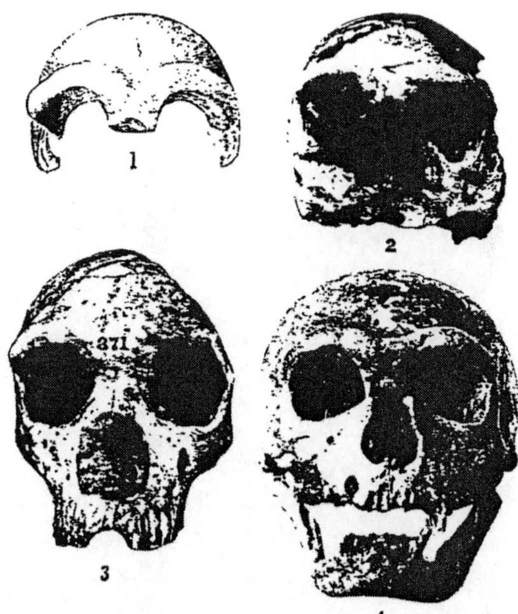

Plate X.

1. The Neanderthal Skull, with the heavy, massive supra-orbital ridges.
2. The Spy Skull. These Skulls are those of Nilotic Negroes.
3. The Gibraltar. In the absence of any knowledge of the implements found with this, I should put it as a primary type of Non-Hero Cult Nilotic Negro, the cranial capacity being 1,180 c.c. Many Nilotic Negroes have a cranial capacity of 1,100 c.c.
4. From the burial position, and the implements found with this Skull, I should place it as that of an early Stellar Mythos Man or the latest exodus of Nilotic Negro.

Plate XI.
A Reconstruction of the Neanderthal Man.

I have placed his height about 4 feet, but probably he was a little shorter.

2. A Reconstruction of Neanderthal Man, by Mr. Trevor Haddon, R.B.A., under my direction (Plate XI).

I give here a reconstruction of the Neanderthal Man from the Neanderthal Cranium; orientated on the same plane as the modern English man, here reproduced, it portrays the enormous development of the supraorbital ridges, showing the frontal lobe, and the greater development of the posterior part of the cranium, in fact, that of a primary type of Nilotic Negro. Remains of him have been found widely distributed over some parts of Europe; up to the present time, none have been brought to light in the British Isles (except the Sussex man), Italy, America, or Asia proper, although as these remains form part of the Osteo-anatomy of those primary Nilotic Negroes who first left old Egypt and followed the Pigmies, it is possible such relics lie hidden in these countries (see Plates VIII to X).

The type is still extant in Africa, and in Australia types of them may be found amongst the Arunta and other tribes to the present day.

Although the remains have been found in a more recent formation, geologically, than those of the Galley Hill and Ipswich men, yet the original prototype left old Egypt thousands of years before the Galley Hill man ever existed. He is much lower in the scale of evolution.

It appears to me probable that the whole of Europe and Asia were populated by this type immediately after the extermination of the Pigmies, whom they drove before them and killed as in Australia. The reason why we do not find their skeletons in the lower strata (Pliocene) is probably due to their burial customs, and disintegration. That we have found part of these skeletons in isolated places, in a more recent formation than other higher types of the human, who left Egypt thousands of years after, is not surprising. The primary leaving their old home in Egypt would spread over Europe and Asia travelling North, until driven back by the Glacial period. They would then return South until the cold ice period had passed away, when they would again advance northwards.

How many times these migrations occurred before a higher race came up from Egypt and fell upon them we cannot say, nor is it possible to form an accurate estimate of the time that must have elapsed before this type was exterminated, probably many hundreds of thousands of years. For these Neanderthal men would doubtless retreat from their enemies in isolated groups, and when not molested, would propagate a true breed until the last man was extinct, except, of course, when they had taken Pigmy women in marriage. They would not be allowed to take the women of the superior race; the superior race on the other hand, while exterminating the males, would take the females of the inferior race. The Implements found with the skeletons (or part of skeletons) and also their cranial capacity and other features prove them to be of a higher type in evolution than the Pigmy. These men would be driven away into isolated groups, in more inaccessible places, by the superior race which followed, the latter would take and occupy the best hunting grounds and best fishing places, and both might, therefore, live in the same country for thousands of years at the same time, but would not mix, like the Pigmy and Papuan at the present day. *This is a representative instance of misjudging the type and class of man by the geological formation his remains have been found in,* and in which the present Geologists and Anthropologists have gone wrong in their deductions, from finding these osteo remains in a strata of more recent formation than those remains found of a higher type of the human. I cannot let Professor Keith's statement pass without the greatest protest, namely—"*The Galley Hill man, who is one of the modern build of body, lived in England long before the Neanderthal man lived in France and Central Europe.*" The Neanderthal Man was born and lived in Europe probably over 300,000 years before the Galley Hill Man. The proofs of the same he will find in this work. The Neanderthal Man was evolved from the Masaba Negro and was the first type of the Nilotic Negro that left old Egypt, whereas the Galley Hill Man is a Stellar Cult man and certainly 300,000 years later in evolution.

Professor Keith thinks that this branch of the human,

Neanderthal type might have sprung from the gorilla and the others from the gibbon, on account of the great difference in the Osteo-anatomy found in this primitive man—but why? Does he suppose that a cart-horse and a thoroughbred horse were evolved from the same original or from two different sources? To me his statements (*supra*) are as absurd as Sir Harry Johnston's ("The Uganda Protectorate," p. 474) where he states: "Because white races may have risen twice or thrice or four times independently from Mongol, Negro and the Neanderthal-Australoid type"!

Then from whence did they come and how evolved? And what facts can he bring forward for such a theory? None.

Such a statement is too unscientific for any argument or consideration.

In the cranium (Plate XII, Ipswich Man), the intercranial mass forms a cast of the brain on which the third frontal convolution is well marked, *there is a greater supraorbital ridge than in the modern type of man but much less than in the Neanderthal man. The frontal lobe is not quite so well developed as in the modern man, but much more so than in the Neanderthal man.*

Probably from the Implements found near, this is part of the skeleton of a Nilotic Negro, but of much later exodus, and higher in the scale of evolution than the Neanderthal Man.

Besides Totems and Totemic ceremonies this man would have "Hero-Cult," whereas the Neanderthal man would not.

3. A Reconstruction of the Ipswich Man (Plate XIII).

If this Ipswich man is a Nilotic Negro, it can be so proved, not only by the worked flints found with or near him, but also by his cranial capacity and buried position. He was buried in the thrice-bent position, but it has not been stated if he faced North or South, and I have not been able to obtain any photographs of the flint implements, or the description of these, found near, as Mr. Reid Moir, to whom I wrote, had none he could send me. Therefore, without these details, I cannot positively say whether this was a man of late exodus of the Nilotic Negro Hero-Cult, or one of an early exodus of the Stellar Cult people.

Originals of this class are still extant in Africa and many other places. He is an earlier type of homo than the Galley Hill man, whom we must classify as belonging to the Stellar Cult people, from the instruments found near him.

I reproduce here (Plate XIV) the photograph of the skull found in one of the cists at Harlyn Bay, in Cornwall (with many thanks for permission to reproduce the same to Mr. Reddie Mallett). This human was buried in the thrice-bent position facing North, and over the cist was a triangular stone with apex pointing North; therefore, a Stellar Mythos man at the time that Horus was Primary God. The whole form and convolution of the cranium and jaws show a "modern type of man."

The position of burial proves him to be a Stellar Cult man, and therefore he lived here before the Druids of the Solar Cult came to this country—he came after the Nilotic Negroes.

He was here in Cornwall at the time and built those temples, the remains of which I have found in the West of England—*with Two Circles*, representing the division of Heaven into two Divisions, North and South, which are many thousands of years older than those of Three Circles (Solar Mythos), a good example of which may be seen at Rough Tor, in Cornwall, and represent the division of the Heavens in three twelves, i.e. thirty-six divisions, and we can gauge the time by that of the division of Egypt into thirty-six Nomes.

Professor Keith appears not to understand—

1. Why the existence of the tall Cromagnon race should have appeared and died out.

2. Why the Neanderthal type have been found in a higher or more recently formed strata than the Ipswich and Galley Hill men, whom we may term modern men.

3. Why the tibia in the Ipswich man differs from the others that have been found, and

4. Where the big brain type came from.

These questions, and many others, can never be solved by our present geologists and anthropologists as long as they continue to believe in the present fallacy which is still taught by Professors in great Universities.

Yet the solutions of these questions are quite easy when you learn to read the true alphabet of the past. The tall race of Europe were those Hero-Cult Nilotic Negroes who came up from Egypt and whose descendants are still found in Africa (see Nilotic Negroes). The Turkana and Bor-Jieng are a fine race of men, some of them over 7 feet in height, very muscular and powerful, have high foreheads, large eyes, rather high cheek-bones, mouths not very large and well-shaped, lips rather full. These were followed by another race from Egypt of shorter stature, like the Ipswich man and after the Galley Hill man (see later), yet we often see "throwbacks" of these men and women at the present day; personally I know several—bodies not long, but very long legs with small amount of muscles below the knees, and broad shouldered, and if placed side by side with a Suk or Turkana and painted the same colour would pass for one of them.

I cannot follow Professor Keith in his idea of the differences found in the tibia being any point to be taken into consideration of the evolution of the human race, except by the posture they adopt in "sitting." Any one who has been amongst the natives in different parts of the world, and who will observe the different modes of sitting that these natives have adopted, must feel convinced that it cannot have any other effect upon this bone than that which we find, and it has that; if Professor Keith will examine the bones of these living types (photographs reproduced in Plate XV), and I might also include the Sakais, I am sure he will be as convinced as I am (who have examined many of them). But he must take the typical types. By this I mean, when from infancy one tribe adopts a particular position and from generation to generation they have always done so. The study of photographs reproduced, I think, will convince him, as an anatomist of the first order, that the constant strain and action of the different muscles of the leg, in the posture adopted, as well as the pressure upon the bones themselves from childhood always on the same side, will produce the condition he finds.

I give here photographs of two Nilotic Negroes (by kind permission of Sir Harry Johnston and Messrs. Hutchinson & Co.) of the "big-headed" or "big-brained" people whose ancestors have left their "osteo remains" in Europe (Plate XVI).

As in the case of all other Nilotic Negroes, there was no progressive evolution of them outside Africa and they died out in Europe and Asia, as a distinct race. I might have remarked that the size of the head is not always the indicator of a clever man, or a clever race, it is the amount of "grey-matter" which is the representation of clever men. You may have a large brain with the "grey-matter" very "thin" and no depth in the "sulci" and another man with smaller brain, possessing greater depth of sulci, and thicker grey matter.

The latter man would be much the higher type of the two intellectually, although the other would be much stronger physically.

For further note see Appendix, p. 497.

CHAPTER III

BURIAL CUSTOMS

THE burial customs, rites and ceremonies, from the remotest times, were founded in the faith that the departed still lived in the spirit. They were buried for rebirth. *The corpse was bound up in the thrice-bent position* in the fœtal likeness of the embryo *in utero*, and placed in the earth, as in the mother's womb—it did not denote a resurrection of the body corpus—but was *symbolical of rebirth in spirit.*

In the Stellar Mythos many symbols of reproduction and resurrection were buried in the tombs as amulets—and fetish figures of a protecting power. Elaborate preparations for the spiritual rebirth were made.

Amongst the Nilotic Negroes and even the Pigmies "Little Spirit Houses" were and are built; food is placed here as a propitiation to the departed spirits, so that they shall not return and do harm. Also food and implements were often buried with them.

There is no doubt that primitive man practised many modes of sepulture. Burials took place in various places, as we find remains of man buried in caves, in barrows, tumuli, and under dolmens, principally in the thrice-bent position, either sitting or laid on their side in this way; sometimes the corpses were placed in pots, and sometimes only slightly covered with earth. This custom we find in Africa, South America, and other parts of the world. Seldom is any attempt made to care for old people, and often they are left to die of hunger. Moffat tells us that he once came across an old woman thus exposed whom he was able to succour. She apparently was a Bushwoman. Often old people are killed or encouraged to commit suicide, this more especially amongst some of the Eskimo. Mr. E. W. Hawkes states that "amongst some tribes, when a man or woman gets so old that they can no longer enjoy life, they express a desire to leave this mortal life. Then one of their children or relatives assists them with a rope or knife." Artemidorus states, "The prototypes, or primitive savages of Egypt, bury their dead by fastening the legs of the corpse to the neck with withes of buckthorn; they then joyfully and with laughter pile stones upon the corpse, until it is hidden from view. Then they set up a goat's horn on the pile." (This was the deceased's "totem.") This, therefore, must refer to the primitive Nilotic Negro, as the Pigmies and Masaba have no totems or totemic ceremonies.

From many fractured skulls taken from Wiltshire barrows, Dr. Thurnam inferred that on that occasion, the chief's funeral, slaves were sacrificed.

Herodotus, describing the funerals of the ancient Scythians (early Stellar Cult people), states that "on the death of a chief the body was placed on a couch in a chamber sunk in the earth, and covered with timber, in which was placed all things needful for the comfort of the deceased in the other world. One of his wives was strangled and laid beside him, his cup-bearer and other attendants, his charioteer, and his horses were killed and placed on the tomb, which was then filled with earth, and an enormous mound raised over all." This custom survived down to a comparatively late date amongst the Sclavonic people. Julius Cæsar thus writes of the Gauls: "Their funerals are magnificent and sumptuous. Everything supposed to have been dear to the deceased during his lifetime was flung upon the funeral pile: even his animals were sacrificed, and until quite recently, his slaves and dependents he had loved were burned with him."

Prescott states that "when an Inca of Peru died, his obsequies were celebrated with great pomp and ceremony. A quantity of his plate and jewels was buried with him, and a number of his attendants and favourite concubines, amounting sometimes to a thousand, were immolated on his tomb."

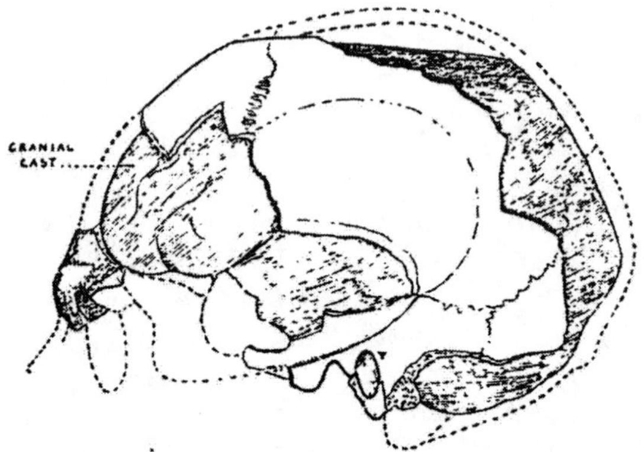

The Ipswich Cranium, profile drawing, after Professor Keith. The intercranial mass forms a cast of the brain, on which the third frontal convolution is well marked. The frontal and occipital bones are fragmentary.

In this Cranium there is a greater supra-orbital ridge than in Modern Man, but much less than in the Neanderthal Man. The frontal lobe is not quite so well developed as in the Modern Man. I have given a reconstruction of this man.

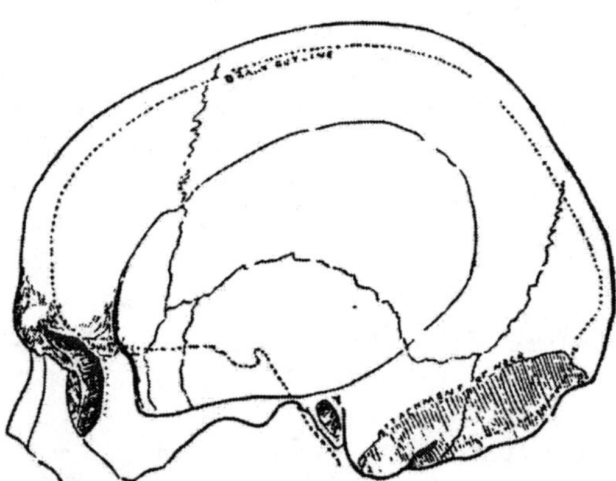

Modern English Cranium (¾ natural size), after Professor Keith, showing the higher development of the frontal lobes of the brain and the lesser development of the posterior part.

PLATE XII.

Plate XIII.

Reconstruction of Stellar Cult (Ipswich) Man.

PLATE XIV.

Stellar Cult Man. From Harlyn Bay, Cornwall.

Plate XV.

Squatters, showing modes of posture affecting the tibia in a different way.

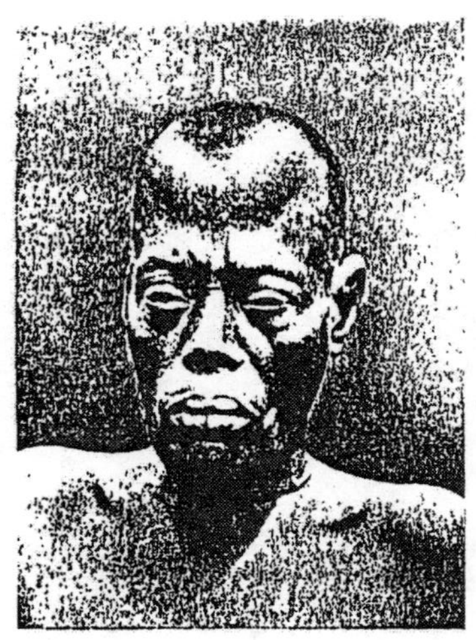

PLATE XVI.

Nilotic Negroes.

Found at Uxmal, Yucatan, body buried in thrice-bent position, but it has not been stated for certain if this faced north or south; if south, then he was buried at the time Sut was primary God—before Horus I.

PLATE XVII.

These were Solar Cult people. Among the Aztec, a throng of slaves was sacrificed to accompany him on his dark journey.

Bodies mummified and preserved, in a thrice-bent position, as in the case of the old Stellar people of Egypt, have been found at Chimu, in Peru, and other parts of South and Central America associated with the Stellar Mythos people, who were long anterior to the Solar Incas.

Killing dependents at the grave of a nobleman, or encouraging them to commit suicide, was also an ancient Japanese custom.

The Fuegians strangled wives, slaves and friends to attend the deceased, and put with the deceased his club and whales' teeth to defend him against the spirits of the other world, just as the Nilotic Negroes used to do. The same custom prevailed in the New Hebrides.

The Dyaks hunt for heads to serve them in the spirit world, and bury food with the deceased for his sustenance. Head-hunting was practised amongst the Muzo, the Colima, and the Panche in South America for the same reason.

The Sandwich Islanders usually bury their dead in the thrice-bent position.

The above are all old African customs, and the custom of eating enemies killed in war or captured in raids was, at one time, extensively practised.

The prehistoric Briton had no objections to eat human flesh, as proved by the discovery at Braintree of human skulls split open to extract the brains, and human bones split from end to end to extract the marrow. Amongst the early Totemic people it was originally the custom to kill and eat the mother of the tribe alive, after she had ceased child-bearing. It was a sacrificial meal: all her children fell upon her at once, and ate her alive, every part of her being eaten, and all taking part in this religious ceremony, so that the mother never died, but always remained alive. After this custom died out, they ate the Zootype representing her—and still do, as a religious rite. When the Spaniards landed in Central and South America they found this custom amongst certain tribes, such as the Nahua, the Carabs, Guaranis and others.

Early voyagers found the practice amongst the Pacific Islanders, where it persisted till quite recently. This was an African custom of early Totemic people, and may be traced all over the world where these Nilotic Negroes went.

Amongst the Matabele the practice is the same as amongst the Zulus, the former being an offshoot of the latter. They bury their dead in the thrice-bent position with the face to the South (Stellar Cult, during the time that Set was primary god), but we find some with the face towards the North, showing that these lived later and were Stellar Cult people during the time that Horus was primary god.

Amongst the Nilotic Negroes we find the practice varies, hence the variations found all over the world.

The Baris simply bury their dead in the deceased's house, with the body in the thrice-bent position and mounds raised over the dead similar to some South African people.

The Madis raise mounds of stones similar to the Bushmen, and the Shilluks formerly practised human sacrifices to appease the spirits of the departed.

Amongst the Niam-Niam the body is laid in a niche excavated in the side of the grave in the thrice-bent position.

The Wanyamwezi carry their dead to the forest and leave them to be devoured by the wild beasts.

Amongst the Nandi the body is taken away at night a few hundred yards to the west of the hut—towards the setting sun—laid on its side and lightly covered. The hands are supporting the head, but the legs are not bent up. Men are laid on their right side, women on their left.

Some Nyasaland tribes bury in the lying position, the body on its back and arms down by the sides. This in all cases represents a departure from ancient custom, and probably is of comparatively recent introduction.

A grave of one of the prehistoric men was discovered in 1907 in a cave at La Chapelle aux Saints.

It was a shallow pit 1·85 metres in length, 1 metre in breadth, and about 30 cm. in depth.

The body was lying from E. to W. I have not found out which way it faced, but I believe North.

Around it were a great number of well-worked implements, fragments of ochre, broken bones, etc.

Professor Sollas states: "This was evidently a ceremonial interment, occupied by offerings of food and implements for the use of the deceased in the spirit world. It was almost with a shock of surprise that we discovered this well-known custom and all that it implies already in existence during the last episode of the Great Ice Age."

In March 1909 another skeleton was discovered in the lower cave of the famous Moustier itself.

A male about 16 years of age, lying on right side with face to the ?, right arm bent under the head, and left arm extended. Burnt bones and implements were deposited about the skull, etc. Flints, carefully *dressed on both sides* and beautifully worked, just within reach of his left hand.

Both these were probably early Stellar Cult people.

The sacrifices offered to the dead are made to propitiate the *living spirit of the dead—not the corpus;* the fear that the ghost might return and manifest displeasure or revenge, dread of the spirit, caused the propitiation.

The form of burial in use at present amongst the Australian aborigines mentioned by Mr. Walter E. Roth in his book "Ethnological Studies Amongst North-West Central Queensland Aborigines," 1897, states that the mode of burial is recumbent, but that the face is always towards the North, although the practice is dying out where they have much intercourse and contact with the whites. They could give him no reason for this, except it had always been the custom of their ancestors, but, as we shall prove later on, the customs amongst the Nilotic Negroes vary a little, although in the majority of cases it is the thrice-bent position, either sitting upright or lying on the side.

With the Stellar Cult people it was always so. I give here a photograph of a tomb from Yucatan (Plate XVII). The carving over this tomb is very distinct. The inscription reads thus: "The Thrice-bent Man. The Altar welcomes the crushed body, lying face downwards, of the man from Uxmal." This in Maya is, "Ta ox uuɔ u Tem Kam uuc noocal Oxmal" (see next page).

"The Thrice-bent Man"—that is, the thighs bent up on

the body and the legs again bent on the thighs. In Cornwall also, at Harlyn Bay, there have recently been discovered Stellar Cult remains in this position, i.e. "the Thrice-bent Man," and those in the round cist show that this is one of the earliest forms of burial which, probably, we shall

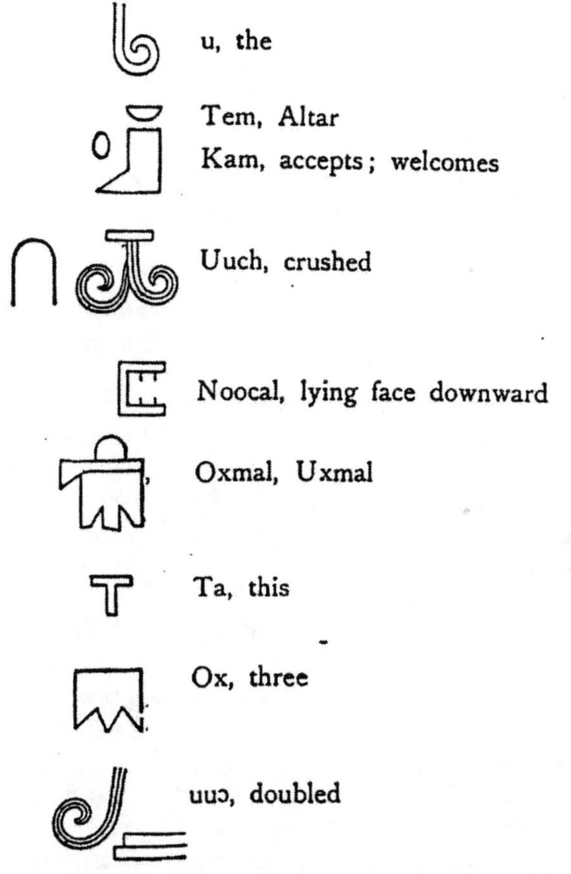

find practised by the human race all over the world—i.e. a round hole and the body as a "Thrice-bent Man" placed in it. In an ancient graveyard, six acres in extent, found in Tennessee, from 75,000 to 100,000 osteo-remains were found buried in this manner—round cists with thrice-bent position. From the length of bones it was estimated that these people were about 4 feet high.

The fact that it is mentioned of the man from Uxmal that he is "*lying face downward*," i.e. to the South, leads me to attribute him to one of the early Stellar Cult people at the time that Set was primary god. The two tombs found in Cornwall are of Stellar Mythos people during the time that Horus was primary god. These all face North. There was a triangular stone with apex to the North over the one in Cornwall.

Forty-nine graves of the Stellar Cult people (Neolithic) were found by Professor Mosso at Pulo, and he states that these tombs are exactly like those found in Egypt from the Delta of the Nile as far as Nubia, with stone weapons in their hands and vases near them—as in the photographs shown here.

Mr. Reisner examined 1,200 of these tombs in Egypt, and Weigall, in his recent excavations in Nubia, extending in the direction of the Sudan from the First Cataract, has found many.

In 1826, near Lausanne, 15 were found—many in Savoy—and in other parts of the world where the old Stellar Cult people went.

Preserving the human mummy perfectly intact was a mode of holding on to the individual form and features, as a means of preserving the earthly likeness for identifying the personality hereafter in spirit.

The mummy was made to preserve the physical likeness of the mortal.

From the first lowest phase of simply preserving the bones to the highest art of mummification, we have to return to the Nilotic Negroes and the Egyptians in Africa.

Consider what De Morgan and M. M. Amelineau found at El-Amrah. There were a number of oval graves sunk in the stony soil to a depth varying from five to six feet, wherein were the skeletons of human bodies lying upon their sides, having their faces to the North, hands crossed before their faces, and knees bent up on the body upon a level with the chest. With them were buried flints, small bronze implements, pottery, stone vases, shell ornaments, etc., the mode of burial practised by the Makalanga, natives of New Guinea, Solomon Islanders, and natives of Australia at the present day, and many other native tribes still extant. From the period beginning with the third and fourth dynasty we find that the body was laid upon its back, as at present with us, and arms laid on the body. Taking all these facts into consideration, we must come to the conclusion that :—

1st. At the earliest date of Pre-Totemic Palæolithic man there are few records of any burial-ground, but from the knowledge of the customs still in practice amongst the Pigmies, it is probable that they buried their dead as the present Pigmies do, either where they died, or in burial-grounds where they placed them under about a foot of earth and leaves, on their sides, in a bent-up position.

2nd. At a late Palæolithic, and all through the Neolithic age they had recognized "burial-places," and buried their dead either in a "hole" or "cist" in the thrice-bent position, or on the side, with face to the South or North, which was the custom all through the time of the Stellar Mythos and early part of the Solar, until the custom of burying with face to the East began. That during a certain period "urn-burial" was practised by the Egyptians and some peoples of South America.

3rd. During the Solar Mythos, dating from the third and fourth dynasty, the body was laid on its back, flat, as at present with us. This we find throughout all the world, wherever remains have been found, not taking into account what may be termed "accidents" or cremation, which came into use later. One of the most interesting points about this find in Cornwall of prehistoric things is that there is evidence of a long occupancy of this district by Neolithic man, i.e. Stellar Cult people.

That this thrice-bent position form of burial was the common practice amongst the early races of man will not admit of any doubt, nor that it had sprung from one original, and was taken from the knowledge of the position of the *fœtus in utero*; they believed that the spirit entered the body of the child in this position, and when the body died it was placed in the same position for the spirit, until it left again. The burial customs, rites and ceremonies, one and all, from the remotest times, were founded in the faith that the departed still lived in the spirit; in the earliest mode of interment known the dead were buried for rebirth. The corpse was bound up in the fœtal likeness of the embryo *in utero*, and placed in the earth as in the mother's womb; it did not denote a resurrection of the body, but was symbolical of rebirth in spirit; not only were the dead elaborately prepared for the spiritual rebirth, but many symbols of reproduction and resurrection were likewise buried in the tomb, such as amulets and fetish figures of a protecting power.

What, then, does this indicate? Undoubtedly that here we have a progressive people; and this custom, like that of hieroglyphic language, was simply the beginning of that higher and more perfect state which we find in both; the one brought to perfection by embalming; the other—those artistic hieroglyphic writings—developed perhaps during many centuries from the time when they commenced to record their "rude marks," and "crude embalmings," to the time when those Egyptians whom we now know so well excelled in a perfect Eschatology and perfect mummification.

We also learn from recent finds in Egypt that during the evolution of burial customs the change in the position of "the thrice-bent man," to that of "straight," was gradual, as many skeletons have been found with the upper parts placed in a contracted position and the lower limbs placed straight.

CHAPTER IV

ANCIENT IMPLEMENTS AND HOW TO DISTINGUISH THEM

THE oldest evidence of man's existence on the Earth is afforded by his implements.

They were very primitive at first, some of them probably only just a big stone, a broken piece of flint or quartz, or a branch of a tree broken off. The latter, of course, would perish, so we have no evidence of it, but no doubt it was one of his primary weapons of defence. Even the Gorilla at the present day will break off a large branch and defend himself with it if hard pressed in a fight with the natives.

The earliest implements we find which we can attribute to the use of man are flints or stones of some kind, rudely chipped into flakes and scrapers, to which Geologists have given the name of *Plateau Implements*, because they are found in gravels capping the high plateaux of Kent and elsewhere.

I have divided and differentiated all the implements found into definite classes or divisions, which can be easily distinguished the one from the other, and have demonstrated to which exodus of our forefathers each class belonged, so that Geologists and others, when finding one or more of these implements, can with confidence say at once to which type of humanity it can be attributed for its production.

The terms *Palæolithic* and *Neolithic* are bad terms to use in a *chronological sense*, as many of these instruments are still manufactured and used by various natives throughout the world in isolated places at the present day; therefore in using these terms, I wish my readers to understand the term

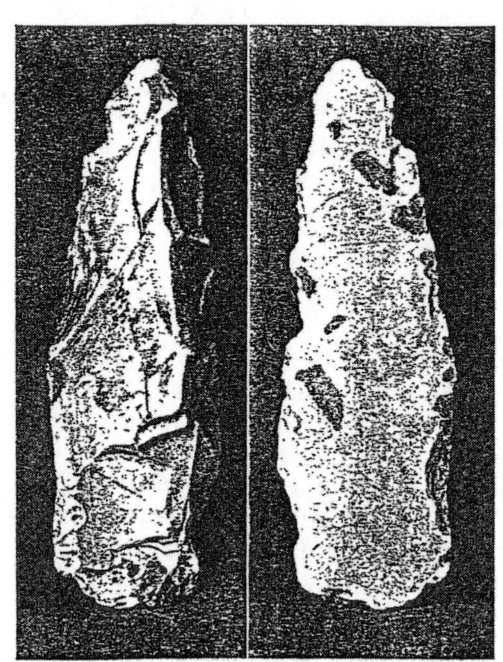

Plate XIX.
Palæolithic.
Pigmy implements, chipped on one side only.

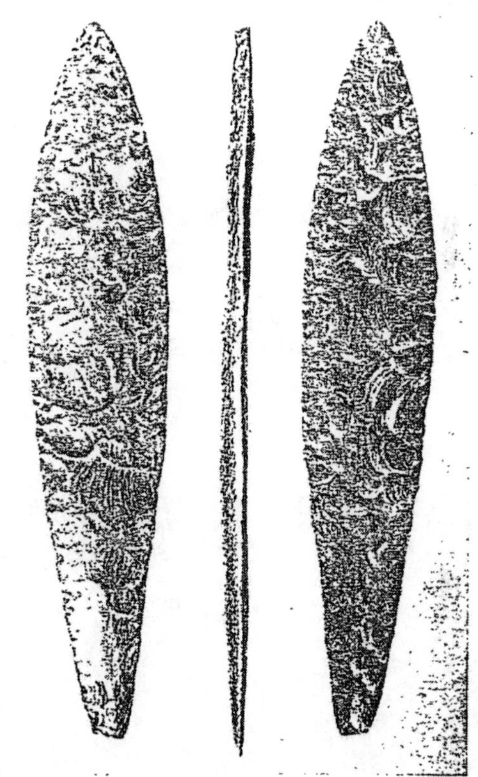

Plate XXV.

Neolithic. Stellar Mythos People.

The most perfect and beautiful specimen I have seen of flint dagger, or large double-edged knife, beautifully worked on each side and polished. The above illustration represents each side, and the edge is portrayed in the central figure.

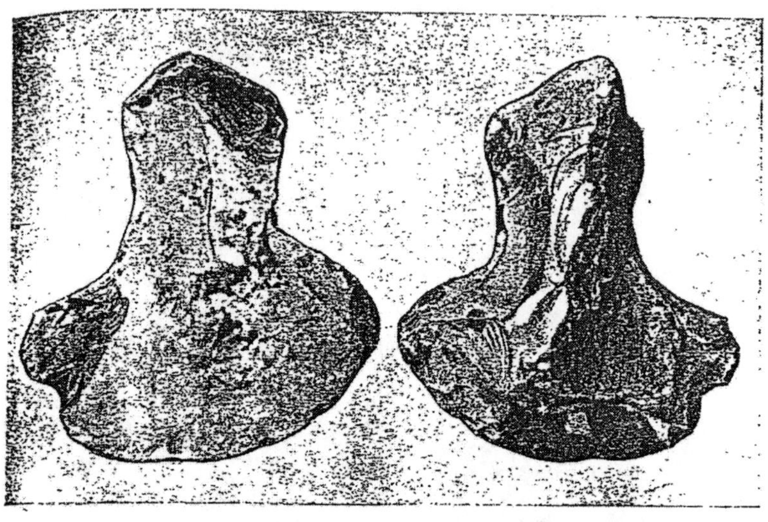

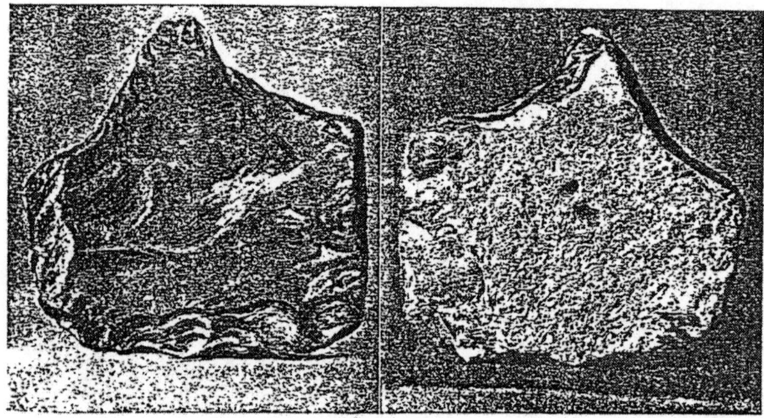

PLATE XVIII.

PALÆOLITHIC.

Pigmy instruments, the so-called Plateau Implements, found all over the world. A flake chipped on one side only, or just the end chipped off on each side, as a scraper, not hafted, and associated with the Pigmy people only, or Pre-Totemic or Non-Totemic people. M. Rutof states that he has found them in the Oligocene formation. These have certainly been found in the Pliocene formation over 1,000,000 years old.

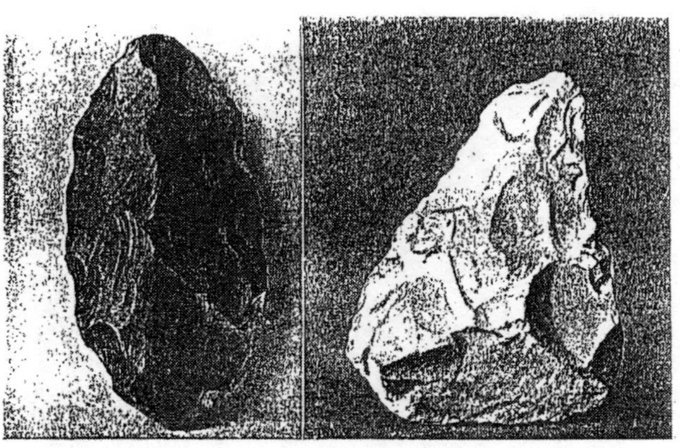

Plate XX.

Palæolithic.

Nilotic Negro instruments, chipped and worked on both sides roughly, found all over the world and associated with the Totemic people, i.e. those that have Totemic Ceremonies.

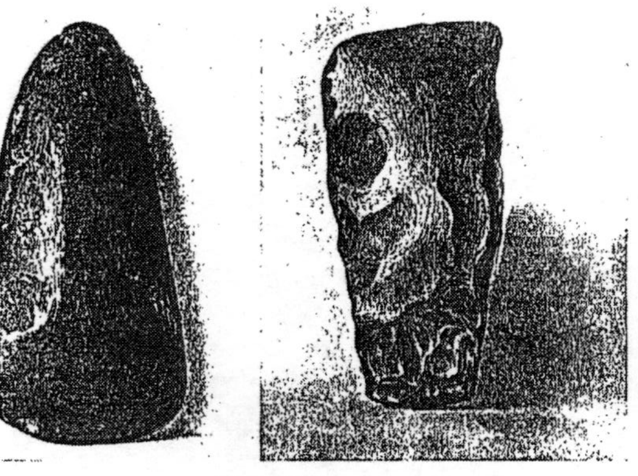

PLATE XXIV.

IMPLEMENTS OF THE STELLAR MYTHOS PEOPLE OF THE FIRST EXODUS.

These are flaked, chipped and fashioned on each side, some very beautifully, into knives, spear-heads, arrow-heads, etc., and are only found in Europe, Africa, Asia, and some parts of Central, North, and South America. *They are not found* in Australia, Tasmania, or Oceania, and are only associated with the Stellar Mythos people, who never went to these latter countries. But I am of the opinion that they will be found in Java and the Caroline Islands, as there are other remains and proofs that these people went there.

Palæolithic is used as indicating implements *not polished*, and *Neolithic* as those *with polished surfaces.* This will obviate the fallacy of describing the manufactures of these Palæolithi, as those of the Neolithi, living ages before, which may be found in some parts of the world; at the same time it will enable the reader to identify which are Pigmy and which Nilotic Negro implements, and distinguish and differentiate those of the Stellar Cult people.

The Plateau Implements.

These are the implements of our forefathers, the Pigmies. Mr. Rutof states that he found them in the Oligocene formation. These have certainly been found of Pliocene age—I have found them myself in these strata, as have also other Geologists whose authority is unimpeachable.

I give here some photographs of the Pigmy implements (Plates XVIII and XIX). These are found all over the world, a flake chipped on one side only, or a stone with *just the end chipped off* on each side, as a scraper.

No hafted implements can be associated with the Pre-Totemic group—they had not attained the knowledge of hafting.

Some of these implements probably date back 1,000,000 years or more.

In the next development, we find these flints and stones roughly chipped on both the sides, some hafted into sticks, thus forming primitive axes. These were not polished at all, only roughly chipped on each side. I have found some of these in Pliocene strata.

These were, and are, the implements of the Nilotic Negro race, the Totemic people who followed and drove out the Pigmy. I give here some photographs of good specimens (Plates XX to XXII).

In the next stage, early Stellar Cult people, we find an advancement, a higher finish and perfection in the formation of these implements. I wish to draw particular attention to these, because they are very important. The implements *are polished.* In these cases, the flint or quartz is flaked off on each side so as to form a stone knife, sometimes with a double edge, and polished. These are found nearly all over the world, in Africa, Asia, Europe, North and South America, but there are places *where they are not found.*

Exceptions.—They are not found in Australia, Tasmania, New Zealand, New Guinea, and some other islands of the Pacific, where the Stellar Mythos people did not go.

This is a very important fact to notice and observe because their absence from places where the Pigmies and Totemic people still are existing, proves that these implements were associated first with the earliest Stellar Mythos people, who never reached Australia, Tasmania, and some other parts above mentioned.

So that even by their implements in the Tertiary and Quaternary periods we have geological evidence that different races of our forefathers travelled over the world, and that these races still exist, as is proved by extant tribes of the present day, and the implements they use found amongst them.

These implements would live long after the Osteo Anatomy had disintegrated, and we must, therefore, take the above as objective evidence of the existence of these types of the humans where we find their implements. It is perhaps difficult, in some cases where we have a rough semblance of "polish," to say if it was done by a late Nilotic Negro type or early Stellar Mythos man, because in the transition of the cults there must have been the same gradual higher standard of work in implements, as in the evolution of religious ideas, and as in the overlapping of cults which we find later.

I give here some photographs of typical examples of the Stellar Cult people's implements (Plates XXIII to XXV).

Stone implements are found all over Africa in large numbers, and of every type and variety, showing that all the "manufacturers" of them existed there.

The Rhodesian Museum possesses a very creditable collection, some made of jaspery slate rock, others of quartzite. Many very large axe-heads have been found, as well as a number of small ones. Length of one, 15·3 c.m.; width, 9·5 c.m.; depth, 4·9 c.m.; weight, 771·28 grammes. Many are well trimmed, flaked work on each side (Nilotic Negro implements). Some were used by the Stellar Cult people (polished and flaked on each side). Many implements correspond quite closely in the matter of size, shape, and workmanship to those found in Europe and America; also many roughly shaped scrapers, chipped on one side only (Pigmy instruments), and flakes of epidiorite are exhibited in this Museum.

A notable point is that *very large axe-heads* have been found in South America, which correspond in every particular to those found in Africa—and were probably used by the same type of people and for the same kind of work.

In the Gran Sasso d'Italia and in the Valle de Vibrata and on the Promontory of Monte Gargno, Professor Angelo Mosso has found abundance of large, roughly chipped flint weapons which are identical with those found first in France, then in Belgium, in England, all over Europe, and many in North and South America, all of which must be classed as Palæolithic Nilotic Negro instruments. These are precisely similar to the same kind of Palæolithic Nilotic Negro implements found in Egypt. Moreover, we find that the strata of prehistoric generations lie superimposed in the same order in the Nile Valley as in Europe. The votive axes of Stone which are found in many parts of Europe and America correspond in all particulars to those found in Egypt. The next distinctive implements we find wherever the old Stellar and Solar Cult people travelled are those of Copper and Bronze. Egypt was the progressive centre of knowledge of the arts and cults, and here we find the first implements formed in Copper and Bronze, which art was carried all over the world where the Stellar and Solar Cult people went.

The present Geologists, Anthropologists and Ethnologists have divided all these into two divisions—namely, Palæolithic (non-polished) and Neolithic (polished).

By my division, we can divide and differentiate all these into the Palæolithic, which are again subdivided into the Pigmy implements, that is, those of the Pre-Totemic people, and those of the Totemic people—Nilotic Negroes—then the Neolithic, those of the Stellar Mythos people.

Palæolithic, not polished.

1st. Pigmy, chipped on one side only, not hafted—Pre-Totemic people.

2nd. Nilotic Negro, chipped on both sides roughly, not polished, and hafted—Totemic people.

Neolithic, polished.

Polished and worked on both sides, forming sometimes a double edge—the early Stellar Mythos people.

Copper and Bronze.

Stellar and Solar Cult peoples.

Thus we see the cults bear out and prove in a most remarkable and perfect way their co-relation with the geological records found.

Some Geologists give or reckon from 300,000 to 400,000 years for the Quaternary and Recent periods, and many think it is uncertain if man existed during the Tertiary period. But it has been fully confirmed and demonstrated by Professor Sergi that the skeletons found in Lombardy were in Tertiary strata, and that these were similar to the present types of man, and I maintain that they were Stellar Cult people, on account of the implements, etc., found with them. That man existed in the Tertiary time there cannot be a possible doubt, because of the numerous implements found in the strata of this Period. But Stellar Mythos people living during that time must place the origin of man a million years before this at least, on account of the time that must have elapsed to accomplish the evolution from the Pigmy.

We should not expect to find many human remains in the lower strata. Primitive man only buried the corpus a few inches or a few feet under the earth, and the result must be that the whole body, including the Osteo Anatomy, would soon disintegrate, except in special cases where the bones might be preserved for an almost indefinite time.

The implements found associated with the old skeletons correspond with and are precisely of the same kind as we still find in use amongst the Pigmy and Nilotic Negro races (types of which still exist in Africa and other parts of the world) that left Egypt thousands of years ago, carrying with them the primitive knowledge of certain Totemic rites and ceremonies. All these rites and ceremonies are identical in each race, and the tribes are anatomically and physiologically the same in whatever part of the world we find them; and original types are still found in the old home, Africa.

They all had a primitive monosyllabic language, and many words at the present day are identical with the primitive Egyptian hieroglyphics; this shows that the late Egyptians had obtained these ideographs, which had been carried on from the originals, from the Pigmies and Nilotic Negroes. As regards the sounds or how they originally pronounced these ideographs we have no means of knowing for certain, but we do know how these primitive people pronounce them at the present day.

I have not used the terms which are at present in vogue amongst Anthropologists and Ethnologists, i.e. Magdalenian, Solutrean, Aurignacian, and Mousterian, as these are only names of places, and might be very misleading. There is no locality with which a definite form of these flint implements can be solely associated.

Hair.

Professor Sollas' division of Man *by his three different kinds of hair* would be most amusing if he was not a Professor at Oxford; as he is, it is very, very sad to see such statements. But then the present Anthropologists follow the same line of thought, and so one cannot be surprised.

He divides the humans thus into three groups—

1st. One in which the hair is without any twist—that is, it is perfectly straight.

2nd. That in which it is twisted to an extreme, as in the Negro or Bushman (which are two quite different classes).

3rd. Those in which the hair is only twisted enough to be wavy, as in many Europeans.

And he finishes up with this remark: "The Tasmanians, like the Negro and most other races with dark skins, all belong to the same race or class."

He never mentions anything about the *difference in section of hair*. Compare the Australian with the European; then make a section of these, and tell me where he finds the missing link.

Let him make a section of the hair of the Pigmy, and then compare it with that of the Chinese and Japanese, and tell me where he finds the missing link.

He evidently does not know that an existence in a cold climate for prolonged periods produces straight, coarse hair. If he took a section of the straight hair of the Eskimo and American Indians, could he tell me where he would find a corresponding section? I think not; *but he would find all the Nilotic Negroes the same*—including the Australian, Ainu, Eskimo, North and South American Indians, as well as the Nilotic Negro in Africa. And as these must have travelled North from Africa to cross to America, the thousands of years in a cold climate accounts for all the Americans having straight hair.

The Chinese and Japanese and all the earlier Stellar Mythos people are different in higher formation, and the White Race still different.

Then he is erroneous in his description of the character of the hair of those he quotes. *The Tasmanian Pigmy race have a discrete peppercorn hair.*

That of the *Nilotic Negro is not—it is frizzy and curly*, sometimes almost straight; *and that of the true Negro who never left Africa, but came after the Pigmy and was evolved from the Masaba Negro in Africa*, may be described as woolly, or *kinky*, and quite different from that of the Pigmy or Nilotic Negro.

Nearly all the Nilotic Negroes who travelled North during the thousands of years that have passed have developed a straight hair, except the Ainu. Some of these are frizzy or curly; but climate and mode of hair-dressing, to be described later, account for the change that has taken place during the thousands of years that man has existed here.

We can see from the Nilotic Negroes how they lengthened and dressed their hair in the same way as the old Egyptians. We observe also from unbaked clay figures that the Mediterranean races in the Pre-Mycenæan and Mycenæan ages had hair of the same character and arranged in the same style as the Egyptians at that time.

Professor Angelo Mosso, in his "Dawn of Mediterranean Civilization," gives examples of the similarity of hair arrangement, which, as he states, "shows another point of contact between Crete and prehistoric Egypt," but adds, "and it is a noteworthy point that in dynastic times there is an end of resemblance in the adornment of the head between the Egyptians and the Minoans." The reason for this must be obvious. The prehistoric Egyptians were first Totemic people, then Stellar and Lunar Cult people; the Minoans were Stellar and then Lunar Cult people.

The Egyptians were a progressive people, and when they evolved to Solar Cult, they not only changed many forms and ceremonies, but changed the fashion of dress, as well as many other things. They wished to blot out all forms and everything connected with the former Cult. But the Minoans remained Stellar and Lunar Cult people until a much later time; then they adopted the primary Solar Cult.

CHAPTER V

PRIMARY MAN

ALL the Pigmies, some of the Nilotic men, and some of the early Stellar people, were mostly hunters from necessity, winning a precarious existence from the chase of wild beasts, and the collection of food, yet they were superior to all others in nature; they could make fire, and form clubs from sticks and stones, etc.

The Stellar Mythos people from hunters became shepherds. They found that they could tame animals (e.g. the Masai) and that wild plants could be cultivated. They first settled into camps or villages, generally on the hills, and surrounded them with ditches and mounds to keep off the wild animals, etc. These people were Neolithic. They then began to build houses and temples, fortresses, etc., until they acquired a knowledge in building that none since have surpassed—as, for example, the Pyramids and those beautiful and stupendous remains we find in Central and South America, Asia, Africa, and other places. The temple in Egypt recently discovered is another example.

Palæolithic man must be divided into three distinct classes at least, both anatomically and by his primitive implements, as well as into *Pre-Totemic* and *Totemic.*

The first and primary are the Pigmy race—***who are Pre-Totemic or Non-Totemic, and have peculiar and distinct anatomical features***, while their implements are characterized by a very primitive form.

The Nilotic Negro race are also Palæolithic, ***but are Totemic***; and there are at least two distinct types of evolution in their Totemic ceremonies, the first without "Hero

Cult," and the second with it. Anatomically, one is superior to, and higher in evolution than the other. From their implements we can trace a great advance, not only in a higher art of manufacture, but also in the knowledge how to make others; this, the primary race had not acquired. Their Anatomical features prove also how much progress the human race had made in evolution up to this point. Following these came the first Exodus of the Stellar Mythos people, who were Neolithic.

Now, the only man that I know who has observed the difference of these two types of Nilotic Negroes is Dr. Haddon, Reader of Ethnology at Cambridge, who, I believe, was the head of an Expedition that went out to Torres Straits, and he has published the result of his observations in his reports of the Cambridge Anthropological Expedition to Torres Straits.

He has distinctly described the difference in a great many particulars between these two different exodes of the Nilotic Negro, and his observations are true and accurate, although he does not recognize them as the descendants of Nilotic Negroes.

If you mix these up and make no distinction between them, as all writers hitherto have done, you have not found the key to unlock the mysteries of the past; and I am certain if Professor Sollas and other writers on this subject took into consideration the evolution of the "Cults," with Anatomical and Physiological conditions, which they now treat with scorn, not only would they be able to trace the history of man on a firm and definite basis, but they would also discover that their geological researches would agree critically on every point.

The original Pigmy was "born" in Central Africa and spread throughout this world over a million years ago, and remnants of this first race are still found in the forests of Africa, in the forests of Bolivia, South America, in New Guinea, the New Hebrides, the mountains of China and the Philippine Islands (particularly the North Island of Luzan). With the Pigmy religion dawned, by the propitiation of Elementary Powers, propitiation of departed spirits and a belief in a Supreme Being. Theirs was the first articulate language; from the Pigmy, the human race has gradually developed in body and mind up to the present White Man, and the Christian doctrine, by evolution, from the various cults preceding it. From the Pigmy and Bushman and Masaba Negroes, who have no Totemic ceremonies, we gradually pass to the Nilotic Negro, and various tribes of a higher type of the Nilotic Negroes, in whom we find Totemic ceremonies with Hero Cult finally established and practised. That these Nilotic Negroes followed the Pigmy all over the world there can be no doubt, because these Nilotic Negroes have the same signs and symbols, the same Totems and Totemic ceremonies everywhere, which must have been derived from a common centre, and could not possibly have been developed from their own ideas and own surroundings in each case, and yet be so widely distributed as we find them at the present day. By these Nilotic Negroes in Egypt the old Astro-Mythology was first developed, followed by the Stellar Mythos, and carried throughout the world (except in Australia, Tasmania, Patagonia, and Oceania). After that came the Lunar and Solar Cults. *The Solar Cult was not universal, and only in a few countries, comparatively speaking, can remains of the same be found, viz., Egypt, Europe generally, except the extreme North, the South of Asia as far as Japan, but not the North of Asia, in America only in Yucatan, a little North and West of this, and South as far as Peru.*

This, I contend, is what I have shown and proved in my work, as well as that many of the signs and symbols used by the present generation were brought on from the earliest Stellar Mythos, and that the Rituals were obtained and carried down from the Eschatology of the Ancient Egyptians. I have also shown and proved that the whole of the ancients of the Mexican and Central American States and South American buried Cities, obtained their learning and doctrines, as well as their pottery and implements, from the Egyptians, as is proved by my decipherment of antiquities found there, and now in museums, and others still extant.

I have answered some critics of my book ("Signs and Symbols of Primordial Man") in this country in the *Pall*

Mall Gazette and *British Medical Journal.* To those in America, in the *Masonic Sun*, of Toronto, Canada, April 1911, and to others, I reply, with de Balzac: "Fierce critics, gleaners of phrases, harpies who pervert the ideas and purposes of every one!" I fear nothing for this book from them, because it is written in accordance with decipherments of existing monuments (which are hidden mysteries to the masses owing to their having not yet learned to read the language) and from the knowledge of the Anatomical conditions of these old people. Notwithstanding, real gleaners of truths will know that I am correct in my translations. I have found by much research, studying so-called "traditional histories," tracing by analysis and synthesis, the real truth about the doctrines which were so strongly imbued in the minds of many past generations, through ignorance, and false teachings, and "advanced theories" which will not stand the least critical examination and analysis, that the computation of time by years of the world, even for pre-Christian history, being as absurd and irrational as it is for the epochs of the earth and the universe, must be abandoned as the unscientific assumption of rabbins and scholastics which has grown into wilful, mischievous falsehood, in the face of the annals of nature and of mankind. I must strongly insist upon and proclaim what I bring forward to be the truth in these matters. The Middle Ages, so called, were retrograde in the advancement of real knowledge and learning. From the time of the fall of the Old Egyptian Empire not much advancement was made in true knowledge for many hundred years. There was a set back. The Greeks and Romans, although better known to the present generation than the old Egyptians, because scholars have been able to read their literature, were no advancement on the old Egyptians in science—how much in art it is difficult to say; they have been much overrated. I have a sculptured head and bust of Isis, fashioned out of Syenite Granite, more beautiful as a work of art than anything I have seen in any gallery or museum in Europe. The Greeks and Romans copied and gained most of their knowledge from the Egyptians, did not improve upon it, but perverted much of the truth, and the Roman Church has done more than anything else to keep the people down in a dark and degenerate age. It is only in the last hundred years or so that evolution has again shown advancement, more especially this last fifty years. But there are so many false dogmas and theories founded on misbeliefs, still taken for gospel truths, in opposition to truths now founded on scientific facts, by the masses who follow the teachings of men as ignorant as themselves, that it will be many years before these truths filter through to their brains. And the present "Press" does not help much ; rather, on the contrary, I think much harm is done at the present day by eulogizing works by some Professors because they are Professors of Universities—men who have a perfect right, as men, to think and believe anything they like, but certainly, as Professors, have no right to disseminate false information on these subjects which scientific investigation has proved to be incorrect. Because they are ignorant of the key to read these facts, they have no right to ignore and treat, or profess to treat, with scorn that which they have never learnt, and relying on Germany and other foreign countries to "give them a hand," will not assist England to keep to the front in this branch of science. Much harm is done also by the so-called "Yellow Press" and low-class literature. Perhaps one should take consolation in the fact that the lower classes are now taught to read and write, whereas a hundred years ago they could not, and thus by and by the future generations will become "thinkers," and so attain a higher degree of evolution ; and as each generation attains a higher level and greater development of brain cells, the old dogmas and false beliefs will be swept away, and a truer enlightenment and higher doctrines will take their place. How far each nation will rise, and how high a point of evolution will be attained before its fall, it is difficult to say. The past history of the rise and fall of all nations has been very much the same. No one nation has attained a progressive standard and lived beyond a certain point without dissension occurring amongst the priests, followed by Socialism ; and Socialism has been the cause of the downfall of every great nation without exception, and it will be so again if any nation becomes entirely Socialistic.

The masses generally are too ignorant ; others have but one object, i.e. to obtain wealth, which, when obtained, is in the majority of cases squandered on self-gratification, and degeneracy ensues. But it is ridiculous for scholars and philosophers to think they can with impunity ignore hieroglyphical discoverers and the monuments deciphered by their aid. The epoch of Menes, or the beginning of the *Imperial history* of Egypt, is nothing but the beginning of the *last stage* of the religious and social development of the Egyptian nation, and although every nation that has existed owes to *science* its advancement in evolution to a higher standard, yet the same history is found in each fall of a nation. The men who have gained wealth and position through *science*, directly or indirectly, and are ignorant of scientific truths themselves, ignore the very men whose brains and work have placed them in the exalted position they occupy. Many of the British wealthy manufacturers at the present time have used the brains of poor men of science, robbed them without remuneration, and stolen their knowledge with impunity. These men would hold their hands up in pious horror at taking a penny from any one, but without the least compunction will steal, and have stolen, the result of many years of hard brainwork, which has advanced the evolution of the nation in some way or another, and have left the poor scientist to starve, or forced him to seek a living or recognition in foreign countries. The nation becomes poorer by the loss of such men, and degenerates. Nevertheless, what I state is an absolute fact; I know many who have thus been treated.

If polygenism is not yet dead, few, I believe, there are to keep the dying embers alive, and these few can have no knowledge of the comparative anatomy and physiology of the lower human races, such as the Pigmy, Bushman, and Nilotic Negro, and other aboriginal natives in various parts of the world allied to them, or perhaps they do not believe in the evolution of the human race. These men, no arguments, nor facts found and brought forward, would convince. Zoologists may think that they are finding evidence of "convergence," and may think and believe that morphological similarities do not of necessity mean phylogenetic kinship, and that similarity of natural surroundings has given birth to similar elemental ideas in the primitive mind, leading to comparable rites and ceremonies in different parts of the world. Professor Frazer has come to this conclusion, which critically, in my opinion, is opposed to all facts found.

Totemism and Totemic rites throughout the world, and amongst all tribes past and present, are fundamentally the same, have the same meaning, "*are sacred ceremonies,*" have signs and symbols and passwords, or sacred words, all identically the same. Professor Frazer could give no explanation, that could be satisfactory, why in New Guinea you find one tribe with certain characteristic anatomical conditions with a Totem of a Sow or Pig, and in Africa and South America you find tribes identically the same in anatomical conditions, with the same Totem and Totemic rites and ceremonies; and why another tribe in New Guinea, with different anatomical conditions, have the Crocodile for their Totem, and why you find tribes in Africa and South America identically the same on every point. (Dr. Lorentz and his companions, who met some of these people in New Guinea, did not understand their sacred rites as admitting them as members of their Totemic tribe by the sacrifice of a Pig, and sprinkling the blood of the same upon them. The Sow or Pig amongst these people, as a Zootype of the "Great Mother," is quite understood, however, by those who have learnt Totemic ceremonies and Sign Language. The form of Zootype would prove at what date these people left Egypt, like the Bear of the Ainu, the Lizard of the Australians, the Beetle amongst those Tribes around Panama and the Tortoise in North America.) Then he omits any notice of the *Primary race with no Totemic rites or ceremonies*, who were driven back into the forests and mountains by these descendants of the Nilotic Negroes as they came up from Egypt bringing their ceremonies with them; we find that this occurred in every part of the world. In the same way those Nilotic Negroes were driven out and exterminated in Europe, North America, and most parts of Asia, by a higher type of the human race having its centre of progressive development in Egypt and the Nile Valleys, and with the fully developed Stellar Mythos. These again were driven out or exterminated in some places by those who came after them with the old Solar Mythos and later Eschatology in its final development; this we only find in Europe, Yucatan, some parts of Central America, and South America as far as Peru, and among the Hindus and Japanese in Asia. In Europe and North America and some other parts of the world, these doctrines have also disappeared and given place to a higher type of Eschatology, viz. the Christian doctrines, now in a more perfected form, as originally developed by the Egyptian Copts. The cult, however, is the same as that founded by the Egyptian Copts from their old Solar Eschatology.

Then, again, the adaptation to environment was a necessary condition to primary man in whichever country he settled, if he wished to continue his existence and to prosper. He would die out if he could not comply with the conditions of existence imposed on him, and this law holds good to the present day. Some white men cannot live and prosper, i.e. marry and rear children, in some tropical climates; neither could some natives who now live in a tropical climate exist and prosper in the Arctic regions. Yet we find many of the human family existing and thriving in each of these districts, and as they all spring from one original, and from one original centre, it is only by a process of evolution that has taken place, that the adaptability of man to various and different surroundings has been accomplished.

There are two men who stand out before all others in attempting to solve the mystery of this question—Darwin and Bergson.

Personally the writer does not agree with Bergson in his entirety, but is of an opinion that the solution lies between the two theories.

That adaptation to environment is the necessary condition of evolution is certainly true. It is quite evident that a species would disappear should it fail to bend to the conditions of existence which are imposed upon it; but these conditions—i.e. environment—alone, do not and will not produce an evolution from one type to a superior and superior again; there are other causes to assist to a much greater extent or degree.

If we take the first man—the Pigmy—as an example and proof, what do we find? This, that although he has spread all over the world and now still exists after more than a million of years in various climates and different parts of the world, i.e. different environments, yet when he has "kept his class" and in the same environment he has not developed to a much higher type of evolution, but we find that in some places and under different environment, from the "original class" he has developed into a higher type, a distinct second class.

Further, when we find that the best developed men of the second class have married women of the first class, they have produced a third class somewhat of a different type, sometimes developing into a higher type. Well-developed men and well-developed women at certain ages, as a rule, reproduce a fine type of child, but if the parents are of the same class, so will their offspring be always of the same class in primitive man; in latter cases you may have "throws back," but not in primitive man; there are only throws back towards the Anthropoid Ape—not the human, if I may use this expression.

It is by environment and by *intermarriages* that a different type is produced, and very often we find a higher type is thus reproduced in the primitive races. This is the only way, then, that a higher type in evolution has taken place; i.e. it is owing to two causes, viz., intermarriages and environment.

Man, as he went North and was exposed to the cold and rigorous climate, and compelled to fight harder for his life and existence, would become better developed, anatomically at least; the weak ones would die out and only the fittest would remain when he was driven back South by the Glacial Period; and then, when the higher type came up and intermarried with the women he found, another type would be produced.[1]

[1] What effect on the functions of the Pituitary Body environment, the different foods and conditions to which our Primary Ancestors were subjected, and the resulting effects on the osseous anatomy—if any—is a question which has not been solved up to the present time. I am under the impression that some change took place in the Pituitary Body during the evolution of man, commencing with the Masaba Negro and reaching its maximum change at the time of the Non-Hero-Cult Nilotic Negro. The development of the Supra-Orbital Bone in the Non-Hero-Cult Nilotic

Bergson looks upon *life as something apart from matter*, something which permeates, modifies, and controls matter, and thus he views evolution from this side of life, whereas Darwin looked upon evolution from the more materialistic aspect, *the adaptation of the living structure and functions to meet the requirements of the environment.* Therefore, in my opinion, the true solution is the modification of the views held by both these great men. When Bergson uses the term "matter," we must ask, What is matter? We must define matter as that which occupies space and possesses weight. "Paut," a handful of earth, is matter, inorganic, without life. "Bodily" matter is organic, i.e. it has life, for a time at least, and then returns to "paut" again. But this original life, a part of the organic matter, *passes its life on to another*, and this again to another; yet the Spirit of the one does not pass on, and does not die: it passes into what is understood as "Spirit form" and lives another and different life from the material life that humans have here. In this second life there is no matter, but there is "Energy," and no passing on, but also there is no cessation of the existence of the Spirit: it is eternal. We have here, then, two different aspects of *life* to deal with, *life* contained in matter, passed on from one to the other, and Spirit life without matter, not passed on, but always existing as the same; yet both exist for a time in the same matter.

The question of the first "cell of life,"[1] which in my opinion is composed of corpuscles of "Energy," incorporating itself into inorganic matter has never been solved by humans, past or present, beyond than that it was the will or creative

Negro is very pronounced. It commences to develop in the Masaba Negro, reaches the highest point in the Non-Hero-Cult Nilotic Negro, and then declines again gradually, being less pronounced in the Hero-Cult Nilotic Negro, and much less so in the Stellar Cult people, and may be considered to have reached its normal state in the Solar Cult people. Not only are the Supra-Orbital Bones much enlarged and thickened, but the whole of the bony structure of man is of a "coarser type" in the Non-Hero-Cult Nilotic Negro than in any others. This is a subject upon which future workers may throw great information, and I would suggest the study of the Pituitary Body and the Hormones of the same, in the Non-Hero-Cult Nilotic Negro, to solve the question.

[1] See my "Origin and Evolution of Freemasonry connected with the Origin and Evolution of the Human Race" (George Allen and Unwin Ltd.).

act of a divine Creator; it is too incomprehensible to be analysed by science. Life, therefore, is something apart from matter (paut), although bound up with and incorporated as part of it, at least for a time, as organic. In that passage of life from one body of matter to another, life is also bound up with matter, which Huxley rightly termed protoplasm—"as something that life fed on"—and in the passage it reproduces more living matter for a time. Now the conditions or environment in which this life has to exist do alter the future life to a material extent, either to a better or to a higher state, or to a lower or degenerate state, a great deal depending upon these corpuscles of Energy.

Then one class of cell of life meeting another class of cell of life, in which it incorporates itself, reproduces a third and different cell, if the above cells are of a different class or differently arranged corpuscles of Energy; or, if of the same class, it reproduces the same kind of life, which, however, may be modified by some inherent power existing in the cells inherited from previous generations, being a different grouping of the corpuscles of Energy; and the higher or lower standard produced depends not always upon environment, but upon a higher type or class of cell of life, or upon a degenerate or diseased cell, as we may express it, but which physiologically are only different groupings of the corpuscles of Energy. In the one case you have strong, healthy people produced, in the other case a degenerate people; but environment will affect either for good or ill after birth, if not before, and have a modifying effect one way or other. My belief, therefore, is that Life is inorganic matter impregnated with Energy, this Energy being transitory in matter, but ever existing as Energy.

Energy assumes, or may assume, various forms, composed of corpuscles differently grouped, as we know that these corpuscles are $\frac{1}{1000}$ size of a hydrogen atom and can penetrate through the hardest block of steel and have a velocity of 90,000 miles per second. We know that these corpuscles group together in various forms and numbers, all having different properties. These exist in Body-matter as "Life," and when one dies these corpuscles leave the

Body-matter and regroup themselves into Spirit-form. These corpuscles can exist free from matter.

The "luminosity" we sometimes see, when watching a person dying, immediately after he or she has ceased to exist, is composed of corpuscles of Energy in Spirit-form, and this spirit never dies. What is the Law which these corpuscles follow in causing the grouping together to form a first "Life Cell" and again to form the Spirit, at present is only known to the Divine Creator.[1]

Until medical and surgical science came upon the scene, Nature or natural laws eliminated the degenerate and unfit; it was a case of the survival of the fittest only. But since the science and art of medicine and surgery, with sentimental philanthropists, have arrived and taken part in the "welfare" of the human race, we have been encouraging the unfit and degenerate cells of humanity to perform the functions of breeding to populate the world, to the detriment of the human race.

As Sir James Barr states, "Nature is lavish in the production of life and prodigal in its destruction; her efforts are devoted to the improvement and perfection of the species. She does not look after the individuals except to adapt them to the environment, and thus the unfit are eliminated. Those that survive are the ones suited for their surroundings. They in their turn produce a vigorous, intellectual, and self-reliant race."

We at present, therefore, are working against Nature's laws, and this natural law will in time eliminate the race if we so continue. It rests with the Government and the people to say if they intend to continue in this sentimentality and bring degeneration and destruction to this nation. Or will they become sane, take action, and use the means to obviate this progressive disease that is eating into the very life of the nation? The remedy and means are obvious.

Another important point as to the future degeneration of the nation is that *Heredity must be taken into account.* One cause of degeneracy in this country is the increasing number

[1] It is called the Periodic Law. See my "Origin and Evolution of Freemasonry connected with the Origin and Evolution of the Human Race" (George Allen & Unwin Ltd.).

of its dependent class, largely composed of the mentally unsound and the socially incompetent.

The mentally unsound are lunatics, idiots, imbeciles and feeble-minded, and those of low mental resistance.

The socially incompetent are principally paupers and the habitual criminal class, where the inherent weakness of will and moral fibre and an innate defect of character inevitably take the line of least resistance, probably owing to the abnormal arrangement of the corpuscles of Energy inherent in the Body-matter.

The defect is a germinal one—it is inherited and transmissible ; the cause often is hereditary disease.

The danger lies in the healthy members of the community mating with these degenerates, and thus lowering the aggregate vigour of the nation, as well as these unfit breeding amongst themselves.

The remedy must be to arrest the propagation of the insane unfit, and to raise the conditions of environment amongst the poorer classes, giving them "healthy food and healthy homes." Unless this is done the nation will not live as the predominant Power for long; and if the nation loses its inherent vitality, strength, and character, and becomes degenerate, whatever wealth it may possess will not save it from destruction as a nation.

But not only Heredity must be taken into account if we wish to be the "greatest nation" and to continue so. Good food, education, sanitation, temperance reform, the clearing of the slum areas and "open air" life must be taken in hand and carried out; because, even supposing that all the imbeciles, degenerates and criminals were put in a lethal chamber and removed, yet with these insanitary conditions existing many would degenerate again, and so our last state might be worse than the first. Syphilis must be stamped out (as smallpox, typhoid, and other diseases have been), for there is no disease which produces such dire results to the future race as this syphilis.

The training of the young during development is also most important, and General Baden-Powell has done a greater service to future humanity than almost any other man of the present age in his institution of the "Boy

Scouts," the best school that there is to make the boy *a man.*

Therefore, in taking into consideration the racial decay of this nation, we cannot look upon it from one point only. We must consider the stage of evolution that the White Man has arrived at, and for our ideal standard must ever remember that we have passed the stage of socialism. We cannot level down, and can only level up to a certain point; that is to say, some will always be of a higher type and superior to others, whatever we may do.

In taking into consideration the Antiquity of Man, and the proof thereof, it is not of any use to put forward the fact that the Nilotic Negroes' remains, or even the Pigmies', and their implements and skeletons, have been found in recent geological formations, because some of these Negroes and Pigmies exist at the present time.

We must go back to the oldest geological formations in which the present type of man—the Stellar Mythos—has been found, and that is the Tertiary. The human remains found at Castenedolo, near Brescia, in Lombardy, by Professor Rajazzoni, between 1860 and 1880, are described by Professor Sergi *as of modern type*, and I may add that Professor Sergi is a very reliable and acute observer of facts; moreover, he is quite sure and convinced of these remains *being modern, and in Tertiary formation which had not been disturbed.*

What does this imply? It proves our contention of the vast ages of time that have elapsed since the first human was developed on this earth. The fact of Stellar Mythos man having been found in the Tertiary strata is conclusive evidence that he existed on earth at that time—600,000 years ago at least—and he was of *the present type of the human.*

The question next is, How long did it take to form this type from the original Pigmy in evolution?

Certainly it would require another 1,000,000 years to produce these Stellar Mythos people from the original Pigmy. This time is a low estimate of the ages that must have passed before this evolution could be possible.

Therefore, it is useless for scientists and Anthropologists

to try to shut their eyes to these facts, because they still exist; and why should they not be believed?

The only reason why there are doubts and disbeliefs is because the present generation, like those for the last 1,000 years, have been so influenced by Biblical dogmas that they have been blinded to the great progressive knowledge that has been discovered and brought forward by a *very few workers for Truth*—knowledge of which our old forefathers, the Wise Men of Egypt, were quite cognizant.

Zaborauski attributed to Sergi a statement that the Egyptians are diffused through Asia Minor, Southern Russia and elsewhere, which Sergi does not claim to have stated; nevertheless, it is quite true that it was so. Sergi, however, does not go back far enough for the Exodus direct. The Pigmies went first, after them the Nilotic Negroes—Palæolithic men; after them the early Stellar Mythos people, Neolithic. Many skeletons of these Nilotic Negroes and Stellar Mythos people have been found, and Sergi states that he has found Pigmy skulls in Russian Asia.

The ethnographic observations of Herodotus and other classic writers are much too late to be taken into serious consideration when proving the origin of the human race, although interesting as to the facts they observed about the customs of people they visited.

Professor Sergi agrees and proves by anthropology "that the origin and migration of the African racial element took place in ancient times from South to North."

The types of Cro-Magnon, Homme-mort, and others in French and Belgian localities, as well as the oldest remains in these islands, bear witness to the presence of an African stock in the same region in which we find the dolmens and other megalithic monuments erroneously attributed to the Celts. All these types were widely distributed over the world, and had the same primitive funeral customs, shaped skulls, long forearms, etc. The Celts came a long time after, and modified their funeral customs by introducing cremation. One reason why blondes are found in North Africa may be determined from the retrograde movement of the Northern people, through being driven back again during the Glacial Period.

That only a small percentage of blondes are found in North Africa would prove that only a few "crossed back again" from the South of Europe to Africa during this intensely cold epoch, when life could not exist further North than about the centre of France.

The skulls and physical types of these North African blondes have so many characters similar to the Finns, that it amounts to positive proof when joined with others brought forward in this work.

We see in the evolution of the human race many changes.

1. The Homo is evolved as a little man, with great muscular development and a very small brain in comparison with the present human—about half the size; this brain is flattened, the greatest development being posteriorly. He has a long body, short legs, and long forearms.

2. He grows taller, his legs become longer, but not so muscular in proportion as those of the little man; his chest development remains great, and his brain increases in size—a greater development anteriorly. His body in comparison is not so much out of proportion.

3. In stature and size the whole body is well developed, with tall limbs and bone more robust than in the present generation. The brain is smaller and still more flattened than in the present generation; it is not so much developed anteriorly as posteriorly. One branch assumes greater development of bone and strength and dies out—Neanderthal type. The other progresses to present type.

4. The present state shows great development of brain anteriorly, less posteriorly. Teeth, limbs and bones show a finer and higher development, type "as quality" but not "quantity"; in fact, the difference between a race-horse and a cart-horse. Arms become normal.

5. The primary race are of a primitive type, i.e. long thin upper lip with sub-nasal projection, gradually losing this, in the next form, from infancy, the lower lip becoming more projecting in adult age; then both lips getting "thicker." One form of human, viz. true Negro, developing thick everted

lips; this type dies out, that is, no further evolution from this type, and remains in Africa, and does not develop anything beyond a true Negro.

6. The hair in primitive man is discrete peppercorn; one type assumes woolly or kinky hair (Negro with thick everted lips). He dies out, i.e. there is no further evolution in this type outside Africa.

The other branch gradually assumes a longer type of hair—frizzy, curly, and then into the straight, and evolution takes one branch on to the white man.

7. The forearm in primitive man is long, with a wide difference between Ulna and Radius; it gradually assumes the present type.

8. The nose in primitive man is depressed at brow and wide at the root, very broad and wide at the alæ; it gradually assumes the present type.

9. The colon in primitive man is larger, and of a different shape, than those of the higher races; the change is gradual in the evolution, but there is a great difference between that of the Pigmy and the white man. The bacilli contained in the colon of primitive man are different from those found in the white race.

10. Language began with primitive monosyllabic words with gestures, signs, and symbols; gradually more words are added, until primitive man speaks an agglutinated language, followed by further development into sub-inflected and then inflected; he gradually loses his gesture language, signs, and symbols, which he has replaced by words, of which primarily he had not many.

11. In his primitive form he has no Totems or Totemic Ceremonies, but believes in a Supreme Being and life hereafter and elemental spirits.

In his next development he has Totems and Totemic Ceremonies.

In his next development he has Hero Cult.

In his next, he has Mythology and Astro-Mythology, developing into Stellar Mythos, followed by Lunar and then Solar and finally, with the white races, the Christian doctrines.

12. In primitive form he has a long body and short

legs comparatively. This is gradually reversed; he has a short body and very long legs, as seen in some Nilotic Negroes. Gradually he assumes the form of a "perfect man," and one often sees "throws back" in the present types of men.

These are a few of the most prominent salient points for the student to notice. There are many other points and differences of minor points to be taken into consideration, which we shall describe in detail.

It is difficult to understand why, to certain minds, the admission that these "primordial men" were our ancestors is humiliating. To us it would seem to be something of which we are entitled to be proud, that our species, in the extraordinary evolution of our race, so closely allied to the animal only a million of years ago, should be so superior to it to-day, especially the white man, the highest type of the human race. Surely that is vastly more comforting than the belief that we have always remained the same as we were in the far distant origin, and that we have not progressed with the lapse of centuries. Although, as a matter of fact, life affords us daily proof that there still remain in the midst of the vast humanity, struggling ever onwards to a higher type, beings in whom the rudimentary psychism of primordial man is practically unchanged.

FIRST HOMO.—The Pigmy : *non-Totemic : non-Anthropophagous.*

Height.—Averages from 1·15 to 1·37 metres.

Cranial c.c.—850 to 1,100.

Hair.—Discrete peppercorn in Africa.

Abdomen.—Large, and long forearms.

Implements.—Primitive, chipped on one side only.

Language.—Limited monosyllabic: has sign and gesture language; words about 200.

Religion.—He believes in Elemental Spirits, a Supreme Spirit and life hereafter; propitiates Elemental and departed Spirits.

He has no folklore tales.

He has no magic.

He has dances in which he dances his past history of tribe and descent from Animal Kingdom.

Left Africa and spread over the world.

MASABA NEGROES.—Bushmen and Hottentots.

The same as Pigmy, *supra*, in every way, only a higher developed Human.

Never left Africa; are there now; went South and cut off from coming North again.

PART II

THE NON-TOTEMIC OR PRE-TOTEMIC AND NON-ANTHROPOPHAGOUS PEOPLE

CHAPTER VI

CHARACTERISTICS OF THE PRE- OR NON-TOTEMIC RACES

It is a cardinal tenet in the present work that man originated in Africa and from thence spread all over the world, and that the various types of his progressive evolution are still extant.

I shall therefore take the Pigmy as a type of the original "homo," and try to trace his evolution and the various types found, as far as is known, up to the present time, anatomically and physiologically, his ideas, beliefs, and language, etc., from the Pigmy and Pigmy tribes to the Bushman, Masaba, and Nilotic Negro, comparing each point so that the whole may be critically correct in the final solution of the inquiry.

It has been stated:

1. That these little people are a degenerate race.
2. That they have no language of their own, but only words they have learnt from the surrounding tribes of negroes.
3. That they have no belief in a future life or Great Spirit.

I hold that all these three points are incorrect.

First Point.

They are not a degenerate race; for this reason. You cannot degenerate Anatomically or Physiologically.

We find them as *little people*, following the well-known law of evolution that animals at first develop as small species and then grow large. For instance, in its first form the elephant was quite a little creature in comparison to the present; and so man was evolved as a little man.

His height averages from 1·15 to 1·37 metres; his weight from 3 stone to 7 stone.

His Osteo-anatomy is more primitive that that of the earliest skeletons or parts of skeletons that have been found in various parts of the world, except the Pithecanthropus of Java, which, in my opinion, was an early Pigmy. I contend that the remains found correspond to the Nilotic Negro and to the Stellar Mythos people, with the above exception. And here let me say that there are lower and higher types of the Pigmy, as one would expect to find after all the thousands of years that they have existed, although they have been cut off from the world and all intercommunication by being driven into forests and mountains where the Nilotic Negro did not follow. Those Pigmies who have not intermarried with the surrounding tribes are always alike, except when climate, etc., has had some effect. But there is no "throwback," as you sometimes find amongst the later tribes.

There are six distinct tribes well known in Africa alone.

1. The Akka, average height of 38 given by Deniker 1·378 metres, and 30 measured by Emin Pasha averaged 1·36 metres.
2. Balia.
3. Tikki-Tikki.
4. Batwa.
5. Wambutta.
6. Bazungu.

Marriage.—As a rule they have only one wife, and rarely more than two or three children.

The Pigmies say that they never intermarry with the negroes or others, only with themselves, i.e. with other Pigmies. From information that we have received it is probable that some Pigmy women have been given or sold to some of the negroes in their near vicinity. Dr. Arnold Schultze mentions one authentic case. Mr. Tony Cozens has informed us that their practice in this respect with regard to strangers who come amongst them and who gain their confidence is the same as with the Australians and other Nilotic Negroes. It has been stated that the Pigmies marry at about eight years old and do not live much over forty, but I do not think there is sufficient data to prove the ages of these little people one way or the other. The people themselves have no idea; and when asked by Mr. Geil, one old man said "Many moons," and opened and clasped his fingers many times—so Mr. Geil thought his age to be about fifty years.

Outside Africa they still exist—in New Guinea, New Hebrides, forest of Bolivia in South America, mountains in China, Philippine Islands, and the Sakais in the Malay Peninsula.

As regards proof of the existence of these Pigmies outside Africa, I give later photographs of some of the New Guinea Pigmies. Mr. Geil states in his book "Travels in China" that he saw some there. Mr. W. G. Churchward, F.G.S. (my brother), was some time amongst the Sakais. The report as to seeing and living amongst the Pigmies in the forests of Bolivia was printed in one of the Journals by two travellers.

Captain McDonald stated that there were Pigmies in the New Hebrides. The others are well known.

Negroid Pigmy races are also found in the Andaman Isles; average height 4 feet 7 inches, hair discrete peppercorn. They live in small encampments and in dwellings rudely constructed of branches and leaves of trees, use bows and arrows, chips of quartz and bamboo for knives, and hammers of stone. The Veddas of Ceylon and Kols, or Kolarians, Santhals-Kurkus, Bhels, etc., aborigines in the Indian Peninsula, are all of the Pigmy tribe and of the same stock. Some have altered a little, as might be expected owing to the different environment and probably intermarriages, from their African brothers, but anatomically and physiologically they are the same, all speaking a limited monosyllabic language and all having a sign language. All are fond of dancing and ignorant of letters, but all have signs and symbols, many identically the same, though not any have totemic ceremonies nor magic. They have been exterminated in Europe, North America, Japan, Australia, and Tasmania, but I have no doubt they will be found in other places not yet explored.

They are not negroes, but negrillos or negritos.

Mr. Edward Grogan states that while travelling in Africa

in 1898 he encountered between Khartoum and the mouth of the Zambesi "ape-like creatures" with "long arms and pendant paunch and short legs, pronouncedly microcephalous and prognathous. The stamp of the brute was so strong on them, he further states, "that I should place them lower in the human scale than any other natives I have seen in Africa. Their face, body, and limbs are covered with wiry hair, and the hang of the long, powerful arms, the slight stoop of the trunk, and the hunted vacant expression of the face made up a *tout ensemble* that was a terrible pictorial proof of Darwinism. Mr. Albert Lloyd also mistook one of these men for a monkey. The 'dwarf' was perched high up in a lofty tree, and when he saw he was observed he swung himself from branch to branch with the ease and rapidity of an ape. Mr. Lloyd had his gun up to shoot him, when his native guide cried out that it was a man." These are the lowest forms of the Pigmy. I sent to some friends in Africa to try to get a photograph of some of these, but my friends have not been able to do so.

In a number of the *Quarterly Review* (July 1901), a reviewer of certain books of travel relating to "Negro Nileland and Uganda," after describing the black and yellow dwarfs of the African forest, with their "arms somewhat long in proportion to their body, their furry skins, and, in the case of the yellow dwarfs, their long beards," goes on to say:—

"These Pigmies easily fly into violent rages, very much after the style of apes and monkeys. They have a strong sense of humour and a great power of mimicry. Their mental abilities, though apparently so undeveloped in their natural lives, become considerable when brought out by kind treatment at the hands of Europeans, this producing a strange contrast with the apishness of their appearance and their actions. The pranks they play, half in malice and half in fun, on their full-grown neighbours, their remarkable power of concealing themselves rapidly in vegetation, reminds one over and over again of the descriptions of gnomes and elves in European legends. And, like the elves and gnomes of legends, the dwarf folk of Africa can be kindly to those whom they like, often performing some friendly little service unseen, or leaving some gift during the hours of darkness."

According to Gudmund Andreas, the dwarfs of Scandinavian tradition were so very primitive that they could not be called human. "The Duergar are described as being of low stature, with short legs and long arms reaching almost to the ground when they stand erect."

The traditions found in various countries of Europe point to the conclusion that the primitive race in these countries was the Pigmy.

Sir Harry Johnston, in "The Uganda Protectorate," divides the Pigmies into two different classes—yellow and black. He states that "the yellow Pigmy has a fine downy hair all over the body, a poor development of the buttocks, is bird-rumped, and of more simian appearance than the black Pigmy. This light-coloured downy hair is not peppercorn except on pubes and armpits."

He describes some as of "dirty-reddish colour," and some of blackish colour, with discrete peppercorn hair all over the body. In some tribes the gluteal region in the women is not large, but in other tribes the gluteal region is largely developed.

These details indicate two distinct stages of evolution in the Pigmy. Sir Harry further states that the Pigmies never wash. A peculiar odour is exhaled from their skin; the smell is something between that of a monkey and a negro.

He characterizes them as "having merry impish ways, little songs, little dances, with mischievous pranks, unseen spiteful vengeance, quick gratitude, and prompt return for kindness."

The latter characters correspond to the tales of gnomes and elves found throughout Europe and the British Isles.

Dr. Arnold Schultze, in his "From the Congo to the Niger and the Nile," vol. ii, gives the following description of the Pigmies of the Southern Cameroons. These men were small, stunted, and muscular, with a yellow-brown skin, large, wide-set eyes, bushy eyebrows, big fleshy noses, and very long arms, prominent brows, and protruding mouth with thin lips; and a few individuals, both men and women, were unusually hairy, one man with a great mane and long beard.

He further states that the Pigmies of the Southern Cameroons never make use of bows and arrows, only spears. He convinced himself that these Pigmies had kept their language from foreign taint, whereas the Bagielli dwarfs of the coast spoke only the language of the surrounding tribes. He gives one word only of true Pigmy language—namely, *Mokbea* = "Do not be afraid." It is to be hoped that he will publish more later on. He distinguishes the dances of the Pigmies from the dances of natives by stating that the Pigmies *dance around in a circle* whenever the natives *form a circle of onlookers*, and the dancers dance down the centre.

Take the anatomical features of these Pigmies.

Some of them are broad-chested, short-necked, strong looking, muscular, and well made.

Colour.—This varies from a chocolate-brown to a rather reddish, and in some a yellowish tint.

Arms.—Their arms look longer in proportion to those of the European, but it is only so in the case of *the forearm;* this is at least 1 per cent. longer than the European's—in some a little more.

The bones of the ulna and radius are wider apart than in the European or yellow races.

Hair.—This is important to notice. It is not woolly or kinky, like the negro's, but is rolled closely into little balls—discrete peppercorn. It is the same in the male and female all over the body, both on the pubes and axilla, and is the same in length, colour, and section. On those that I have seen the hair has not been black, but rather of a peculiar very dark greenish grey, of a lustreless appearance.

Colonel Harrison states: "Hairdressing is a great art among them; nearly as many patterns are cut on the head as there are grades of colour." The bristles of the red pig form their favourite headdress and earrings; these they twist up into small bunches and tie in tufts to the hair.

Mr. Cozens stated to the author that he has seen them shave "all over" with pieces of glass, and they told him it was done to protect them from the "Matakania."

The transverse section is ovoid-elliptic.

In the photographs given here (Plate XXVI), Bohani alone has a beard; it consists of a small stubbly growth of short, close-rolled hair on the chin and upper lip, and the hair is the same on the chest and abdomen. They shave part of the head at times, as may be seen in the photographs, sometimes on one side and sometimes on the other, and others again only to leave a fringe. This, probably, is connected with some sacred ceremony unknown to us at the present, with our limited knowledge of their habits. (Pigmies with long hair are noted later.)

Nose.—The root of the nose is exceptionally flattened and very broad, from 40 to 49·5 millimetres. The tip is broad and flattened and the alæ greatly expanded, although this varies in some to a greater extent than one could believe without seeing. The characteristic Pigmy type of nose is well seen in the illustrations.[1]

Eyebrows are well formed and different from those of the negro.

Teeth are well formed and rather large for such little people; they crunch up bones with ease. Mr. Cozens states that he has seen some Pigmies file their teeth, which custom they have probably learnt from the Nilotic Tribes.

The palate is elongated.

The lips are quite characteristic. They have not the massive fleshy, everted lips of the negro, but rather long, narrow lips. The upper has a noteworthy sub-nasal projection, and when the Pigmies are drinking or speaking, the combination of the flattened nose and the lips gives a striking simian appearance. The lips of one woman have been perforated in three places (see photograph of Kuarhe).

Steatopygia.—In some young women there is very distinct enlargement and development of the gluteal region—as in the Bushman, and in some no two types are distinctly indicated.

Ears are small, the *lobules not well developed, and in some the pinna is crudely formed.* Some have the ears pierced, but not all. The chin is small in all, and the mandible is narrowest and inclined to be pointed.

Brain.—The average weight of brain in a healthy Pigmy is *900 grammes.*

[1] There are some more photographs of Pigmies in "Signs and Symbols" and "Origin and Evolution of Primitive Man."

I noticed several scars on the upper arms, on the left side, but could not get any definite answer as to their meaning. The general contour of the head is well depicted in the photographs. I give Dr. Smith's measurements, for I had not a suitable opportunity to measure them myself, as stated in my article, "A Study of the Pigmies," in the *Standard*, July 7, 1905. Dr. Smith's able article in the *Lancet*, August 12, 1905, is very accurate and important as regards measurements: "All the heads are ovoid in norma verticalis, and the length-breadth indices are respectively 76·5, Mongungu; 77·7, Bohani; 78, Mafutiminga; 78·7, Amooriape; and 79·1 both in Matuaha and Kuarhe. The heights of four of the crania are almost identical—135, 137, 139, and 139, the two old people having much the loftier heads, the woman Amooriape's cranial height being 144 millimetres (head length, 174), and the man Bohani's 155 millimetres (head length, 184). The horizontal circumference of the heads is 514, 544, 533, 518, 514, and 505 millimetres, respectively. As a measure of their prognathism, the indices expressive of the proportionate length of the nasal and prosthionic radii are 111, 104, 155, 110·5, 107, and 108. Bohani is the least prognathous and Mafutiminga the most."

Their average cranial capacity is 900 c.c., very low, and the sutures close early, thus preventing the further development of the brain and retarding learning. As a result, progressive evolution would be slow.

Physiological Conditions.

Abdomen.—In all the Pigmies and Pigmean races there is an extraordinary prominence of the abdomen—which has been attributed to overfeeding and a large liver. I am sure you will agree with me that, if this was so, the Pigmy might be one of the late humans in evolution, instead of the primary. But this prominence is caused by no such reasons. It is a primal condition of man, and proves his near affinity to the Anthropoid Ape. It is owing to the cæcum being placed high up in the lumbar region; from it the colon is bent downwards to the right, and the iliac fossa, thus becoming largely distended, pursues a sigmoid course across the abdomen, thus giving this appearance.

Now the reason for this is apparent if my contention is right.

In the animal, important herbivorous processes of digestion occur in the large intestine. A very large proportion of the food, namely, the proteins and carbohydrates, are shut up within the cellulose walls of the vegetable cells, *and since none of the higher vertebrata have the power of producing a cellulose-digesting ferment in their alimentary canal*, these foodstuffs would be lost to the body were not some other means provided for the solution of the cellulose. Requisite means are provided by the agency of bacteria. The mass of vegetable matter which accumulates in the cæcum is a breeding-ground for bacteria, and more especially for the bacteria which feed upon cellulose. These bacteria dissolve the cellulose, and ferment it, with the production of fatty acids, methane, carbon dioxide, and hydrogen, and by this means the contents of the vegetable cells are set free, and can be absorbed by the walls of the large intestine. *Thus we see the importance of the large intestine in the lower types of "homo" who live in low conditions* and depend for their nutriment on coarse vegetable foods; *whereas in civilized man this condition is not found. It is not required, for he has risen, by evolution, to that state in which he selects the most digestible foodstuffs* and prepares them by grinding, cooking, etc., for the process of digestion, separating all the harder parts so that the larger intestine has little or no work to do. *Now, in the Pigmy, we find that the cæcum and larger intestine are more developed than in any other human, and he possesses bacteria in this intestine which do not exist in civilized man.* Moreover, as we have advanced by evolution, so this large cæcum and large intestine have become smaller, as can be seen in the following order: Bushmen, Masaba Negro, Nilotic Negro, Masi, Yellow man, White man. Sir Arbuthnot Lane, Surgeon to Guy's Hospital, has proved by excising the whole of the colon that the white man is just as well without it as with it, and in many cases better.

Music.—Their music is similar to the Soudanese negroes. Mr. Geil, who visited them, says: "They are fond of dancing, very merry, laughing, singing, and continually at jokes. Their voice is really very melodious." Lieut.-Colonel Harrison says that he has "seen natives dance in all parts of the world, but nothing to surpass the agility of these little people, who are taught to dance almost before they can run." He states that "it is wonderful to watch seventy or eighty men, women, and children circling in and out, round and round, always keeping the tom-toms in the centre, every foot along the whole line moving together, and every toss of the head and every twist of the body being executed with military precision." Colonel Harrison adds: "How interesting it would be to follow them, in their own dialect, through all the plays and scenes their dances depict! There are funeral and wedding dances, the hunting and wrestling dance (a curious dance, in which they try to throw each other by the interlocking of legs)." I have seen this and drew attention to it in my articles for the *Lancet* and elsewhere. Then they have the fetish dance, war dance, monkey dance, executed almost in a sitting position, and many more. Dancing seems to be their only source of pleasure, and many an hour is passed in its indulgence.

The rite of circumcision appears not to have been definitely settled by those who have seen most Pigmies—some state that they do not circumcise, others (Sir Harry Johnston in his "Uganda Protectorate") state that the males of all the Congo Pigmies seen by them were circumcised. If so, it is a custom adopted by them from the Nilotic Negroes surrounding them.

Mr. Tovey Cozens, who lived with them for two years, states that they do not circumcise but have a habit of pushing back the prepuce, and so give one the idea that they circumcise.

Burial.—Their dead are buried on the same day they die with the bark-cloth worn in life covering the body, and with green leaves and earth. The corpse is buried close to, or under, the central fire in the encampment. They mourn for a few days. They sing and weep for three days, but do not dance; then leave and build a new camp. Mr. Tovey

Cozens told me that they had burial-grounds, which, however, they took great pains to keep secret; they only buried their dead there at night, and it was by following, unknown to them, that he discovered this. At the same time he agrees that they do not always bury the dead in their "burial-grounds," but only sometimes, as Mr. Geil has stated. They bury their dead in the thrice-bent position, laid on the side. I think *all this proves absolutely that they are not a degenerate race, but a primary one.*

Second Point

—that they have no *language of their own.*

Their language is not a language of "clicks" only, as I have seen stated; many of the sounds are quite soft, mostly labial and lingual, and not so many lingual-dental. I do not say that the Pigmies have not acquired some words from surrounding tribes with whom they have been in contact, but I have not the slightest doubt that they possess a monosyllabic language of their own, and one of the principal proofs lies in the fact that they speak words which have the same sound as the ancient Hieroglyphics of Egypt, as we pronounce them. My contention is that these are the "originals." They may be only a few words to express their meanings, but we must call this "the first and oldest language of primitive man"; the tribes who use it, and the old Egyptians who have used corresponding words, must have brought on the original Pigmy words. No doubt many more words have been coined by surrounding tribes, as we ever find a progressive people coining words; and so words and language grew, as the mode of articulation for different sounds became easier with the practice of those they had. There cannot be any real argument that these Pigmies learnt these words from the old Egyptian Hieroglyphics and from surrounding tribes, because it would involve the acceptance that long ago some ancient Egyptians, in a much higher state of civilization, had been driven back, lost all the records of the time, and degenerated to the Pigmy, which, taking their anatomical condition alone into consideration, would be impossible. The more we study the facts that are already brought to light, the greater proof is there that the Pigmy is the oldest and first man, and with him language originated, and also, as I shall show later, the first sacred ceremonies. I have collected some of the Pigmy words, and give them here with the old Hieroglyphics, and only regret that I have been unable to obtain all their words. They are written phonetically, as Mr. F. Trussel (to whom I am indebted for them) states that he has used the French etymology, as he could not get the sounds in English. Mr. Geil states that they have a sign language of their own, which all understand. This is very important, as all negroes have not got a sign language, although the aboriginal Australians have. It is also stated that they sometimes stick a small stem of fern, bone, or bamboo through their noses. The Mai Darats nearly always do so.

How many distinct words the Pigmies use I have not yet been able to ascertain accurately, because some of the articulate sounds are so accompanied by gesture signs or sign language that it is difficult to gauge if they are primitive words distinctly or a particular sound accompanied by a gesture sign—as Sign Language; but if not to be classed as distinct, words, these sounds undoubtedly became words after. For example "fifee," or so-called hiss of a snake—the "su" for the goose, etc. Then, again, a different intonation of the same word may mean a different thing.

I have been able to find out more about the Nilotic Negro. The first exodus of these people possesses about 900 distinct words. The second exodus a few more.

These are all monosyllabic; they have no written language, but have signs and symbols. They have no prefixes or suffixes, although some Anthropologists and Ethnologists have added these, because they have not understood the gesture sign.[1]

[1] Walter E. Roth, amongst the number, to whom I am indebted for much information, has always been an accurate observer, but is without the gnosis of the Cults or Sign Language, which is a great pity, because he has devoted much time and study to these people. He is trying to read without learning the alphabet first, and so cannot arrive at a true interpretation of all he sees and hears.

Some Pigmy Words

Pigmy	Egyptian Hieroglyphics	Equivalent in English
O-be	ànb	to dance, to rejoice
Mai	mu	water
A-do-da	āx or za	to sleep
Bacchaté	baak	grain, fruit, bread, food
Massouri	maau	good, right, to be good, to do well, to be straight
Tzi-ba	Xerpu-baa	a bow, a piece of wood, etc.

Pigmy	Equivalent in English
O-be	A Dance
Maria-ba	Pipe of Bamboo to smoke
Tath-bà	Whistling into a reed
Oct-bà	Wood
Di-pé	Spear
A-do-da	Sleep
A-pé	Arrow
Tzi-ba	Bow
Mai	Water or drink
Massouri	Good, well
Kon-pé	Clothes of any nature
Ma-di	Hunting horn
La-gou-ma	Bristle head-dress
Kalli-Kélli	Native bell
Bacchaté	Bread. This word is sometimes used—macchaté, the " ch " hard
A-foie	Dagger
Mokbea	Do not be afraid

Herodotus states that the Egyptians were the oldest people in the world, but that Psammetichus once put this to the proof. Two infants were kept carefully apart from human society, their attendants having their tongues cut out so that they could not utter a word. One day, when about two years old, the children came to their keepers, stretching out their hands and calling "Bekkos, Bekkos," for something to eat. This is very suggestive of a Pigmy word, and if so nothing more is needed to show that the Pigmies were the oldest people in the world.

The Egyptian record tells us that the human birthplace was a land of the papyrus reed, the crocodile, and the hippopotamus, a land of the great lakes in "Apta," at the horn point of the earth, i.e. the Equator, whence the sacred river ran to cover the Valley of the Nile with plenty. The African legends tell us that the Egyptians, Zulus, and others looked backward to a land of the papyrus reed as the primæval country. The Zulus told the missionary Callaivary that "men originally came out of a bed of reeds." The birth-place in the reeds was called "Uthlanga," named from the reed. The Basutos have the same traditions. The Japanese and Pueblos traditions are similar to the Egyptian. The Egyptian name "Kami" identifies them by name as the reed-people. The goddess Uati is the African Great Mother in the bed of reeds.

The Pigmies can count up to 100, having a particular sign for 50, also for 100. They reckon time by "Lunar time," have no days, but have "seasons" and "moons."

They build their camps in the form of an ellipse and their huts in the form of a half-moon, all facing inwards, with a large central fire and small fires in front of each hut. Mr. Geil states that "the Pigmies live in little huts, to the boughs of which large leaves are attached with fibre. Inside the leafy huts are little couches of the same large symmetrical leaves, laced together, making elfin beds fit for a fairy goddess. Unlike so many native dwellings, the Pigmy hut is clean, and the Pigmy is clean also in person and is a great hunter."

Lieut.-Colonel T. I. Harrison, who penetrated into the great Ituri forest and made friends with the Pigmies, states that "the whole Pigmy nation seems to be split up into small tribes of say sixty or seventy, each obeying its own chief. They live an entirely nomad, wandering life, like our gipsies, and their average stay in one place is about three months. Their huts, made of branches and leaves, measure about 7 feet in diameter and 4 feet high. They are hidden away in almost inaccessible places, so that no one could ever approach them unawares. They never steal from each other, but rob the negroes and Belgians with impunity. They have the most wonderful knowledge of poisons and counter-poisons, and the negroes around come to them if poisoned by bites of snakes or other reptiles. They live on roots, insects, animals they kill in hunting, and largely on snakes, and fruit when they can get it."

Dances—The Best Dancers in the Old World.

The Dance is Sign Language and was used and performed as sacred mysteries in *various modes*, by which *the thought, the wish, and the want is expressed in act*, instead of, or in addition to, words, which they did not possess. They were too poor in articulate words, and therefore these ceremonial rites were established as the means to memorize facts, in Sign Language, when there were not any written records of the human past. In these dances the knowledge was *acted, the ritual was exhibited and kept in ever-living memory by continued repetition, and the mysteries, religious or otherwise, were founded on the basis of action.* These were acted symbolically and dramatically, more to impress them upon the minds of the individuals concerned, having only this Sign Language to enable them to transmit from one generation to another their then ideas and beliefs as regards the departed spirits, the Spirit World, and life hereafter. How many thousands of years did man exist working out this Sign Language and formulating articulate sounds?

Sign Dances are those danced and taught in all the mysteries of gesture language, dramatized to show the evolution of the human race from the pre-human condition that was actual in nature, by which means the unwritten past was communicated and photographed on the brain by ceaseless repetition of the acted drama from generation to generation.

Signs preceded words and words preceded writings. When man began to observe and think, he would see different forces at work throughout nature; he must have noticed the changes of the moon, the movement of the stars, seasons of the year, and different forces of the universe, celestial and terrestrial. Each of these powers, these forces in nature, would appeal to him, and having only a limited number of words to express his thoughts, he would naturally adopt an animal, a bird, a fish, or something that he could see, apparently associated with one of these powers or forces, which he would thus represent by a sign or drawing of the animal or bird or object so conveyed to his mind—sometimes a compound object, as we see in the Egyptian later on.

Thus we see death and darkness represented by a crocodile's tail, because it was the last thing to disappear below the water at night when the beast retired to the bed of the river, and death, because it struck its prey with its tail to kill before eating.

Its head and eyes, for the same reason, represented light and day, etc.

The serpent was a representation of regeneration, because he changed his skin and became, so to speak, a new serpent.

Mr. Geil states that Pigmies believe in a good spirit and a bad one, and some in the finger of fate; that they have "a weird and mysterious air," which he attributes to their religious belief. He mentions the name of one tribe of the Pigmies as "Ti Kiti-Ki," which would mean in ancient Egyptian "men of the double cavern." It is important to note that the Egyptians later used to think and believe that the Sun-god—"Osiris in mummified form"—or the flesh of Ra, was reborn into the life of a new day *only after he had been drawn through the body of a serpent*, and that he came out in the Double Cave on an island in a lake in spiritual form; or rather, they used the same symbolic language to express the belief of the spirit living and "rising up" after the "corpus" was no more. Thus we see that the Pigmies' idea of the spirit first inhabiting the body of a large serpent was brought on and symbolically made use of in their Solar Mythos. It was a resurrection of the spirit; they probably adopted the serpent as a type because they noticed that it cast its skin once a year and so came forth as a "new serpent," and thus they associated it with the regeneration of the spirit. We feel that to obtain the truth of all the past origins of mythology and all that followed, some capable person should go and dwell amongst these Pigmies, gain their confidence, and study them as Spencer and Gillen did the Australians; then many points would be undoubtedly cleared up, and the "book of the past" could be read and would be open to those who studied and understood. How much of the "original past" of the Pigmies remains at the present day or how much is "lost" is an open question, even when we gain all the knowledge that is possible now; because the ages and ages these little people have lived in their great forest would naturally tend to efface or alter some of their ideas and customs, even if only at first in a modified form, as we see in the case of the Australians.

I cannot agree, therefore, with Dr. Wallis Budge in his statement in "The Gods of the Egyptians," vol. i, p. 27, that: "There is no reason whatsoever for doubting that in Neolithic times the primitive Egyptians worshipped animals as animals and as nothing more: the belief that animals were the abode of spirits or deities grew up in their minds later, and it was this which induced them to mummify the dead bodies of birds and beasts and fishes, etc., in which they thought deities to have been incarnate." Now the Pigmies, whence all sprang, do not worship animals, but they say in Sign Language they believe that the spirit inhabits the serpent for a time after death. People who do not understand have called it reincarnation instead of regeneration of the spirit.

This is, then, *only one mode of expressing their belief that they have future life in the Spiritual World* and *not reincarnation*—as we understand the term; but they do not worship it. The serpent is only a sign, a type or symbol expressing regeneration. It *simply represents a type*, as in like manner the Apis Bull and the Ram of Mendes were *representative types of the attributes of the One Great God*, and they were only looked upon as this, not worshipped as beasts and animals. They believe in a Great Spirit, as is seen and expressed in their care for one who is "defective."

One of the most interesting and important points I

found was that when I drew the oldest hieroglyphic sign for Amsu ✱ they recognized it. In the Book of the Dead, Chapter 164, we find that Mut [hieroglyphs], the Great Mother, "Mother Earth" of us all—Ta-Urt of the Egyptians in another form (a form of Isis)—is associated with two Pigmies, each of whom has two faces, one of a hawk and one of a man. Each has an arm lifted to support the symbol of the god Amsu, and wears upon his head a disk and plumes of the same form as we find worn by "Bes" and others associated, according to Dr. Budge, with Central Africa, but which we associate with those of the "Lakes of the origin of the Nile and Nile Valleys." All authorities are agreed that they are able to recognize photographs and pictures, and will themselves readily draw them, if given a pencil and piece of paper, and can also draw figures to express what they have no words for. When we drew this sign ✱ and asked the chief if he knew what it was, he smiled and pointed to Matuha and Mafutiminga, and told them to tell us what it was. They answered, "That is him," pointing to their chief, mentioning his name. That they recognize this sign, the most ancient sign we have for Amsu—or Horus I risen—is very important to Egyptologists, because it was the first sign used by primitive man to represent the chief—a great one—and it was the sign of the Chief of the Nomes. It is found depicted on the oldest Australian boomerangs and on the ivory tablets found in the tomb of Naqada. A full description and explanation of it is given later on. I cannot agree with Professor Flinders Petrie that it is the ideogram for Neit.

Sacred Ceremonies.—It has been stated that they have no religious ideas at all, but this is evidently a mistake, as before they "dance" they take off their ordinary head-dress (La-gou-ma) of light feathers or that which they have to represent it, and put on the representation of the "Horus lock," also a leopard's skin, with the tail hanging down behind; and they tell you this is *part of their sacred ceremonies.* This dance, which is "part of their sacred ceremonies," consists of a series of more or less zigzag movements in more or less of a large circle; the chief in front, and one following close behind imitating all his steps and actions. Suddenly the chief stops, faces round, and wrestles with the other with his feet and legs. The one behind having disengaged himself, the chief resumes his dance around in a large circle with fantastic movements. Again he faces around as before, still followed closely by the other; again he stops and repeats the former action. This occurs three times, and on the third occasion, he, the chief, throws the other, and the dancing ceremony is ended. They do not use their hands or arms, but only the right leg, and it is a "good throw." Whether this is the first and typical representation of the "fight between Horus and Sut" I am unable to say, but it is very suggestive; the chief representing the day and the other the night, and the chief overthrows his enemy, who is following him. Weird music is being played by those around, similar to the "Soudanese tom-toms," and as soon as it has ended they take off the leopard's skin and tail and Horus head-dress and put on their ordinary head-covering. Naturally, one does not expect to find all the ceremonies of "ancestor worship" amongst these negrillos; we can only find the *commencement of the originals*, but there is a great deal more to be learned and studied yet; therefore, to say that these little men and women have no idea of a future or any religious ideas at all, is, in my opinion, incorrect. I believe that as we are able to gain their confidence and speak to them in their own language my view will be found to be right. We must remember that all native tribes guard most jealously anything pertaining to their sacred ceremonies and often pretend not to understand. Spencer and Gillen have proved this, perhaps more than any other men, in regard to the Australian aborigines. The Pigmies consider that a man who is foolish has been sent down by the Great Spirit, and that he must be cared for and respected or the Great Spirit will punish them. They believe, too, in the Spirits of their Ancestors, build little spirit-houses about 18 inches high, with a small opening, inside of which they sometimes place fruit. They have one side of the head shaved during certain sacred ceremonies. These facts must all be taken as a further proof that they have some religious belief. They also believe in charms, make marks down the middle of their foreheads and cheeks with a red substance obtained from trees, and sometimes smear black liquid on the face. The belief that the Spirits of their Ancestors are supposed to inhabit rocks, trees, stones, and so on, we find has been carried all over the world. The so-called worshipping of spirits in trees, as amongst the gipsies of Germany, or spirits about the trees, as with the Shans in Western China, or spirits occupying houses, as prepared by the Wanandi, the Australian aborigines, the Ainu, the natives of New Guinea, the Indians in Mexico and North and South America and the Esquimaux, must all have originated primarily from these little people. How long ago the first exodus from Pigmy-land took place it is impossible to say; if you ask a Pigmy how long he has been in the forest and where he came from, he answers: "Always here; came from nowhere; always have been in the forest."

Age.—It is difficult to say to what age they live. When asked, the reply was "Many moons," as they reckon by Lunar time only.

Third Point.

That they have no belief in a future life or Great Spirit is not correct.

They take great care of any one who is defective, because they say if they do not the Great Spirit will come and do them harm.

They build little spirit-houses and put food there for their departed friends—a propitiation of the spirit.

They told Mr. Geil they believe that when they die their spirit enters a great serpent, and this serpent comes to see them, remains a few days, and then goes away. The serpent does not harm them and they do not injure it.

This is their only way or mode of expressing their belief that they have a future life in the Spiritual World. It is an expression in Sign Language, *and not a belief in reincarnation,* as some have considered it.

They propitiate elemental power.

Major P. H. G. Powell-Cotton states as follows, which

is a further proof of their belief in a Supreme Spirit: "It was during a forest storm that I received my first inkling that the Pigmies believed in a Supreme Spirit. One evening, about five, as they came to fetch me, after lying throughout the day in the forest, a wind sprang up and dried twigs and leaves came rustling down, while every now and then a dead branch or limb crashed to the ground. With quick glances to right and left at the tree-tops, my head tracker hastened his steps; then, uttering a shrill whistle, he placed his left hand to his mouth, made a sneezing sound into it, and threw it above his head in an attitude of supplication. As the storm grew and the thunder came nearer, I saw him darting anxious looks on either side, till he espied a little shrub, with leaves like a willow. Gathering a bunch of these, he pressed them into the palm of his hand, sneezed over them, and again extended his hand in supplication over his head. Presently a tremendous thunderclap burst overhead, whereupon he hastily plucked a larger leaf, wrapped the other up in it, and tied them to the top of a stick, which he then held aloft, and every now and then, to the accompaniment of shrill whistles, waved it round his head. On return to camp I obtained from him an explanation of these strange proceedings. The first part of his ceremonies was an *appeal to the Supreme Spirit* to send away the tempest, but, as the storm continued, he besought protection for us from falling branches torn off by the wind." *Here we have the Pigmy "offering propitiation to the elemental power"* —the *first origin of religion*. The earliest mode of worship recognizable was in propitiation of the superhuman power. This power was elemental of necessity, a power that was objectified by means of the living type; and of *necessity the object of propitiation, invocation and solicitation, was the power itself, and not the types* by which it was imagined in the language of signs. If we use the word worship, it was the *propitiation of the power in the thunder* and the storm; *not the thunder or the storm itself.*

"Again, during our wanderings in the forest we came across many curious little structures—diminutive dwellings, which we were told were ghost-houses. *These were built to propitiate the shades of departed chiefs*, who, until a resting place is provided for them, nightly disturb the Pigmy villages. *There the people sacrifice and place food for the spirits of the departed.* We obtained much interesting information regarding the existence of religious beliefs, even amongst these Pigmies, and learnt that in some spot in the innermost recesses of the forest an imposing religious rite takes place on certain occasions, in which an altar is erected, whereon offerings are laid while the Pigmies arrange themselves in a semicircle and perform their devotions."

Dr. Schmidt, in his recent book on the Pigmy, believes that they received "a gift" direct from the Supreme Being. I contend that they obtained their religious ideas direct from observation of the elemental powers. Certainly I agree that the Great Spirit might have, and probably did, intend that their brain cells should be capable of observation and thinking; it would only be the natural law of creation of evolution. In this way I agree with Dr. Schmidt, but in no other.

With the Pigmy, the elements of life themselves are the objects of recognition, and the elementary powers were, and are, propitiated. *This was the beginning which preceded the Zootype, and this they expressed in Sign Language*, and this preceded Mythology and Totemism. There are no legends or folk-lore tales connected with the Pigmy, no Mythology, for he was, and is, Pre-Totemic. *He has no Magic. Sign Language is far older than any other form of sociology.*

Pigmies possess no Magic, but they do possess religious ideas. This proves that Sir James Frazer's opinion of this is as erroneous as his opinion "that all tribes have developed all their religious ideas, Totems, and Totemic Ceremonies, from their own surroundings." I have a profound admiration for Dr. Frazer's erudition, as well as for his wonderful literary gifts, but I cannot always agree with his views.

Speaking of this evolution of thought, he states that he believes "*those that practise Magic represent a lower intellectual status, and that it has preceded religion.*"

Now, Magic is very late in comparison with religious ideas, probably thousands of years, and we do not meet this evolution until we come to the Nilotic Negro. Even then amongst the most primitive it is questionable if it is used—not until the time of the second exodus do we find it. It was first used when the elementary powers were born of the first mother—the old Mother Earth, Ta Urt of the Egyptians, of no sex, who reproduced no children, male or female, but was a provider of food and life, as represented by the seven elemental powers.

Magic is the power of influencing the elemental or Ancestral Spirits.

Magical words are words with which to conjure and compel; magical processes were acted with the same intent, i.e. magical incantations which accompany the gesture signs. The appeal is made to some superior superhuman force—i.e. one of the elemental powers in Mythology which became the gods and goddesses in the later cults—or to the Ancestral Spirits.

The amulets, charms, and tokens of magical power that were buried with the dead afterwards became fetish, on account of what they imaged symbolically, and Fetish-Symbolism is Sign Language in one of its ideographic phases.

The practice of Magic was fundamentally based on spiritualism.

The greatest magician was the spirit medium.

The magical appeal made in mimetic Sign Language was addressed to superhuman powers as the operative force. The spirit appealed to might be elemental or ancestral, but without these powers there was no magic.

It was in one phase a mode of soliciting and propitiating the superhuman elemental powers or anamnestic spirits; this might be called the most primitive phase.

In another form it was the application of secret knowledge for the production of abnormal phenomena for the purpose of consulting the Ancestral Spirits.

Telepathic communication of mind with mind, directed by the power of will even without words, was a mode of magic practised by the primitive spiritualists.

All that is now effected under the name of hypnotism, mesmerism, and the like, was known to the ancients as "Magic."

In Egyptian the word "Heka" means to charm, enchant,

or ensnare, therefore "Magic"; also thought, rule, etc., therefore the ruling power of thought.

A. B. Ellis ("The Tshi-speaking Peoples") states that in time of war the wives of the men who are with the army dance publicly stark naked through the town, howling, shrieking, gesticulating, and brandishing knives and swords like warriors gone insane, and from head to foot their bodies are painted of a dead white colour. Dancing in a state of nudity was a mode in which the women showed the natural magic of the sex. Being all in white, they danced as spirits in the presence of the powers, whether sympathetic or not, whilst soliciting aid and protection for their men engaged in battle.

CHAPTER VII

SPIRIT WORSHIP

SPIRIT Worship is of two kinds—

1. *The so-called Spirit Worship of elementary forces, or the propitiation of elementary spirits or powers.* This arose in the mind of man when he observed the various powers and attributes of the forces of nature—water flowing—trees growing—darkness and light and all associated with it—the heavens as the Great Weeper, and light, which was considered the source of life to man and all else. From these powers of nature man imbibed his spiritual ideas, and so was laid the foundation or the beginnings of the later Mythos. Each at first was given, and recognized with, a Sign and Symbol, and afterwards a name was attached or connected with each power or attribute; *one greater than all* became *The one Great Power or Spirit*, and the others were attached as attributes or powers of *The One*.

2. We must be careful to distinguish these *Spirits* or *powers* from the second class, or *the Spirits of the Ancestors*; the propitiation of the latter, which has been called Ancestral Worship, is one of the typical leading features of the religion, or religious idea, of the Japanese of the present day.

This was the dawn of religion, because—

1. Religion proper commences with and must include the idea of, or desire for, another life.

2. This belief in another life is founded on the resurrection of the Spirit.

3. The belief in the resurrection of the Spirit was founded upon the faculties of abnormal seership, which led to Ancestor worship in all lands at one time.

4. It was a worship or propitiation of *the Ancestral Spirits, not of the body corpus,* which died and disintegrated.

The Egyptian Religion was founded on the rising again of the human soul emerging alive from the body of dead matter. The Corpus could not, and never did, come back or make its appearance again in any form, but the Spirit that arose from this was seen by seers.

The seven elemental powers were afterwards, at the time of the Hero Cult and the second exodus of the Nilotic Negro, divinized and represented by seven gods, two at first, Horus and Sut, then Shu, and afterwards the other four. There are two lists of these gods in the Ritual. These were given stars on high, and later represented the seven Pole Stars, and were called *The Glorious Ones*—The Khuti in Stellar Mythos. The first three were the Heroes; they play a great part in the folk-lore tales found all over the world. The old Earth Mother, Ta-Urt, was now divinized and represented as of both sexes, and was depicted by the Egyptians as the Great Mother Apt, who gave birth to the Heroes. So we have the *Heroes* and *Glorious Ones* of Manetho, as well as the *Glorified Spirits.* The latter had once been men; the former were Mythical, and were not nor ever could be human. The Glorified Spirits were the Spirits of Ancestors who had lived here on Earth and risen again after the death of the Corpus.

The Heroes and Glorious Ones were the Mythical Gods, or *divinized elementary powers, who had never lived* or reigned as men or women, and never could do so; and Dr. Wallis Budge, misunderstanding the gnosis, in all his works on this subject has mixed up and confounded these divinized elemental spirits with humans. The Hindus and others have gone wrong in their decipherment of the Ritual of Ancient Egypt; so have the Hebrew translators of the Bible, when they mention that the "Sons of God saw the daughters of men that they were fair; and they took them wives of all which they chose" (Genesis), and "the Sons of God came in unto the daughters of men, and they bare children to them." The translator thus mixes the Mythical gods with the human women!

It is in the religious part of man's brain that the greatest stability of the past is photographed. Primitive man's occupation was principally taken up in two things—

1st. *Hunting to obtain food.*

2nd. *Religious ceremonies,* enacted over and over again from generation to generation, in accordance with those sacred tenets which his forefathers had taught and impressed upon him; for he had no books or writings as we have. He used Sacred Signs and Symbols common to all, and danced the history of the past in various forms and ceremonies which all understood, so that if a stranger came to the camp it would be known whence he came by his dance.

They lived by hunting and fishing. They used bows and poisoned arrows and little spears and clubs.

Let us now take into consideration of our argument—

What was the opinion of the old Egyptians of the Pigmies, during the time they were working out their Mythologies; and how have they left a record of their ideas and beliefs? This is answered in unmistakable language: *they represented the first type of homo by the Pigmy—at the creation of the Solar Cult.*

Hitherto, all through the Stellar Cult the representation was by Zootypes. In the Lunar Cult the Great Mother was represented by the part-human figure of a woman in the two forms or phases of the Pigmy—namely, "steatopygous" and "non-steatopygous."

In the Solar Cult man was represented, and the Pigmy was thus represented and portrayed as the God Bes, thus recognizing the Pigmy as primary homo.

The Pigmies—not giants—represented the earliest human form of the seven primal powers. The giants were the Zootypes of the superhuman powers and were not human, but the Pigmies are human.

That the Egyptians knew the Pigmy as the Pigmy is proved by De Rougé, "Mémoire de Tombeau d'Ahmes," where the Egyptian name is written, *Nemma,* i.e. Pigmy (see Bunsen's "Dictionary of Hieroglyphics," p. 444).

I shall now add to this a description of the Sakais found in some parts of the Malay Peninsula. The description is given in Mr. Clifford's writings and by my brother, Mr.

W. G. Churchward, F.G.S., who went amongst this tribe, and although neither of these gentlemen is much versed in anthropology and ethnology—hunting having been their pastime—yet their account is sufficiently clear to prove the identity of the Sakais with the Pigmies. I am also indebted to Signor C. B. Cerruti's excellent work, "The Sakais."

"Sakais" is a Malay scornful appellation, which signifies "a people of slaves," for the aborigines of the Malay Peninsula—"The Mai Darats."

These are little people living in families and nomads. Each family lives separate, having an "Elder," who exercises, however, no authority except as consultant. They believe that they came from the West through the forest and have no ideas or traditions of the sea. Probably they are the ancient Benuas of India, who have been driven out, passed through Indo-China from Africa, and travelled on until they settled in the Malay Peninsula. There is a tradition amongst the Malays that they were originally Kurumbs. Dr. Short, in his ethnological studies of India, describes certain characteristics and habits of the Kurumbus inhabiting the forest, which perfectly coincide with those met with amongst the Mai Darats or Sakais.

They have nothing in common with the Malays surrounding them, where we find them "pure," no mixing with the Kampongs, Malays, and Chinese Gedes having taken place. The Sakai is short in stature, 4 feet to 4 feet 6 inches, well formed except in the lower limbs, which are slightly bow-legged. The cause of this is to be found in the habit they have from their earliest childhood of sitting upon their heels, thus leaving the knees wide apart; sometimes they will remain for whole hours in this attitude without showing any fatigue whatever.

Their feet are rather large and properly arched. The big toe is well separated from the others and is very strong. The muscles of their arms are not much developed. The fore-arm is very long in proportion. Their hands are long and slender. Their chest muscles are very well developed. They have large abdomens. They possess a wonderful agility in climbing, etc. They can endure great fatigue. The colour of their skin is between light and burnt ochre,

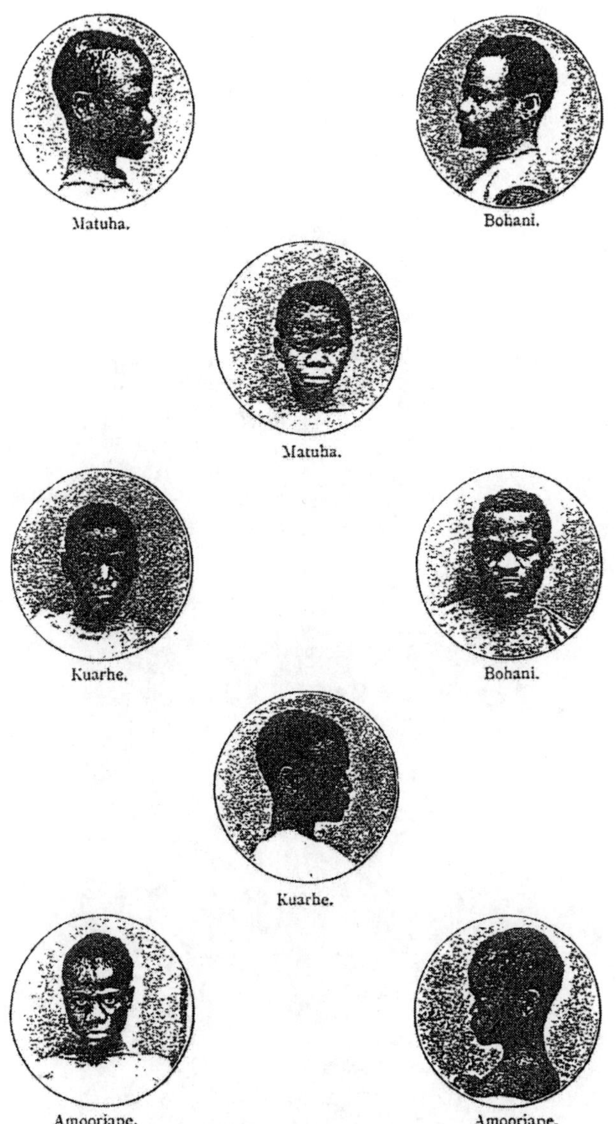

PLATE XXVI.

Pigmies, showing characteristic hair, nose, lips, etc.

A village hut, and group of Mai Darats, called "Sakais."

Making-fire.

PLATE XXVII.

Dancing.

Playing the "ciniloi" (pronounced "chineloy").

PLATE XXVIII.

Three types of tattooed Bretak Sakais.

A Sakai family.

PLATE XXIX.

Young man procuring food, with his blowpipe.

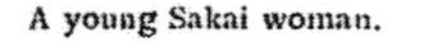

A young Sakai woman.

PLATE XXX.

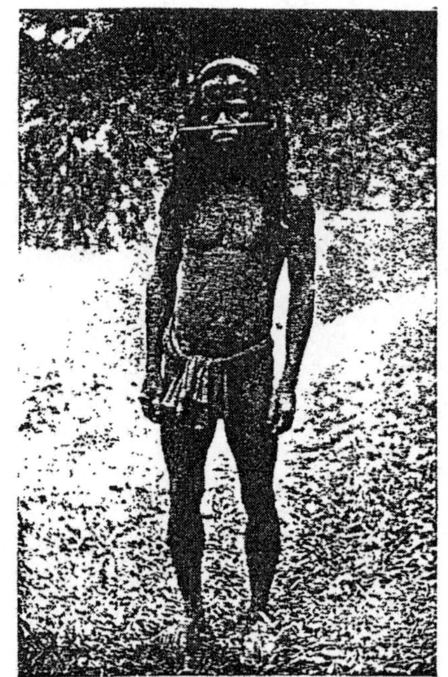

A half-breed Sakai.

Note women suckling young pigs.

PLATE XXXI.

getting darker as they grow older. The children are of a much lighter colour. The women, as a type, differ very little from the men, but are rather shorter. The Sakai's head is of the same form and shape as the Pigmy's. The forehead neither retreats nor protrudes. The mouth is well cut, not too large, with Pigmy lips. The chin is sharp. Their noses are flattish and bulging at the end, and sink in at the top. Hair may be classed as discrete peppercorn, grown long in most cases.

The Sakai women often have beautiful long tresses. They sometimes wear a nose-stick made of bamboo. They play the "nose-flute." As a rule they are almost completely nude. They are nomads and live by hunting. Their food consists of serpents, rats, toads, lizards, beetles, monkeys, deer, wild bear, wild sheep, and fish. They chew tobacco and betel. Once they are weaned they never touch another drop of milk, only water. (See Plates XXVII to XXXI.)

They never marry into their "own family," but always take a mate from another; if they cannot agree they part amicably, the wife going back to her family. The male and female children of the same father and mother are considered brothers and sisters; also the sons and daughters of brothers (who among us would be only cousins) are classed the same and call all their uncles "father."

The children of a woman are supposed to bear no relationship to those of their mother's brothers, and little or no attention is paid to that which exists between them and their uncles.

The wives of brothers call themselves sisters and are known by the name of "mother" by their nephews and nieces, but sisters' husbands have no claim to relationship other than that of cordial friendship.

No tie whatever exists between the parents of the husband and those of the wife, nor between the father and mother of the wife and their sons-in-law.

The Elders of the various villages are upon a perfect footing of cordiality, and never incite or permit the shedding of blood or even a conflict between their tribes. Their hut (*dop*) is the centre around which all others are erected.

They marry about the age of fifteen years and may live to

sixty years. Sometimes the Sakais build their huts either up a tree or suspended between stout poles, to defend themselves against wild beasts and other animals.

Their bed consists of dried leaves and their huts are made after the style of the Pigmies.

They sometimes paint their faces, the principal colours being red and black, sometimes a little white, but it does not last more than a day and is then scraped off.

They are very fond of dancing and singing, and have for musical instruments "a nose-flute," the ciniloi. They also will take two bamboo canes of 6 to 8 or more inches in diameter, being careful to select a male and female reed. These they beat one against the other, the result being a deep note with prolonged vibrations.

They have also the "krob," a very primitive kind of lyre, consisting of a short but stout piece of bamboo, on which two vegetable fibres are tightly drawn. The plectrum used is a fishbone.

The Sakai possesses only one weapon, the "blau" (for blakoo), called a "sumpitan" by the Malays.

Now, they sometimes procure knives and hatchets, but these are not used as arms. The blau is a cane bamboo from 10 to 12 feet long, not large in diameter, but perfectly round inside, and at one end there is a mouth similar to that of a trumpet. The arrow, or dart, is made of very hard wood, about 12 to 14 inches long, about the size of a big knitting-needle with a little pith cone at one end; the other extremity finishes in an exceedingly sharp point, sometimes of bone, well inserted into the wood, contrived in such a mode that when the dart strikes an object the point breaks off and remains there.

The force of penetration is so great that the body of a man standing 30 metres off may be pierced through without the dart being broken. They can send this little dart 50 or 60 metres.

The points are smeared or dipped in virulent poison, which they are great experts in making.

The Sakai women have a peculiar custom. If a wild pig or a rat (mother) is killed and leaves quite little young ones, the Sakai women will suckle them, and they become pets when they grow up. These animals will come every night back to the hut to sleep beside the women, and, when they move from one camp to another, will follow; these are never killed or eaten.

They bury their dead in a grave 5 or 6 feet deep, the body generally placed in the sitting position, but sometimes on its back, with face always turned to the West. Some tobacco, betel, and other food is placed near and carried to them every day for seven days—the time of mourning; they then forsake the village and march and form another.

The woman who is about to become a mother separates herself from the rest of the family and retires by herself to a hut apart. Nobody assists her at her confinement. Her own husband, the father of the new-born babe, dare not cross the threshold or enter the hut.

For some time after, the presence of a stranger in the village is not tolerated. Great superstition is attached to birth, more so than to anything else amongst the Sakais. The hut in which the poor woman is, is jealously guarded both night and day, and it would be death to any one not belonging to the village to go near.

Language.—Their language is monosyllabic, and I have here added a list of the words commonly used amongst the Sakais, but as their language is totally devoid of every rule of orthography, I have given the pronunciation according to the sounds and rules of Italian.

I have taken the following from Signor C. B. Cerruti's works.

Arm	Glahk
Arrow	Grog
,, (poisoned)	Grog mahng tshegrah'
,, (not poisoned)	Grog pe'm tshegrah
Bamboo	Annahd'
Banana	Tellah'e
Betel-nut	Blook
Bird	Chep
Body	Brock
Born	Egoy (alphabetical sound of *e*)
Blow-pipe	Blahoo'

Brother	Tennah'
„ (elder)	Tennah' bop
„ (younger)	Manang se ne (*e* sounded as in met, men)
Child	Kennon
Cigarette	Rocko
Come soon	Hawl aghit (*a* as in father)
Cover	Tshenkop
Day	E eah top
Dead	Daht
Death	Daht
Dog	Chaw
Ear	Garetook
Earth	In noos
Evening	Danwee
Evil	Ne' ghne' e' (alphabetical sound of *e*)
Eye	Maht
Father	Abbay', abboo', appah'
Father (in-law)	Tennah' amay
Fear	Sayoo neot
Female	Knah
Fish	Kah
Flute	Tshinelloi
Foe	Pay kabaad
Foot	Jehoo
Forest	Dahraht
Fowl	Poo
Fruit	Pla'
Good	Bawr
Good-bye	Abbawr
Hail	Tayho oontoy
Hand	Tahk
Harm	Ne', ghne' e' (like evil)
Head	Kovey
Heart	Noos
Hen	Poo
Hill	Loop
Hot	Baykahk
Hunger	Chewahr

Husband	Carelore
Hut	Dop
Illness	Nigh
Leaf	Slah
Leg	Kaymung
Lightning	Bled
Malay	My gope
Male	Crahl
Man	Sing no
Mandoline	Krob
Mangosteen	Play semmetah
Many	Jeho e
Medicine	Penglie (*ie* as in lie)
Moon	Ghecheck
Monkey	Dak
" (with long tail)	Raoh
Mouth	Eneoong
Morning	Pawr
Mother	Amay, kennen, kenung
" (in-law)	Tennah' abbay
Mountain	Lot, loop
Night	Sin oar
No	Pay neay'
Noon	Dahjis
Nose	Moh
Old	Din grah
One	Nahnaw
People	My
Plain	Barrow
Pond	Tebbahov
Poison	Chingrah
Quiver	Lock
Rage	Roh
Rain	Mahny
Rat	Hay loy
Rice	Bah
River	Tayhoo
Season	Moosin
Sing	Jeoolah
Sister	Kaynah

Sister (elder)	Taynah kaynah
„ (younger)	Mennang kaynah
Sky	Sooey
Sleep	Bet bet
Slumber	N'tahk
Snake	Teegee
Sorcerer	Ahlah
Spirit	Ghenigh nee
„ (evil)	Ahtoo
Star	Pearloy
Storm	Poss
Sun	Mahjis
Thunder	Nghoo
Thunder-bolt	Nahkoo
Tiger	Mah moot, mah noos
Tobacco	Bakhoo
Tree	Jehoo oo
Two	Nahr
Valley	Wawk
Water	Tayhoo
„ for drinking	Tayhoo engot
Wedding	Ba' kaynah
Wife	Kay el
Wild boar	Loo
Will, wish, want	Engot
Wind	Poy
Woman	Knah, caredawl
Yes	Aye aye (I)

They cannot count beyond nine, and this they do by counting three three times over.

They are very hospitable, and you are quite safe with them, when once you have gained their confidence.

Religious Ideas.—They propitiate elementary spirits and the spirits of the departed, and believe in a Great Spirit.

Many of them have intermarried with the Malays and others, and one often meets with a mixed type.

The above description, partly given to me by my brother, who lived amongst them for some time, and partly extracted

Group of Pigmies found on Saddle Peak Range, New Guinea.

A full-grown Negrito woman, mother of two children. The Negritoes are the smallest race of to-day. The average stature is 58 inches.

PLATE XXXII.

Bushman mixed type.

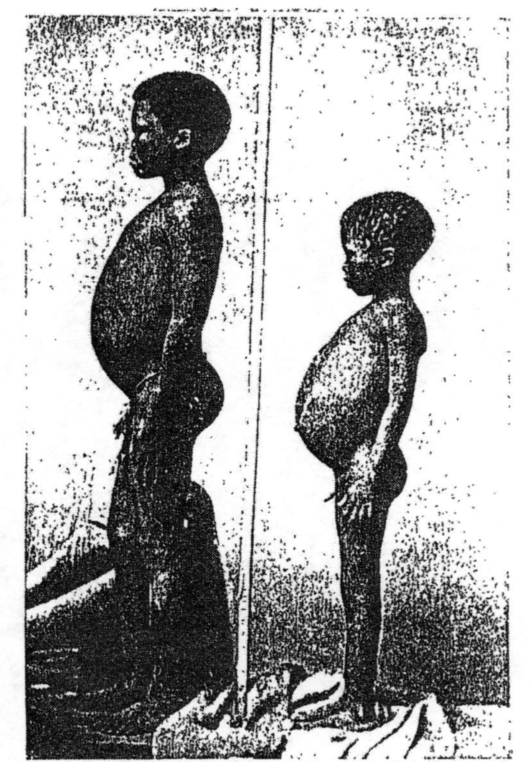
Bushmen, showing large abdomen, peppercorn hair, depressed nose, and long forearm.

PLATE XXXIII.

from Signor C. B. Cerruti's work, is a conclusive proof that these are the same in every particular as the African Pigmy described above.[1]

The occurrence of the Pigmies or Negritoes in the Papuan sub-region has in the past been a subject of much discussion, and the view that they have not been met with there has been widely accepted by anthropologists. Those discovered in New Guinea have all the characteristics of the African Pigmy, some in a modified way. They are nomadic, but some of these Pigmies in New Guinea are said to till the ground, and we see that some of their huts are raised somewhat from the ground, like the Sakais'. It is also stated that they cannot count beyond five.

Information obtained from an article by Mr. W. R. Ogilvie-Grant in *Country Life* (March 4, 1911).

I give here (Plate XXXII) a group of Pigmies found on Saddle Peak Range, New Guinea, taken by Dr. Eric Marshall, sent by Mr. Claude Grant. The measurements of 24 Pigmy men are as follow:—

cm.	cm.	cm.	cm.
148·9	145·5	143·4	136·5
148·7	145·0	140·8	136·2
146·7	144·8	140·2	135·8
146·6	144·0	138·8	134·9
146·2	143·9	138·5	130·3
146·1	143·8	138·1	129·1

—an average of about 4 feet 5 inches.

They have much lighter skins than the lowland natives. Although their skin is of a light-brown colour, some of the men have thick bushy beards. Other characteristics are identical with the Pigmies.

Not only in New Guinea, but also in many islands belonging to the Solomon Group, Captain McDonald found a race of men about 4 feet high, inhabiting the mountains of the interior. They are divided into families, roaming from place to place, having no intercourse with the other natives (Nilotic Negroes). They have for weapons only

[1] I am much indebted to Mr. F. Castle, of Malay, for sending me all the photographs of the Sakais.

bows and arrows, the latter deadly poisonous. These, therefore, are evidently the Pigmy race still existing here, having been driven into the mountains by the later arrivals —viz. an exodus of Non-Hero Cult people from Egypt, the same as we find in New Guinea.

At the time of Homer the Pigmies were not confined to Africa; they were believed to exist in India. Homer thus mentions them:—

> "So when inclement winters vex the plain
> With piercing frosts, or thick descending rain,
> To warmer seas the cranes embodied fly,
> With noise and order through the midway sky;
> To *Pigmy nations* wounds and death they bring,
> And all the war descends upon the wing."

Mention is also made of them by Pliny and Strabo. Pomponius Mela locates them in a certain part of Arabia. P. Jovius says that they are found in the extremities of the northern regions.

Professor Sergi ("The Mediterranean Race" and "Varieta umane della Melanesia," Boll. Accad. medica di Roma, 1892) speaks of "heads so small, though normal in anatomical constitution," that he believes them to be Pigmies. He has seen similar types from the Kurgans and ancient burial-places of Russia, and among the skulls which, in the Mediterranean, go by the name of Phœnician.

He states: "The types or shapes of these skulls are different, for the most part, from those belonging to the great stock, and they often present characters of inferiority in their structure. Many, including all those I have measured and considered to belong to the Pigmy stock, are inferior in cranial capacity to the Negritoes, or Eastern Pigmies."

Hence he concludes: "The low stature, the construction of the head, various external physical characters and peculiarities of the skeleton of the face and its fleshy coverings, led me to infer that in very ancient times there was an invasion of Pigmies from Africa into the Mediterranean, also invading Russia and probably other European regions."

He further states his belief in their African origin and writes: "They must have mixed with the tall stock and followed it in its migrations through Europe, just as they have formed an inferior stratum of the Mediterranean population."

Here Professor Sergi is not correct, in my opinion; for *the Pigmies to follow the tall race* is inconceivable. The Pigmies were the first people who, having left Africa, peopled all Europe, Asia, and America, *and the Nilotic Negroes followed.* We have a living proof of my contention in photographs here reproduced from New Guinea. The Pigmy primary was followed by the Nilotic Negro, who drove him into the mountains and forests, but these still exist here in New Guinea, just as they did in Europe a million years ago and also in the forests of Bolivia at the present time.

Some of these skulls have a capacity of 900 c.c., which proves them to be Pigmies, although a highly developed Pigmy may reach as much as 1,000 c.c. This also proves that the Piltdown Skull, with capacity of *at least* 1,200 c.c., *was that of a Nilotic Negro,* and that Professor Woodward's reconstruction of this with the jaw of a chimpanzee cannot possibly be correct (see *supra*).

On Plate XXXII is a photograph of a Pigmy woman from the Philippines. These have all the characteristics of the Pigmies. The Ilongots in these islands are the next higher type and are the wildest people in the Philippines—forest dwellers, probably a mixture of the original Pigmy with a Non-Hero Cult Nilotic Negro.

CHAPTER VIII

NON- OR PRE-TOTEMIC PEOPLE

PROFESSOR SOLLAS regards the Tasmanians as the oldest of all peoples (chap. iv), and states that they may be called "Eolithic"; at all events, they are Palæolithic, and that is about all that is true of his ideas concerning them. *He classes them with the Negroes!* which they are not, and never were. He also states that their cranial capacity is the lowest of all humans. This, too, is absolutely wrong. Their cranial capacity averages 1,100 c.c. The Pigmies have an average cranial capacity of 900 c.c., a difference of nearly 300 c.c., and their skulls are pentagonal, like those of the oldest Egyptians. The Tasmanian skull is neither long nor short (mesaticephalic), the cephalic index (the ratio of its breadth to its length) is 73·9, and the cranial capacity averages 1,100 c.c. They originally inhabited Australia, but were driven out by the Nilotic Negroes and crossed over to Tasmania, where the Nilotic Negroes did not follow them.

The Tasmanian was a highly developed Pigmy in every sense of the word. The "African Pigmy" has the lowest cranial capacity, and is a class of the human species evidently unknown to Professor Sollas, since he never mentions them as such, although they were the first and greatest of hunters.

Their hair was "discrete peppercorn" and, as far as evidence goes, they had no totems or totemic ceremonies. They used the primitive wooden clubs and spears. Spear-throwers and shields were unknown to them. They used rudely chipped stone implements as knives—scrapers—and had not advanced to the stage of the "Nilotic Negro" (Australians)—that of hafting a stone on to a stick—but held the stone in the hand.

Their stone instruments are the same kind of Palæolithic stone instruments that you find all over the world, and are the "Pigmy Implements" classed by geologists as "Plateau Implements."

Wherever all these Pigmies or Negritoes are found, they make their "boats," if they possess any, after a primitive fashion. These are a kind of half boat, half float, such as are still found occasionally on the Nile and Lake Nyanza—a bundle of leaves and bark tied up large in the middle and tapering at each end. In Egypt these were made out of papyrus or the leaf-stalks of the ambatch-tree. The Tasmanian boats were made of the bark of several kinds of trees (species of the eucalyptus); and the bark, after being removed, was rolled up with pointed ends. Three rolls were required, one large one for the bottom and two smaller ones to form the sides; these were then firmly lashed together, tough coarse grass being used for this purpose. These boats were very often of considerable size, 9 to 10 feet long, 3 feet broad, and over a foot high, with a depth inside of 8 or 9 inches. They would carry four or five persons; poles were used as paddles, and in shallow water they were punted. With these they would go three or four miles out to sea. The Seri Indians use the same kind of raft, which they call the "calsa," but the Seri are Totemic and early Nilotic Negroes. We find the same or a similar kind of boat in South America, on Lake Titicaca.

The Pigmy, starting from Africa, could, in the ancient world, make his way across to Sicily, into Italy, across Europe to Greenland and the East of North America, or he could cross over to Asia and then into Western America by the North, where the Aleutian Islands are now. If he turned South when in Central Asia, it was possible for him to reach India, the Malay Peninsula and the East Indies, united here and there by land connections, and with the help of his frail canoe he would be able to cross over into Australia and then on to Tasmania. Some, no doubt, went North by the Icelandic Bridge—Greenland—others via the Alaskan Isthmus and others South to Australia and Tasmania.

In my opinion geologists have given far too short a time for the Quaternary or Pleistocene age, and "Recent," of the world. Although we have found skeletons as evidence of man's existence in the Pliocene age, or the Tertiary, yet the most distinctive evidence—I might say positive proof—that we have yet discovered is that "he came into being" before the Pliocene age, and I feel confident that we must, on account of this evidence, place the time of this age at over 1,500,000 years. Therefore I cannot agree with Professor Sollas's chronology in any way. The immense ages it must have taken for the Pigmy race to spread all over the world, as we find him, along with the remnants of the original race and their stone knives, and, after these Pigmies, the exodus, on two different occasions, of the Nilotic Negro, before the Stellar Mythos people, would, I think, be conclusive proof of this.

There are no legends or folk-lore tales connected with him—the Pigmy; no Mythology, for he is and was pre-Totemic, and in the Totemic language he must be classed as what the Arunta tribes term "Inapertwa beings," "in the Alcheringa." These were the Alcheringa, or the Unopened or uncircumcised, who had to be transformed into men and women by cutting and opening. Besides, he was not anthropophagous.

In Chap. 164 Rit. there are some ancient mystical names which are said to have been uttered in the language of the Nahsi—the Negroes,[1] the Anti, and the people of Ta-Kenset, or Nubia. These names and words I have proved to be original Pigmy words, still extant—and, therefore, some of the first articulated words ever uttered by the human race; therefore parts of the Ritual of Ancient Egypt had been composed in those languages, and if in those languages, then in the lands where those languages were spoken, including the country of the Nahsi, who were so despised by the dynastic Egyptians.

Gerald Massey[2] was wrong when he wrote that these little people had no verbal language of their own, or any ideas of spirit life hereafter, because the evidence we have brought forward proves the contrary.

[1] The language of the Nahsi (the Negroes), the Anti, and the people of Ta-Kenset or Nubia. These names were applied to Sekket as the Supreme Being, "the Only One."

[2] "Ancient Egypt," Book V, vol. i.

I contend that I have critically proved that :—

The Pigmy was the primary "homo" and that he evolved in Africa, around the Nile and the lakes.

The "Pigmy race" spread throughout the world.

The oldest remains found are those of Pigmy type, viz. "Java man."

They have an articulate language of their own, which is monosyllabic—not written except in signs and symbols, which might be termed Ideographs.

They have a sign and gesture language, each sign representing a word or ideograph (?), and they can draw these signs and symbols.

They propitiate elementary powers.

They believe in a life hereafter.

They believe in a Great Spirit.

They build Spirit Houses.

They sometimes use the "Nose-stick." Mr. T. Cozens states that this is frequently of bone.

They have sacred dances.

They mark the forehead and cheeks with red paint and sometimes smear blood on them.

They believe in Spirit Ancestors.

They have been in the forest from the remotest antiquity.

They state that when they die their spirit enters a Great Serpent, which comes and visits the camp for a little time and then goes away; it never does them any harm, and they never molest it (a mode of expressing their belief that they have a spiritual life hereafter).

Their hair, lips, and anatomical conditions generally, as "primitive man," are not developed by evolution to the same extent as those of the negro.

Their method of existence consists in living by the bow and arrow and the spear, etc., and not cultivating the ground or domesticating animals, as Neolithic man—although it has been stated that some few Pigmies cultivate patches of ground in the forests of New Guinea.

They are nomad, but Palæolithic and not Neolithic.

They do not intermarry with other tribes; this we also find is the custom of the oldest tribes of Mexico, the Seri. They are small of stature.

Their language is monosyllabic and without affixes.

They do not practise magic.

They have no Totemic rites or Totemic ceremonies, but have dances in which they show forth, in sign language, such ideas and beliefs as their limited language cannot express.

We thus see that they are not a degenerate but a primary race.

They have no Totems or Totemic Ceremonies, no Folklore tales.

The Bushmen are very little removed from the Pigmies, and, in fact, claim to be their "first cousins." There are many features to prove *that they are the next type* in evolution, going South; the Masaba Negro comes next, going North, East, and West. (See Plates XXXIII to XXXV.)

They have in common:—

1. The same peculiar odour as the Pigmy, different from that of the Negro.

2. Projection of jaws and lips.

3. Flatness of nose and its characteristic broadness at tip and root.

4. Lobeless ears—ill defined.

5. Elongation of the palate.

6. Large size of teeth for such little people.

7. The same characteristic "discrete peppercorn" hair.

8. A primary monosyllabic language.

9. No Totems or Totemic ceremonies, but they have a sign language and sign dances, and show the same skill in dancing, which the Negro does not possess, since the latter has more articulate words to express his meaning, and so would discontinue the practice of dancing the object, etc.

10. The same convexity of the sub-nasal space.

11. Occasionally, and mostly in young women, steatopygia (greatly developed gluteal region, caused by the primitive mode of copulation practised, which fashion was changed by the Nilotic Negroes).

12. Many identical words.

13. Their sacred signs and ceremonies, propitiation of elementary powers, belief in after-life conditions and a spiritual world.

One of the proofs that the Bushmen are the earliest

inhabitants, and older than the Hottentots, may be gathered from the fact that the Bushmen's implements have been found in the floors of ancient cave dwellings below the deposits of the Hottentots' implements. At Burghersdorp, Bushmen's implements have been found 20 feet below the surface floor of an open cave, and here was exposed a kitchen midden (Dr. D. R. Kannemeyer). In no way do these implements differ from those of the Bushmen of the present day (*Cape Ill. Magazine*, vol. i).

I look upon the Hottentot as a further development of the Bushman through intermarriage and climatic environment.

As regards rock drawings which have been ascribed to Bushmen, I do not think that all those we find have been done by them.

The rock drawings and paintings in the Tuli district (here reproduced), as published by the Rhodesia Scientific Association in their *Transactions*, Plate I, vol. i, are not, in my opinion, those of Bushmen, neither are those near World's View, Matoppos, vol. v. These paintings are all 10 feet above the ground. The men are painted red and the animals black. The Bushmen's paintings are generally 3 or 4 feet above the ground. (Plates XXXVI and XXXVII.)

The figures of these men are not the representations of Bushmen, but of the tall Nilotic Negro, with very broad shoulders and well developed chest, tapering waist and long spindle legs, and we know that many of the Nilotic Negro tribes draw and paint figures of animals, etc., on the inside walls of their huts or houses—the Acholi, for example.

The cave in which these latter were found is about 18 feet in length. The floor of the cave was filled up with a deposit of earth, ashes, etc., the upper layer being about 9 inches think. Below this were several layers of fine ash; about 2½ feet altogether. Rough stone implements of small size were found; there were small pieces of bone and fragments of ostrich-egg shells with abundant stone chips in the lower layers. Near the back of the cave and very close to the bottom lay a skeleton. Dr. Edgar Strong has examined this and is of opinion that it is not the skeleton of a Bushman proper, but rather of an individual about fourteen years of age, belonging to a larger race. Near this lay another skeleton, buried in the "thrice-bent" position. The size of the bones indicates that the defunct was a full-grown specimen. Portions of an oval-shaped, shallow wooden dish, fairly well made, along with iron bangles and a small rusted assegai, were found near the bones. (I have taken the above from the Proceedings of the Rhodesia Scientific Association, vol. v, 1905, and am much indebted to the Association for their kind permission to reproduce them.) All these details point to the conclusion that this was one of the tall races of the Nilotic Negroes.

The natives of South Africa have a tradition that the whole region was first inhabited by the *Abutiva*, and they call it the land of the little people, i.e. the Bushmen, who were originally Pigmies.

The Bushmen and Pigmies look upon themselves as cousins and do not try to exterminate one another, whereas the Nilotic Negroes kill the Pigmies whenever opportunity occurs.

Bushman.

Bushman.

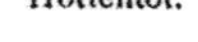

Hottentot.

PLATE XXXIV.

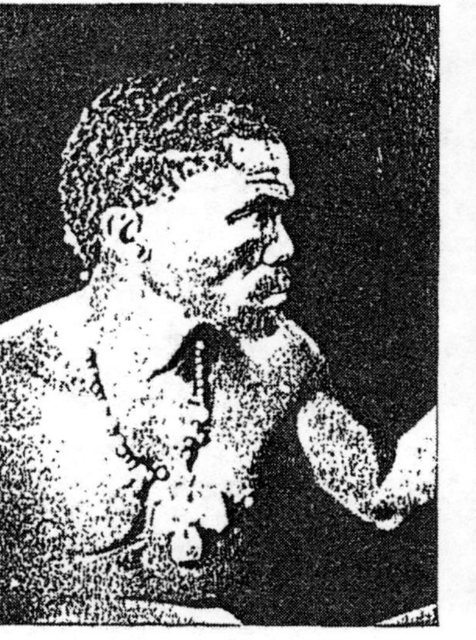

Bushman.

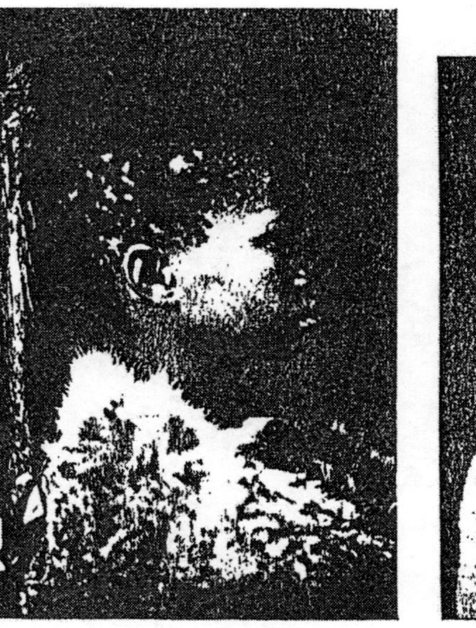

Bushman mixed.

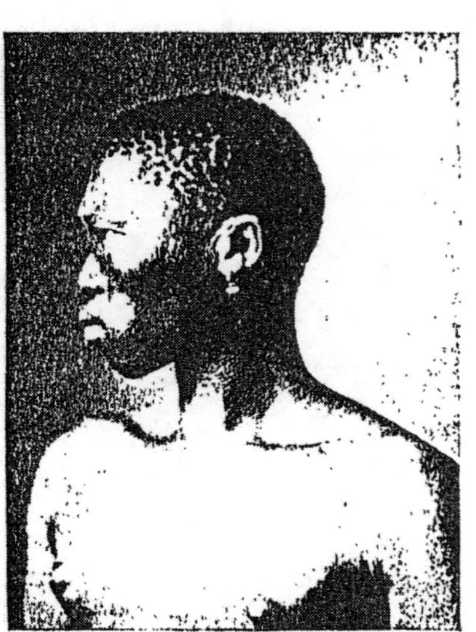

Bushman.

PLATE XXXV.

PLATE XXXVI.

Rock drawings and paintings in the Fuli district, South Africa.

PLATE XXXVII.

Rock drawings and paintings in a cave in the Matoppos, South Africa.

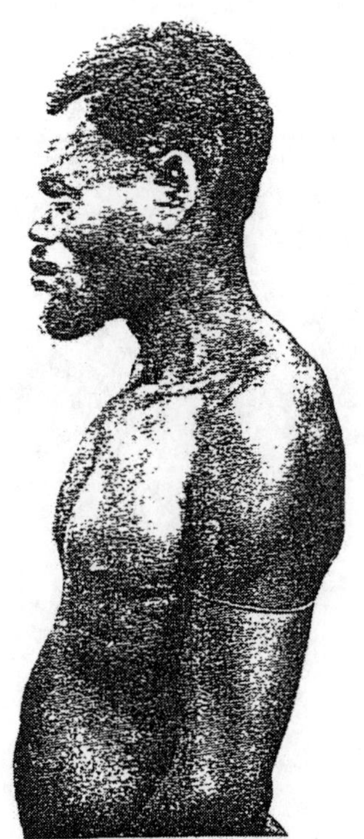

Plate XXXVIII.

Masaba Negro—profile.

Plate XXXIX.

Masaba Negro, showing long forearm.

PLATE XL.

Masaba Negro, full face, showing Pigmy lips, long forearm and large abdomen.

CHAPTER IX

MASABA NEGROES

I HAVE used the term "Masaba Negroes" to denote those Prognathous types of the human who were evolved from the true Pigmy, and from whom we trace the next development of the human race, namely, the low type of Negro to the North-East, and the true Negro to the West and South-West, and which now exist, under various names, in Africa. To the South the Bushman developed, and from the Bushman the Hottentot.

Dr. Robinson, in his travels through Haussaland, remarks on the very ape-like appearance of the wild mountainous tribes in the Banchi country, north of the River Benue.

Sir Harry Johnston also remarks upon these, which he mixes sometimes with Pigmies, although he classes them somewhat of a different type. He does not differentiate the two. Their colour is of a dirty chocolate-brown, and some allow the hair to grow as long as possible; its length is increased by the addition of string intertwined or interwoven with the hair, and the whole is then loaded with grease, clay, and red camwood. Some have short scrubby beards. These are met with in groups and nomad tribes all down Central Africa, north from Bahr-el-Ghazal to the upper waters of the Zambesi, and westward from the Bahr-el-Ghazal to Portuguese Guinea. At the present time examples and types may still be found around West Elgon, and between Lake Keron and Lake Albert Edward on the East, and Congo Forest, countries west of Ruwenzari, mixing with the Bapobo, Banzoke, Bakonde, Bagesu, Bosea, Basokivia, and East Africa, the Doko and Sandawi.

Appearance.—Short, little men with much prognathism,

ape-like appearance, low brows, strongly projecting superciliary arches (see Plate XXXVIII).

Nose.—Flat, with wide, bulging nostrils like the Pigmy (see Plate XL).

Lips.—Long upper lip and receding chin.

Face.—Broad in zygomatic measurement.

Hands and Feet.—Long hands, long forearm; feet large and clumsy. Knee turned in and tibia much bent.

Abdomen.—Large, like the Pigmy (see Plates XXXIX and XL).

Hair.—Peppercorn, sometimes elongated artificially.

Cranium.—Cranial capacity averages 1,000 c.c.

They do not circumcise, as a rule, or decorate the body with any patterns or tribal marks, so far as evidence shows. Next to the Pigmy, they are the most primitive and fundamental race that has not as yet evolved any of those Totemic ceremonies which we find in the Nilotic Negro, the next race who emanated by evolution from this lower type.

Dances.—They have dances like the Pigmy. There are no initiatory rites, but some bore holes in the upper lip.

Language.—Their language is related to the Pigmy language; many words are alike, *and the guttural-labials of these—like the Pigmy, Bushman, and Hottentot Kp and Gb—extend all over Africa, and are not found elsewhere except amongst the Pigmies. The Masaba, however, double their ideographic sounds, as for example, Ba–Baba, Bu–Bubu.* These people are called by the Nilotic Negroes Banande or Munande, and by some *Masaba,* which name I have chosen so that there may be no misunderstanding between "Banande" or "Munande." The resemblance of these *Masaba* Negroes in West Elgon to the Congo Pigmy, is very striking, as also that of the North-East African, the Doko, to the lower Andorobo Nilotic Negro. They have bows and arrows and live to a great extent on raw flesh, wild honey, bee-grubs, reptiles, and bananas, when they can get them.

Their Osteo-anatomy proves the connecting type, or link, between the Pigmy and the Nilotic Negro.

These people never left Africa and cannot be found outside Africa.

From these, the "Nilotic Negro" developed to the East, North, and probably North-West, whilst to the West the true Negro was evolved from them; climate, environment, and intermarriages being the cause of the evolution of the different types.

In the Masaba Negro (reproduced Plates XXXVIII to XL), we see the commencement of the formation of that "beetling brow," so prominently depicted in the skull of the Neanderthal type of man, and in the "Australian Workii man" we perceive its fullest development.

These latter were one of the primary types of the Nilotic Negroes developed in Africa from the Masaba, and were probably the first exodus of Nilotic Negroes from Africa after the Pigmy. He travelled North into Europe, and East, reaching Australia; what other countries he may have gone to has not yet been discovered. In Europe his bones have been found in several places, but as he would not be allowed to marry a higher type of woman he has been gradually exterminated.

In Australia, where we find him to-day, he drove out the Pigmy[1] from Australia into Tasmania, where he did not follow him, and has lived in Australia ever since, because the Stellar Mythos people never came there to exterminate him. Other Nilotic Negroes of a higher type followed him there, and are still there, but as these were all Nilotic Negroes with Totems and Totemic ceremonies they would not exterminate each other. At the same time, I may mention here that before the advent of the white man these Non-Hero-Cult natives were strictly confined to their own territory, which had been allotted to them, and were not allowed to go outside this by the Hero-Cult people. As far as is known at present, this type never reached America. I am of opinion that the lowest type of Nilotic Negro which went to America is represented at the present day by the Ona of Fuegia (see later).

In the forests towards the West of Africa these Masaba Negroes have been met with and described by writers who have visited these regions, and their description coincides with that which I have here set forth.

[1] For photographs see "Signs and Symbols of Primordial Man."

PART III

THE TOTEMIC AND ANDROPHAGI PEOPLE

CHAPTER X

NILOTIC NEGROES, TOTEMIC GROUP

Characteristics

The Totemic People, *Nilotic Negroes*, all originally Anthropophagous.

1. Non-Hero-Cult. ↓	Enlarged supra-orbital ridges, giving very beetling brows. Cranial capacity, 1,000 to 1,200 c.c. Implements roughly chipped on both sides, not polished. Implements first hafted. No mythology. No tattoo, but produce cicatrices on body. Have magic. Made rough pottery.
2. Hero-Cult.	Cranial capacity, 1,100 to 1,500 c.c. All the above, plus Hero-Cult and tattoo in place of cicatrization generally. Grown much in stature—very long legs; less beetling brows; finer type of face. Worked in metals, wood, and leather. Tamed wild animals, made pottery, cultivated the soil, and some became agriculturists. Hair grows longer.

Totemic, or when the sexes were first divided at puberty. In pre-Totemic times they had not been so divided. The Australian Arunta tribes, in their isolation, have preserved some relics of a primitive tradition of the pre-Totemic and prehuman state, in what they term the "Alcheringa," representing a mythical past, which did not commence with those who have no clue to origins. It was a past that was inherited, and never had any contemporary existence for them. These rudimentary beings the Arunta call "the Inapertwa, or imperfect creatures." We know what was meant by the term because it is still applied to the girls who have not been opened and the boys who have not undergone the rite of circumcision and sub-incision. Such beings still remained the same as the Inapertwa creatures, because they had not yet been made into men and women. The Arunta tradition tells us further that the change from prehuman to human beings, and from pre-Totemic to the Totemic status, was effected by Two Beings, who were called the Ungambikula, a word which signifies "out of nothing," or self-existing. Though these two are not designated women, they are two females. There being no men or women in those days, only the rudimentary Inapertwa, it was the work of the Ungambikula to shape the Inapertwa creatures into men and women, with their lalira, or great stone knives made of quartzite. These Two Beings were the primitive creators of men and women from the undistinguishable horde of the imperfect Inapertwa, as founders of Totemism by means of the Totemic rites. They are said to have changed the Inapertwa into human beings belonging to six different Totems:—

1. The Akakia, or Plum-tree.
2. The Inguitchika, or Grass-seed.
3. The Echunpa, or Large Lizard.
4. The Erliwatchera, or Small Lizard.
5. The Atninpirichina, or Parakeet.
6. The Untaina, or Small Rat.

The Two Beings, having done their work of cutting which was to establish Totemism, then transformed themselves into Lizards (*Spencer and Gillen*). Hence it was the Lizard of Australian legend that was reputed to have been the author of marriage, because the Lizard was an emblem of the feminine period. The Arunta state it had been found that many of the young women died in consequence of unlimited promiscuous intercourse with men, who were unrestrained, and women unprepared by the opening rite, when there was no law of Tabu. Therefore it became necessary to protect them from savage treatment (S. & G.). The tradition is one and universal, with many variants, and is to be found in the Mangaina Myths of Creation. It is fundamentally the same in the Mythology of the Californian Indians.

The non-division and classification of the African Negroes of various types found in the evolution of the human race, all collected under the term "Bantu" by Sir Harry Johnston and others, is, in my opinion, one of the most misleading that could possibly be adopted, and here I agree with the late Sir H. Stanley, that it proves the unscientific element of the grouping of the various authors, and their want of knowledge of the evolution of primary Homo, because under this term they have included some of the most highly developed Masaba Negroes, Nilotic Negroes, and some True Negroes, and have not shown the differentiation of these types anatomically, physiologically, or their religious beliefs and works of art. There is a wide difference between the earliest and the latest phase of human evolution in the above. Simply classifying, or trying to classify, the human race under "Language" leads one to about as erroneous an opinion as it is possible to conceive, especially when the original inner African "roots" are not taken into consideration.

Sir H. Johnston, in "The Uganda Protectorate," states that "the Nilotic Negro type extends from the Western frontier of Abyssinia through the Bahr-el-Ghazal region to Bornu, perhaps even to the Central Niger, and from about 200 miles South of Khartum to the North-Eastern Shores of the Victoria Nyanza." There is one feature which specially distinguishes the Nilotic Negroes and their modified offshoots. This is *nudity in the men.* They remain in the primitive condition of "the Garden of Eden." They do not know they are naked. The Greeks before the Roman Conquest were also in this happy condition. We find it amongst some tribes of American Indians and in most tribes of Australia; also the Wankonde, the Central Zambesi and Zulu, the Gala and Bahima descendants of the Nilotic Negro.

There is a very characteristic attitude of the Nilotic people, wherever found, which marks their affinity to that race. This is the posture they adopt when at rest. They stand erect on one leg, and, bending the other, press the sole of the foot against the inner surface of the knee of the leg which serves as a support. This is an attitude in which they will stand for hours (see a Shulluk warrior, Plate XLI).

Some of the New Guinea natives who are descendants of the Nilotic Negroes still practise this custom. They are probably descendants of the Shulluk or Dinka tribals.

The True Nilotic Negroes, at the present day, may be divided into the following, according to Sir H. Johnston :—

The Shulluk or Shuoli
The Dinka or Jañge
The Nuêr
The Shangala
The Chir
The Mandari
The Jañbara
The Dyur or Luō
The Alura
The Acholi or Shuli
The Lango
The Umiro
The Kumūm
The Jardum
The Ja-luo or Kavirondo, as well as many other tribes belonging to branches of these, including—
The Turkana, Masai, Sūk, Bari, Nandi, and Andorobo Groups

The Nilotic Negro was evolved from the Masaba Negro, and performs Totemic Ceremonies. These Totemic Ceremonies originated with the Nilotic Negroes, who were the next to leave Africa and follow the Pigmy throughout the world.

The type is distinct from the True Negro, as commonly understood, the latter being evolved from the Masaba Negro in the West. The True Negroes did not leave Africa except first as exported slaves. They have developed a different type of features anatomically.

These Nilotic Negroes are generally divided into four distinct groups, but as a matter of fact there is a connection between the groups, namely, the higher types can be traced by evolution from the lower. At present they may be classed as follows:—

Group I. Peoples of Bahr-el-Ghazal.
Group II. Peoples of the main Nile Valley and its tributaries.
Group III. The Kavirondo Tribes.
Group IV. The Masai Tribes and their allies.

These last, however, are of a much higher type and have advanced by evolution from the former.

To these must be added also the Niam-Niam, who, as regards anatomical features, are distinctly Nilotic Negroes, and inhabit the watershed of Bahr-el-Ghazal and the North-Eastern Congo.

Some of these are now of a mixed type through intermarriage, but the pure originals are Nilotic Negroes.

These Totemic Nilotic Negroes, both here in Africa and outside Africa—all over the world—proved by their traditions that "their beginning" is immeasurably earlier than the Egyptian tradition preserved in the astronomical Mythology. Their beginning is, in fact, with *Totemism.* This was preceded by a period or condition of existence called amongst the Arunta "the Alcheringa," or the far-off past of the mythical ancestors, of whose origin and nature they have no knowledge, though they have preserved the tradition. These were, therefore, the pre-Totemic people, i.e. the Pigmies and Masaba Negroes, who have no Totemic rites or ceremonies, and who existed here in Africa ages before these Nilotic Negroes with Totemism. The Nilotic Negroes became a higher type from the Masaba by evolution. I contend that amongst the lower type of Nilotic Negroes, Totemism commenced with a primitive form of Sociology, and a mythical mode of representing external nature by Sign language and Sign dances, which they had inherited from the Pigmy. As yet they had no Mythology. Thus we find amongst these in all parts of the world that Earth, Water, Air or Wind, Sun and Rain, are now all Totems, without being put into any kind of Mythology. It was the elements of life themselves that were the objects of recognition, and these preceded the Zootypes. The Zootypes followed immediately after. The next phases were the Nilotic Negroes with Hero Cult. This was the commencement of their Mythology, these powers being divinized and given stars on high.

The Totem has been sometimes called the "original ancestor," as it was given to the girl when she became pubescent and subsequently the mother of the tribe. It is obvious, therefore, that the sole "original ancestor" in Sociology, Totemism, and Mythology must have been *the Mother*, because in the latter case the female Totems of the Mother on Earth became the Totems of the Mother in Heaven, when she was divinized in Mythology.

The two women in Totemism have been handed down in their Mythology, and later carried on in the Eschatology.

According to the African version, the human race originated from Mother Earth in two classes: "The Forest Folk" and the "Troglodytes," born of the tree and the rock. At first, they were the Children of Earth, or "The Earth Mother." The Mother is then divided, or followed by the Two Women, who are distinguished by the emblems of their birthplace—Totems. The Tree and the Rock (or a stone with a hole in it, which is one image of the Mother Earth) correspond to the Wood and Stone Churinga of the Arunta. The Australian Totemic system begins with being dichotomous: a division of the whole into two halves. They erect two Totems—posts or sacred poles, one for the South, one for the North. There are two ancestresses or self-existent female founders, two kinds of Churinga—wood and stone, and two women of the lizard Totem. The Port Mackay tribe divides all nature into these two primary motherhoods, the dichotomous system, founded on the two-fold character of the Mother as Virgin and as Gestator, whom the Nilotic Negroes had divinized as She who conceived and She who brought forth. This, in Africa, was the work of the Nilotic Negroes with Hero-Cult. They had divinized the old Mother Earth as Apt, who, in this phase, brought forth children.

The two divisions were fundamental and universal here amongst the Hero-Cult people as the two Tiruti; they had the two halves, North and South, divided by the Equinoctial line, the two Earths, upper and lower, the two houses of earth and heaven, etc. (see later). But there is always a distinction between the two classes of these Nilotic Negroes, the one with no Hero-Cult, with the old Earth Mother, and the other of Hero-Cult, who had divinized the old Earth Mother and divided her into the two different characters. This creation of man, or man and woman, was mystical in one sense, Totemic in another—in reality, it was *Evolution.* This, however, explains the Semitic version of pre-Adamic and Adamic, which simply means pre-Totemic and Totemic.

Appearance.—The Nilotic Negro varies in height from 5 to 7 feet. Probably environment has much to do with the height of some of the types. An example is afforded in New South Wales and in the United States, where we find that the White Man, after a generation or so, becomes taller.

He has a slight projecting muzzle and retreating chin. The cheek-bones exhibit great breadth, and are particularly prominent just below the outer angle of the eye. The forehead bulges somewhat. In the higher type there is a distinct inclination to be tall and long-limbed. The leg below the knee is exceptionally long, straight, and thin, with very little development of calf. .

They are powerfully built, with good square shoulders. The primary types are ugly, especially the women, whilst in the highest types many are handsome.

Face.—Varies ; as a rule, somewhat oval to round.

Jaws.—Lower jaw prominent, thick, and heavy ; in some cases the chin becomes more prominent and more developed.

Hands and Feet.—Generally small. Arms long, especially the forearm.

Nose.—In the lower types the noses are large and very depressed at the roots, *with supra-orbital ridges strongly marked,* but still more prominent is the great proportionate width of the nose—as in the Pigmy, it is broader than long, and the end flattened. When the nose-stick is worn the width of the alæ is of course increased ; as these progress to the higher types, the nose becomes less depressed and less wide at the alæ, until we reach the higher type, when we find some handsome, well-formed noses.

Lips in Lowest Type.—Until about the age of twenty to twenty-five the upper lip projects beyond the lower—as in the Pigmy; in later years the lips project more and more, and in some cases after twenty the lips are "full," but not to the extent found in the True Negro.

Hair.—There is a very considerable difference to be found in the hair of the various tribes; in some, it is wavy or curly; in others, very distinctly curly or frizzled, but never woolly or kinky, as in the True Negro. It has never attained the true kinky or woolly form. Of course, there are many degrees of fuzziness. Moreover, when intermarriages with the Negroid or Masaba woman have taken place, we find many who are not of the true type.

The Sango women have long hair reaching below their hips.

Section, elliptical in transverse section.

Heads.—The heads are dolichocephalic mostly, but we find amongst the various tribes mentioned above sub-dolichocephalic, mesaticephalic, and some nearly brachycephalic.

The indices vary from 69 to 81.

Cranial capacity averages about 1,200 c.c. (higher type generally from 1,250 to 1,400 c.c.).

Colour.—The children of these Nilotic Negroes at birth are red or copper-coloured, but the colour gradually darkens, and in the adult assumes a chocolate hue.

In some tribes exposed to certain climatic conditions the colour is still darker, but the Masai children when born are yellow. All children have the very prominent stomach of their ancestors, the Pigmies. As they grow older this disappears.

In the lower types of the Nilotic Negroes we find long bodies and short legs, as we do in the Masaba; then there is a gradual evolution until we find in the higher types very long legs and comparatively short bodies—the types varying from the one to the other. All still exist in Africa.

Burial Customs.—These vary amongst the different tribes. Generally speaking, they bury their dead in the thrice-bent position, either lying on the side or in a bent upright position, with food of some kind placed in the grave for the departed spirit—others leave the bodies to be devoured by wild animals. Decapitation is practised by many.

Amongst the Noakkaras at the present day, when a freeman dies all his favourite wives are strangled and buried with him, and in the case of the Sultan these amount to hundreds ("From the Congo to the Niger and the Nile").

Customs.—General promiscuity amongst young men and girls before the latter arrive at the age of puberty is common amongst the Nilotic Negroes. These are customs of their pre-Totemic ancestors. But as soon as the girls arrive at puberty they then become Totemic, as witnessed by the Masai and other Nilotic tribes still extant in Africa.

The pre-Totemic state was one of promiscuity, more or less like that of animals. There were no Totems or Laws of Tabu, nor any covenant of blood, and no magic practised.

Totemic Sociology.—All the Nilotic Negroes have Totems and Totemic Ceremonies. The lowest of these tribes in the scale of humanity have no Myths or Hero-Cult, but in Africa we see that as these advanced up the ladder of evolution, Hero-Cult was evolved.

First, we have the Mother or Mother Totem from which they claim descent, the "Mother Earth"—Ta-Urt, and the belief in the Elementary Powers. These, then, here became divinized, their Astro-Mythology commences; Mother Earth (Ta-Urt) is given a place on high, and is now represented by Egyptian Apt, and two children are born of her—twins—one, Set, the first-born, the other, Horus I (Kavirondo tribes and some others); and as these people developed in evolution we find more children born of the Great Mother in all, with various attributes, until we come to the commencement of the earliest Stellar Mythos people (the Masai, Sūk, and Muhima tribes).

The lower types mark their bodies by cicatrization and the higher types by tattoo. All, or nearly all, knock out their incisors, upper and lower, which custom was first practised with those of the Snake Totem, so that originally the Masaba Negro did not knock out his front tooth, although some are found with a tooth knocked out, but this is of recent date.

They all believe in their Ancestral Spirits, and propitiate these as well as the Elemental Powers or Spirits.

The cicatrices made to pattern, or otherwise, were found in the primary Nilotic Negro in the earlier stages before he mastered the art of tattoo ; the tribal badges found amongst these primitive people sometimes represent their Totem, therefore the pattern varies in different tribes. The art of tattooing may be called a Totemic mode of Sign Language ; some, such as the Masai, the Ainu of Japan, and the Chukchi of Siberia, only tattoo their women, but amongst others the Totem is sometimes tattooed on the person of the clansmen, such as the Iroquois, the Ojibways, and other tribes of the Red Men. The Masai paint it on their shields. The Esquimaux indicate the particular Inoit tribe by different ways of trimming their hair, and the women by the figures tattooed on their faces. Amongst the Seri tribe only the women paint their faces and tattoo, to indicate to which clan they belong.

Some of the cicatrices mark the period of pubescence in the earlier Nilotic Negroes, and in the later, certain parts of the body are tattooed only at that time.

These Nilotic Negroes, who, as I have said, came next in evolution to the Masaba Negroes, have in many cases killed off the lower types of men and taken their women. Thus we have in some cases a modified mixed type, but one easily recognized when the true type is known and the reason considered.

Some customs and characteristics of the various tribes must be here separately mentioned.

I will take the lower types first.

The Sandaiwi

are a type which might be intermediate between the Masaba and the Nilotic Negro. I class them as the earliest Nilotic Negro.

They speak a language resembling that of the Pigmy, the same gutturals, and some clicks. Probably they were the forefathers of the Nandi. As they have no industries, they live by hunting.

Shulluks repairing their boat.

Shulluk warrior standing in characteristic attitude.

A Shulluk of the White Nile.

PLATE XLI.

AFRICAN NILOTIC NEGROES.

I am much indebted to my friend, M. Julius Lundsberger, of Omdurman, for the Nilotic Negro photographs reproduced in this work.

Shulluk women.

Native village, Upper Nile.

Women porters, Upper Nile.

PLATE XLII.

TYPES OF NILOTIC NEGROES.

PLATE XLIII.

ARTIFICIAL DEFORMATION.

All the women of the Sara have this artificial deformation of the lips as a sign of beauty. The effect is produced by piercing the lips and gradually enlarging the holes by inserting wooden discs, the size of which is increased as the lips get distended.

Bari family, Lado enclave.

Nyam-Nyams at Maridi, Bahr-el-Ghazal.

PLATE XLIV.

The Landu

are a very low type, little removed from the Masaba.

Character.—Long-bodied, short-legged, long forearms, Pigmy noses. Feet large, with long toes; large abdomen.

Industries.—They make baskets and unglazed pottery.

The face is not so simian-looking as the Masaba Negro.

The hair is allowed to grow as long as possible, and its length is added to by intertwining string with it, forming a mop of little plaits, with grease, clay, or camwood to straighten and make it longer.

They have a scrubby beard.

They bore the upper lip with from two to eight holes, into which they put a piece of quartz, small sticks, or a flower. They practise circumcision, but do not knock out their front teeth.

Burial Customs.—They bury the dead in the thrice-bent position.

These are the characteristics of the majority, and they are nearly related to the Masaba Negro—a further development, but many have intermarried, and so we find at the present day that they are a mixed tribe or class. They are closely connected with the Bambuba and the Mourfu.

Babira and Baamba.

These file their teeth and practise cicatrization to a hideous extent. Probably the Bagobo natives of the Island of Mindanao descended from this tribe, as they have identically the same hideous customs.

They practise circumcision, but do not knock out their front teeth. They do not tattoo.

They do not make pottery.

The Babira women have a curious practice of piercing the upper lip with one hole, into which they insert buttons of wood, larger and larger, until the hole is sufficiently widened to admit of a large wooden disc, which stretches out the upper lip in a stiff manner like a duck's bill, sometimes as much as 3 inches.

The Asenza women (the Asenza tribes found up the Luangwa Valley and in the Portuguese territory North of the Zambesi) have the same peculiar custom of causing the lips to protrude sometimes quite 3 inches beyond the nose. Also the Sara tribe as seen in Plate XLIII. A tribe on the Amazon in South America has the same custom.

The Bongo

expand their lips and ears to an enormous extent by inserting pieces of wood in these organs, the lips being thus distended to five or six times their natural size. (There exists a tribe at the present time on the Upper Amazon in South America which corresponds to these in every particular.)

The Musgum women at Maniling also deform their lips in this same fashion.

One peculiar feature of the Bongo women is the great fatness and enlargement of the buttock—steatopygia. Polygamy is allowed, but the number of wives is limited to three.

Head.—Short and round.

Colour.—Reddish.

They are agriculturists and hunters, also great and skilful workers in iron.

Their weapons are barbed jagged lances, bows 4 feet long, and arrows with 3-foot shaft, tips poisoned by the great Euphorbia.

Huts are conical and about 20 feet in diameter; on the top of the hut they form a straw platform, which can be used for a sentry as a lookout post over the stockade by which the cluster of huts is surrounded.

Burial Customs.—The corpse is placed in a sack in a sitting posture, about 4 feet deep in the earth. The site is marked by a heap of stones surrounded by posts, many of which are carved. The women are buried with their face to the North and the men with their face to the South. A similar burial custom to this is found among some Madagascar natives.

The Aluru,

who dwell north-west of Lake Alberta, have no spears, only bows and arrows; their huts are primitive.

Tribal marks are made by raising lumps by cicatrization on the skin of the brow.

These are a branch of the Acholi.

The Bari women raise scars of a herring-bone pattern on the upper arm down to the inner aspect of the elbow.

The Melville Islanders have the same pattern.

Niam-Niam.

The Azandeh or Zandey live about the watershed between Bahr-el-Ghazal and the North-Eastern Congo.

The characteristics of the Niam-Niams are as follows:—

Head.—Round, and face circular.

Eyes.—Almond shaped and sloping.

Nose.—Flat and square.

Lips.—Thick, and chin round.

Colour.—Chocolate-brown.

Dress.—A mantle of untanned leather or undressed skins. The chiefs wear a head-dress of the skin of the leopard or wild cat.

Hair.—Arrangement amongst the men is very elaborate; it is plaited into little plaits and elongated as much as possible.

Tribal Mark.—A set of squares filled with dots, placed on the cheeks and forehead.

Arms.—Lances, two-edged swords, knives, large painted shields, but their peculiar weapon is the throwing axe, made of wood or iron and curved like a boomerang; it is used for killing game as well as in war, and is similar to those found amongst many tribes in South America.

Huts.—Large and well built; about ten or a dozen are found together in a circle round an open space; the roofs as a rule are simply conical, but may be double-pointed.

They are agriculturists and hunters, and used to practise anthropophagy. They work in iron, manufactured weapons, pottery, basket-making, and wood-carving. They are still celebrated for their rock and horn drawings and paintings: probably the ancestors of the Turanians.

The Andorobo.

These dress their hair in pigtails, as also do the Kamasia; they use bows and poisoned arrows and small-bladed, long-handled spears. Their huts are very primitive, like the Masaba and Pigmy.

They are hunters only, eat raw meat, never cultivate the ground, keep no domestic animals except dogs.

The Acholi

tribes have a stone pencil through the lower lip and sometimes another through the upper. This custom extends throughout the Karamojo country.

(The Carabs of South America, who probably descended from these or the Karamojo, practise the same custom.)

Most of the people of the Karamojo country pierce the lobes of the ear and upper part of the rim, and have two or more brass rings through the lobe to make it hang down.

These people propitiate their departed spirits and build little houses or temples for them—a conical roof of thatch over a circle of upright sticks about 18 inches high. Round about these may be found long loops of string from which pieces of green hang down. *They have no Hero-Cult;* they are primary or pre-Hero-Cult people, but propitiate the ancestral and elementary spirits and believe in a Great Spirit.

Their dances are "Totemic" in all its various phases.

The medicine-men are the greatest amongst them. They have several kinds of omens.

Their houses are like huge bamboo baskets, to which they add porches of wattle and daub in front of the doors. The interior is daubed with black mud, the surface being made remarkably smooth. On this grey or black surface bold designs are painted in red, white, or pale grey. These designs are either geometrical patterns or conventional figures of men or beasts, such as the giraffe. These were probably the originals of the Rock and Cave painters, the remains of whom are found in many places throughout the world.

Burial Custom.—Thrice-bent, in sitting position.

These Acholi belong to the Dinka group and have the Snake as Totem.

The Kavirondo people are related to them anatomically.

The Dinka, Acholi, Nandi and the Masai are distinctly related as regards language.

THE MANGBETTU,

who are Hero-Cult people, deform their heads by elongation. They place tight bandages around the head of the child until this is accomplished, thus giving the same shaped head as we find depicted on sculptures, etc., in Peru and other parts of South America, also in Central America, as visibly portrayed in "Signs and Symbols of Primordial Man," Fig. 52. Their women allow their nails to grow long, sometimes several inches, like some of the Eastern people (Siam). Their pottery surpasses that of any of the other Negro tribes, and many of the objects, bottles, etc., are similar to those found in Peru and South and Central America.

THE JA-LUO OR NYIFIVA OR ABANYORO

have the crested crane as their great Mother Totem; shave their head in the form of the Swastika; have Hero-Cult, and believe that the spirit of the dead goes up into the sky to the North. They reside in fixed villages.

These live much by agriculture—sorghum, sweet potatoes, peas, beans, pumpkins, etc. They eat all kinds of meat and fish, except the flesh of the hyena and the crested crane, which are sacred (Totems).

They make rafts of ambatch wood and have poor small "dug-out" canoes.

Amongst these people there is a prejudice against "blacksmiths"; they obtain their pig-iron from the other tribes, who smelt it.

THE KAVIRONDO

forge pig-iron, and the blacksmiths are called "Yothetth," in some of the tribes, constituting a separate caste called "Uvino," interesting to Freemasons. Amongst the Gemi tribe "the Blacksmiths" are a secret society possessing deep magic called "Jamkingo," and are said to possess the power of prophesying. The Chief Odua, when quite young, prophesied the coming of the white man.

They believe in magic and that people can be killed by it. They have several omens. If a hailstorm occurs, no one goes to work in the fields on the following day.

They bury the body on *the left side* in the thrice-bent position.

They smear themselves with white clay for mourning, indicating in Sign Language that the departed are now spirits (Plate XLV).

The Kavirondo Tribes

(there are about sixty tribes or clans in this group), generally, all have Hero-Cult. They believe in the Great Mother, and in two gods, Awafwa, who is chief of all the Good Spirits—cattle are frequently sacrificed to him = the later Egyptian *Horus I*; and Ishishemi, who is a sort of devil and typical of evil, and is the chief of all Bad Spirits and the source of all that is evil = *Set* of the later Egyptians.

They pull out their two middle incisors, have tribal marks, and bury their dead in the thrice-bent position.

The Tinguians, who live in the rugged mountains of North-Western Luzen, are very skilful ironworkers, and correspond in almost every particular to these.

They are Hero-Cult people and bury their dead as these Kavirondo people do. Anatomically they correspond to them. The main industries are cattle-breeding and agriculture.

The Madi

have three or four parallel longitudinal cuts raised on the cheek as tribal mark.

They dig deep trenches round their villages, throwing up the earth on the inner side into a parapet, along the top of which is planted a stout stockade of poles. They enter by a narrow gateway, easily closed by a heavy beam. They are agriculturists and keep goats, sheep, and cattle, just as our forefathers did here in this country at the commencement of the neolithic period. Probably they originated from these. Amongst many ot the American tribes we find the same customs.

They knock out their incisors.

Sometimes a disk of wood is inserted in the upper lip, like the "peleli" of the Babira and Nyasaland natives.

Some of the Andorobo and Kamasia dress their hair just like the Masai, in pigtails; others cut their hair short and wear over the head a cap of leather, like the undress head-covering of a Norman knight. There is a tribe in South America which have the same customs.

Burial.—Generally they bury their dead in the thrice-bent position in shallow graves, near the centre of some grove of trees; along with the dead body they place a calabash of milk and a packet of tobacco, or some other food-stuff.

Omens.—The Madi and Kavirondo have the same omens. For a bird to cry out on one's right hand on starting a journey is a bad omen; it is a good omen if the bird sings on the right hand on returning, and vice versa, It is a bad omen if a black snake crosses the path; a good omen if a rat crosses in front of one. We find the same beliefs in omens and the same customs amongst Borneo natives.

They are a Hero-Cult people. They believe in a Central Deity, as do the Masai.

Totem, Crowned Crane.—They believe in ancestral spirits and circumcise, and excise the clitoris in women. They knock out two front teeth.

They pierce the lobes of the ears, insert pieces of wood, and extend them enormously.

They are agriculturists as well as great hunters. They semi-domesticate the wild bees by placing bark cylinders on trees for them to build in. They also make wine from the sap of the wild date palm, and beer from the grain of eleusine and sorghum. They have tribal marks.

They keep dogs, cattle, sheep, goats, and donkeys.

The cattle are marked by their respective owners.

Superfluous bulls are castrated.

The neck of the big breeding bull of the herd is hung with an iron bell. They used to live in caves, driving out the former inhabitants, namely, the Pigmy and Masaba Negroes; some of these latter, Pigmy and Masaba, still exist here, a proof of my contention, that the earliest man (a Pigmy) may exist in the same continent, or country, at the same time as other men (Nilotic Negro descendants with Hero-Cult and Non-Hero-Cult men) who are more advanced in evolution by probably many hundreds of thousands of years, and also the white man (present type of man).

They are closely allied to the Sikisi or Sotik, Eljunono, El Tùkĕn Muti, Japluleil, and Andorobo.

The Nandi or Nandick and Allied Tribes.

The men are quite nude; they use bows, poisoned arrows, and spears, and generally stab with the spear instead of throwing it.

They smelt iron and do clever work in leather; their women make pottery, which is rough and unglazed.

A higher type which has been evolved from the lower, and which we still find here, must now be mentioned, as some of these very tall races travelled North and have left their remains in many parts of Europe: the Cromagnum Man, and the big-headed people, who came up from Old Egypt, an exodus of a particular type and race which became extinct in Europe, or, by intermarriage with others, formed another race. These latter types still exist on the eastern side of Ruwenzori.

All these higher types have mythology and the commencement of the Stellar Mythos, the same religious beliefs as we find in Old Egypt, and even, in fact, with the old pre-monumental and early monumental Egyptians. They are the Nyemprias, Turkana, Sūk, Laluka, and Masai, and the fine-featured, handsome Elgum or Wamia.

The Dinka.

The Dinka are the most northern of the Nilotic Negroes, living in the basin of the Bahr-el-Ghazal. They are a muscular, well-built people.

Colour.—Very dark brown, although they often appear quite black, as they cover themselves with powdered charcoal, mixed with oil.

Mourning Custom.

Women of the Upper Congo smear their bodies with white clay upon the death of their husband. They remain husbandless for about a year, and are then distributed among the dead man's brothers and children. A child thus often inherits many wives.

Grinding Durra for Merrissa, Kordofan.

Notice the peculiar characteristic Nilotic Negro hair.

Plate XLV.

Plate XLVI.

The Jieng—the tallest race in the world, over eight feet high.

Plate XLVII.
Masai woman.

Plate XLVIII.

"The Great Mother: Lunar Cult Figure with serpent showing on front of figure with jug with water of life."

(From Cypriote vase from British Museum and Burroughs Wellcome & Co.)

Head.—Long and narrow, contracting at the top and back.

Jaws.—Powerful and prominent.

Lips.—Thick and projecting.

Hair.—Generally scanty, often shaved, a single tuft being left. Some, however, train the hair into longish stiff tufts, which stand out from the head like spokes.

Clothing.—Women have aprons of untanned skin, which cover from the hips to the ankles. They wear iron rings in ears and lips, and heavy iron rings around their legs and arms. Men go completely nude.

Both men and women knock out the incisor teeth in the lower jaw.

Tribal Mark.—A series of raised lines radiating from the top of the nose, over the forehead and temples.

Weapons.—Clubs are their favourite weapon and a bow-shaped instrument for parrying the blows from their enemy's club. They use spears also, but no bows or arrows.

Houses.—Large circular and conical huts about 40 feet in diameter; the roofs are made of straw and thatch, supported by a central tree-trunk. The huts are not grouped in villages, but in small clusters near the sheds and tethering-ground for their cattle, of which they have large herds; they have also sheep, goats, and dogs.

Totem.—The snake is their Totem; they are protected and allowed to live in the roofs of the houses. This was one of the Totemic types that was afterwards divinized and became the symbol of Rannut, the great Earth Goddess (see later).

The Dinka are famous as cooks, and their food is prepared with great care.

The Nyemprias.

An older tribe than the Masai, taller and thinner, but possessed of the same general features, with oblong eyes.

The Masai took their origin from these.

The Laluka.

A very tall and fine race, average height 6 feet, very muscular and powerful. They have high foreheads, large eyes, rather high cheek-bones, mouths not very large and well shaped, lips rather full (the Tehuelche of South America possess similar characteristics and correspond to these). The Turkana are very similar to these, only taller, frequently exceeding 7 feet in height (types of the Cromagnum Man of Europe). These Turkana dwell to the West of Lake Rudolf; they are among the tallest race known. Captain Wellby considered that the men in one district presented an average of 7 feet in height.

The Rev. C. A. Lea-Wilson, who has been working as a missionary in a district of the White Nile which is a thousand miles South of Khartoum, has just given to Reuter's some interesting facts about a race of giants in the Sudan. "Mr. Lea-Wilson and a party penetrated to the West of the river into the Bahr-el-Ghazal, and reached eventually a neighbourhood inhabited by some 8,000 jet-black typical negroes known as Jieng, who are among the tallest tribes in the world. Amongst the customs of these are the following: All adults have six of their lower teeth removed; they have a habit when at rest of standing on one foot like storks, characteristic of the Nilotic Negro wherever found; they believe in a Supreme Being, to whom they sacrifice through their chiefs, or witch-doctors; they are by way of being dandies; and they hunt lions and hippopotami with spears alone. Elephants they kill by dropping weighted spears upon them from the branches of trees."

These Jieng are Nilotic Negroes with Hero-Cult like the Sūk and Turkana (see later) (Plate XLVI).

The Sūk

are another gigantic race.

Nose.—Both the Sūk and the Turkana men and women tattoo, and pierce the septum of the nose.

Hair of Women.—The women of the Turkana do not shave their heads.

The Masai and Sūk women do.

Circumcision is practised.

Colour.—They are chocolate-brown in colour, and in feature somewhat of the Masai appearance, but better looking.

Hair of Men.—The men cultivate their hair and wear it long, in the shape or form of enormous chignons.

They pierce their underlip and insert in this hole a bird's or porcupine's quill or the long sharp tooth of an animal.

The Sūk carry about with them a small stool to sit upon, three-legged, made of tree branches; it is about 8 inches long. They are one of the first people who formed seats to sit on, instead of sitting on the ground or on a mat (see also the Masai).

Their religious beliefs are the same as those of the Masai.

The Sūk and Turkana were once evidently one people, and probably originated from the intermarriage of the Masai with another tribe of Nilotic Negroes.

The Wamia or Elgum People

are also an elegant, fine-featured people. All these (*supra*) have much affinity in language, religious beliefs, customs, habits, and to a great extent, bodily appearance, with the Masai.

They worked out the beginning of the old Stellar Mythos, and should be classed with the Masai as the oldest Monumental and pre-Monumental Egyptians, with the earliest Stellar Mythos and the latest Mythology and Totemic Sociology.

The Masai

are a splendid type of man as regards development of both body and brain.

Face.—Somewhat Mongoloid in appearance.

Eyes.—Slanting, and somewhat narrow.

Cheek-bones.—Prominent, with pointed chins.

Nose.—In many cases well formed, with high bridge and delicately chiselled nostrils.

Heads.—Rounder and broader than those of other tribes.

Hair.—The young man of the Masai allows the hair of his head to grow as long as it will. By tugging at and straightening it as far as he is able, he plaits into it twisted bast or thin strips of leather, and coats these with mutton fat and red clay, which he gathers into pigtails to form queues, the largest of which hangs down over his back.

The women shave their heads, and from girlhood to old age are clothed from the body down to the knee with dressed hides and a kind of leather petticoat underneath. They adorn themselves with beads and wire. The young warriors wear ostrich feathers in the head-dress, as did the old Egyptians, and white hair around the knee of one leg. The Zulu does the same (see Plate XLVII).

The men wear *capes over their shoulders only*, and sandals of hide when travelling.

Ears.—Large, and much deformed in both sexes by piercing holes and inserting blocks of wood, etc.

Lips.—Not thick. The custom of filing a triangular space between the upper incisors appears to have been frequent.

Colour.—A dull chocolate-brown, but babies when born are yellow.

Burial.—Bodies placed in thrice-bent position and covered with stones.

There are two classes of the Masai. One moiety live a pastoral, semi-nomad existence. The other moiety are agriculturists, keeping cattle and living in permanent villages.

I here give the form of tattooing practised by the Masai women.

Circumcision.—They practise circumcision and excise the clitoris in women.

They have the Snake Totem—as also the Dinka.

The Snake dance is still danced by the Masai and Dinka in Africa, the Mogui Indians of Arizona, and the Warramunga in Australia.

It was the Nilotic Negroes—whose original Totem was represented by a Snake—who first knocked out their incisor teeth, but amongst many tribes at the present day this practice has fallen into disuse. In Totemic Sociology the Snake represents the Goddess Rannut, one form of the Mother Earth as the renewer of life and vegetation. Amongst the Masai her name is Angai, "The Great Goddess of the Great Firmament," thus proving that she had been divinized (The Great Mother Apt, Egyptian).

Sir H. Johnston states that "*the Sky Goddess* is sometimes invoked when a severe drought threatens ruin to their pastures; on such an occasion as this, the chief of the district will summon the children of all the surrounding villages. They come in the evening, just after sunset, and stand in a circle, each child holding a bunch of grass. Their mothers, who come with them, also hold grass in their hands. The children then commence a long chant." Thus we see that in their distress they still turn to supplicate and propitiate the Goddess in her primary form of Earth Mother, Rannut, as the renewer of *vegetation and life*.

This Totemic Snake, which is good, and in a later phase represented Tem, must not be confounded with the Great Evil Serpent, Apapi.

One must not confuse the Mother of the Totem and the Mother in Mythology.

In the Snake Totem here, as in the Snake Clan of Arizona and those of the Dinka in Africa, the Mother of the clan, or tribe, is represented by a Snake totemically, i.e. the Snake is her Totem and her children are all "snake children." But there was also a Mother in Mythology who did give birth to totemic animals and is often confounded with the human Mother. This was Mother Earth, who, in one form, was represented as a Snake, the renewer of vegetation and life, typified by the Egyptian Goddess Rannut.

That the Serpent was a symbol representing not only the Totemic Mother, but also the Great Earth Mother, is proved by this drawing from a photograph taken in the tomb of Khopirkerîsonbû. The inscription behind the uræus states that it represents Rannut the August, lady of the double granary. This was the original, but we find the symbol representing Rannut carried to many parts of the world. I here produce Rannut with this symbol clearly portrayed, although it is much later—in fact, a lunar form or type (Plate XLVIII), as is proved by the half-human figure of a woman. It was taken from a Cypriote vase. The design is painted in light yellow or light green on red pottery, and I am extremely obliged to Messrs. Burroughs Wellcome & Co. for the photograph. The original, I understand, is in the third vase-room at the British Museum. And although we find many varied phases of this figure and symbol, the original was the Egyptian Rannut the August, lady of the double granary, the great Earth Goddess.

As in Totemism, so it was in the Mythology of Egypt.

TAKEN FROM THE TOMB OF KHOPIRKERISONBÛ.
Open-air offering to the Goddess Rannut symbolized as a Serpent.

Drawn by K. Watkins.

In the beginning was the Great Mother, because the first form recognized in Totemism was the Mother, who was human in the first place, and Mythological in the next. The primary woman in Totemism was divinized as the two Mothers in Mythology, and these were continued in the Stellar, Lunar, Solar, and Christian doctrines. Both types were represented by the Zootype Serpent.

At a remote period Egypt was divided into communities, the members of which claimed to be of one family and of the same seed—which under the Matriarchate signifies the same Mother-blood, and denotes the same mode of derivation on a more extended scale. This, therefore, bears out my contention that the ancestors of the Masai group, who have the Snake as their Mother Totem, were the first to settle into communities here and commenced to form "Old Egypt proper."

Later, in Egypt, the Zootypes of the Motherhood had become the Totems of the Nomes. Thus we find the Nome of the Crocodile, the Nome of the Cow, the Nome of the Tree, the Nome of the Hare, the Nome of the Gazelle, the Nome of the Serpent, the Nome of the Ibis, the Nome of the Jackal, the Nome of the Calf, the Nome of the Siluris, and others, showing the continuity of Totemic Signs.

It was during the status of Totemic Sociology that the artizans and labourers worked together as the Companions in Companies. The workmen in the Temple were the Companions. These Companions are the Ari by name, and these can be traced to the 17th Nome of Upper Egypt, the Nome of the Gazelle, when Ariu is the land of the Ari.

The Chief Priests of the Masai are called Ol-Aibon (which should be interesting to Freemasons, plural El-Aibon).

One of the legends is that when their race began there were four Dictours ruling the world, 1, who was black, full of kindness towards humanity; 2, a white one, who held herself much aloof and was, in fact, the Goddess of the Great Firmament; 3, a grey God wholly indifferent to the welfare of humanity; 4, and a Red God who was thoroughly bad, the type of all evil.

The Black God was very human in his attributes, and in fact was nothing but a glorified man and the ancestor of the Masai = 1, Horus; 2, Apt; 3, Shu; 4, Sut.

The legend amongst these people runs that they believed the Black God originally to have lived on the snowy summit of Mount Kenya, whereupon the other Gods, pitying his loneliness, sent him a small boy as a companion. When the boy grew up, he and the Black God took themselves wives from amongst the surrounding Negro races, and so procreated the first Masai men. Afterwards the Grey and the Red Gods, becoming angry at the increase of people, punished the world with a terrible drought and scorching heat (Sir Harry Johnston, "Uganda Protectorate," p. 831).

Thus we see that Hero-Cult was well established at the coming into being of the oldest Masai, for the above represents the Old Mother Earth, Ta Urt, divinized as Apt, the Great Mythical Mother, who had now brought forth her third son—Shu (see later), the earliest Stellar Mythos people.

One of the tokens of friendship and peace is to hold up some green grass or green leaves in your right hand.

Dead men are never mentioned by name.

Weapons.—Besides the two maces we shall describe later the weapons of the Masai consist of spears and shields. These shields are decorated and adorned with the marks of their "Totem," so that each man is known as to his clan. They also have bows and arrows, knobkerries, swords from 1 foot to 18 inches long, called "Sime." These are of a peculiar shape, and are characteristic of these Nilotic tribes, narrow towards the hilt or handle, and broadest close to the top; they are usually worn over the right thigh in a scabbard of leather. The boomerang was used also (see p. 172). Their main weapon is a huge, heavy throwing spear: the head is long and lance-shaped; the wooden handle is about 18 inches in length, and the head is balanced by a long 4-foot spike at the lower end.

Food.—Milk and meat are the main food of the Masai. The warriors are never allowed to touch vegetable food, and they acquire the necessary salts by drinking warm blood from the living cattle, to obtain which they stun the beast and open a vein. To boil milk in the Masai country is a deadly offence. They live on milk at one period and on meat at another, fasting between the two periods.

They possess considerable knowledge of drugs, and practise a rough and crude surgery.

The smelting and forging of iron and other metals is usually done for the Masai by a tribe of smiths (which we have mentioned above), related to the Andorobo and Nandi, and generally called Elgunono. These people not only smelt the iron (which is usually obtained as a rubble of ironstone from the river bed) by means of clay furnaces, heated with wood fires and worked with the usual African bellows, but they beat out the pig-iron with hammers into spears, swords, tools, and ornaments. As we stated (*supra*), these tribes, or secret brotherhoods, of blacksmiths are very noteworthy, because "Horus Behutat" was their chief Hero, and amongst these tribes originated that part of the Ritual of Old Egypt.

These Masai groups were the earliest Stellar Mythos people, and these and the Madi groups, as Turanians, spread over Asia to America, and probably all over Europe, at one time.

Houses.—The style of houses must be especially noted, for the reason that two Egyptian ideographs took their origin from these.

The first and earliest was in this form. The hut is circular in shape, sides made of reeds, with a great uniformity of thatching in a series of flounces—"rounded beehives built of reeds," and daubed with mud and dung. It has a *high peaked roof.* There are a few exceptions to the *flouncing*, however. The Egyptian ideographic hieroglyph  took its rise from this: a forked stick supporting the house in the centre.

The other form is still more important, because it settles the question of the origin of Pa, the Egyptian ideograph for house, which has never altered, or been used for any other expression throughout the world, and I have found this ideograph in many lands, including Central and South America. *It was taken from the form of the houses of the pastoral Masai.*

Here we find a proof that the Masai were the Old Egyptian pre-Monumental and early Monumental people (find proof later), because the two ideographs one square, and one oblong, picture copies of the houses which were built by these pastoral Masai only, at that time, and even to the present, and the hieroglyph for Pa, house, is the shape and form of their houses. *Therefore, this ideograph could not have been found earlier than these pastoral Masai, and must have been formed by them.* Another hieroglyphic taken from the Masai is *ikm* = shield  with crossed arrows  *ikm nt.* It also proves how ancient these hieroglyphic writings in signs and symbols are, and how many hundreds of thousands of years ago the art of writing commenced. It is also a proof that the Old Egyptians were the first people to found the "written language."

The dwellings of the Masai are here shown to be of two distinct kinds.

The agricultural Masai build their houses as round huts, made with walls of reeds or sticks, surmounted with a conical, grass-thatched roof.

The cattle-keeping Masai build dwellings of quite peculiar construction, *unlike those of any other Negro tribe.* These are low, continuous houses not more than 6 feet high, which may go round or nearly round the enclosure of the settlement.

They are *flat-roofed,* and are built of a frame-work of sticks with strong partitions dividing the continuous structure into separate compartments which are dwellings, each furnished with a low, oblong door. A good deal of brushwood is worked into the sides and roofs to make a foundation which will retain the plaster of mud and cow-dung which is next applied. The mud and cow-dung is thickly laid on the flat roofs, and is not usually permeated by the rain. All the doors are on the inside of the circles made by the continuous houses. They have cups to drink out of and small three-legged stools to sit on. Even at the present day this kind of house may be found amongst the lower class native Egyptians.

Here amongst the Masai Group we find a higher development than in the other Hero-Cult Nilotic Negroes, as is seen by the fact that they assign Stars to their Heroes.

Every village elects a head-man, who settles all disputes and acts as leader of the warriors in case of fighting. They are called here, in Old Egypt, Ropâitu, the guardians, or pastors, of the family, and in later times this name became a title, applicable to the nobility in general; still later they were "Princes of the Nomes." They capture the fawns of gazelles and young buffaloes and tame them. They obtain fish by trapping and spearing, and have as *domestic animals —cattle, sheep, goats, donkeys, and dogs.*

A higher type, still further developed in the ladder of evolution, is found in the Muhima of Nipororo.

Sir H. Johnston thus describes them :—

"In physical appearance both sexes incline to be tall and possess remarkably graceful and well-proportioned figures, with small hands and feet. The feet, in fact, are often very beautifully formed, quite after the Classical European. Under natural conditions there is no tendency to corpulence, nor to the exaggerated development of muscles so characteristic of the burly Negro. In fact, they have the figures and proportions of Europeans. The rather round head, with its almost European features, rises on a long, graceful neck, well above the shoulders, which are inclined to be sloping. The superciliary arch is well marked, though not exaggerated. The nose rises high from the depression between the eyebrows; it is straight and finely carved, with prominent tip and thin nostrils. The lips are somewhat fuller than a European's. The mouth is often small and the upper lip well-shaped, with no great distance between it and the base of the nose. The chin is well developed. The ears are large, but not disproportionately so, compared with Europeans'. The colour of the skin is pale reddish-yellow."

So we see that one of the first phases of the development of the body of man was that he became a taller type, with longer legs but poorly developed calves, rather prominent cheek-bones, not so repulsive in physiognomy, or with a less degree of prognathism. The supra-orbital ridges gradually became much less marked, "the beetling brow" disappearing entirely in the highest types, and the nose and lips assuming almost European phase in the Muhima. As evolution proceeded, the disproportions of body and legs began to disappear and to assume the present type of man. The largest heads are amongst the Masai, Karamojo, and Bahima and allied tribes, the smallest amongst the Acholi and lower tribes; the longest or dolichocephalic are the Lender (ind. 69), the shortest are the Sũk (ind. 84).

The custom of piercing the lower lobe of the ear and hanging a heavy ornament therein until it has expanded downwards to an enormous extent—a custom which is practised by the Andorobo and Masai—is also found amongst the Visayans of the Southern Philippines, the Lirong in Borneo, the Incas of Peru, and the Polynesians of Easter Island. It was a noteworthy custom in early Peru, and is kept up to-day by one of the forest tribes of the Peruvian Amazon region, the Orejones.

Amongst most Nilotic Negro tribes, names are not given to children when young, i.e. for the first few years. They are simply called a rat, a kid, or anything, so that the evil spirits may not be able to distinguish them individually and do them harm. This custom has been carried all over the world where we find their descendants at the present day.

Later in life, when four or five years old, he or she receives a name.

Flattening the heads used to be common amongst some tribes of the Nilotic Negro, but now the custom is dying out. It is still practised by the Klemantans of Borneo, and used to be practised amongst some tribes in North and South America.

Mr. Henry Savage Landor in his most interesting work, "Across Unknown South America," gives the following particulars, with regard to the Bororo Indians of Brazil, of one who came to visit his camp.[1]

He states: "I was greatly impressed by the strongly Australoid or Papuan nose he possessed; in other words, broad with the lower part forming a flattened, depressed, somewhat enlarged hook with heavy nostrils. In profile, his face was markedly convex, not concave as in Mongolian faces. Then the glabella or central boss in the supra-orbital region, the nose, the chin were prominent, the latter broad and well rounded. The cheek-bones with him and other types of his tribe were prominent forwards, but not unduly

[1] I have to thank Mr. H. S. Landor for his kind permission to use this extract from his valuable book.—A. C.

broad laterally, so that the face in front view was, roughly speaking, of a long oval, but inclined to be more angular—almost shield-shaped. The lips were medium-sized and, firmly closed, such as in more civilized people would denote great determination. His ears were covered up by long, jet-black hair, perfectly straight, and somewhat coarse in texture, healthy-looking and uniformly distributed on the scalp. The hair was cut straight horizontally, high upon the forehead, which thus showed a considerable slant backward from the brow to the base of the hair. A small pigtail hung behind the head. The eyes, close to the nose and of a shiny dark brown, had their long axis nearly in one horizontal plane. They were set rather far back, were well cut, with thick upper eyelids, and placed somewhat high against the brow ridges, so as to leave little room for exposure of the upper lid when open.

"The Indians wore no clothing, except a tight conical collar of orange-coloured fibre encircling the genital organs.

"I noticed two distinct types amongst the Bororos, one purely Papuan or Polynesian, the other strongly Malay. The majority were, perhaps, of the Malay type. Amongst the latter, many had the typical Malay eye *à fleur de tête*, prominent, almond-shaped, and slightly slanting at the outer angle. The nose, unlike that of the Papuan types, was flattened in its upper region between the eyes, and somewhat button-like and turned up at the lower part. The lips were in no case unduly prominent or thick. They were almost invariably kept tightly closed. The form of the palate was highly curious from an anthropological point of view. It was almost rectangular, the angles of the front part being slightly wider than a right angle. The front teeth were of great beauty, and *were not set, as in most jaws, in a straight line, the incisors being almost absolutely vertical, and meeting the side teeth at an angle of about 60°. The upper teeth overlapped the lower ones.* The chin was well developed, square, and flattened in the Papuan types, but receding, flat, and small in the Malay. Both types were absolutely hairless on face and body, which was partly natural and partly due to the tribal custom of pulling out carefully one by one each hair they possessed on upper lip and body.

"The ears almost invariably showed mean, under-developed lobes, otherwise well shaped.

"They displayed powerful chests, and the ribs were well covered with flesh and muscle. Their skins were dark yellow.

"The feet of the Malay types were generally stumpy; the Papuan types, on the contrary, had abnormally long toes and elongated feet, rather flattened. They used their toes almost as much as their fingers. The anatomical detail of the body was perfectly balanced. Their arms were powerful, with fine, well-formed wrists, graceful hands, long-fingered. The legs were well modelled, with small ankles.

"The Bororos divided themselves into two separate families, the Bororo Cerados and the Bororo Tugaregghi. The first claim descent from Baccoron, the second from Ithibori. The descent was from the female or mother, i.e. the children took the name of the mother.

"Eight days after birth, they perforated the lower lip of male children and inserted a pendant. The lobes of the ears were perforated at the age of ten or twelve."

They are Hero-Cult people, having the two brothers Ceriado and Ittary, and believe in a Supreme Being, called Marebba, who had a powerful son. They also believe in a wicked spirit called Boppe's Magic, in good and evil ancestral spirits and in elemental spirits.

They make basket-work and fishing nets. They have the Bull-roarers, which they call "ajie," and which Mr. Savage Landor translates "hippopotamus." This magical instrument here has the same meaning and is used for the same purposes as amongst the Nilotic Negroes in Africa, the aboriginal Australians, and other tribes.

They are nomads, and show great skill with their bows and arrows.

Mr. Savage Landor is of opinion that there are only a few thousands, perhaps hundreds, of the wild Indian tribes left in Brazil, and not 15 or 20 millions, as the Brazilians think and state.

Some of the words in the Bororo language are similar to the pure old Egyptian words, and the plural was formed as in the old Egyptian. They have rattling gourds filled with pebbles, for the purpose of calling members of the tribe, just the same as we find in some parts of Asia amongst the Mongolian tribes. (I have one of the latter made out of a human skull brought from Tibet.)

Their legends can all be interpreted through the Egyptian, but not otherwise.

They firmly believe that formerly the world was peopled by monkeys; the same legend is found amongst the tribes of the Philippine Islands, the Zulu Archipelago, and along the coast of the Eastern Asiatic continent. They say that the monkeys were their ancestors and they learnt from them how to make fire. Monkeys made canoes too. The whole world was peopled by monkeys in those days. The Bororos claimed that men and women did not come from monkeys, but that once upon a time monkeys were human and could speak. They lived in huts and slept in hammocks, therefore Pigmies. If you change the word monkey into pigmy, the hairy race, we have here a real tradition of the human race.

The Bororos called themselves "orari nogu doghe" or "people who lived where the pintado fish was to be found." They spoke of the Ra rai doghe, the long-legged people, now extinct, and two other tribes. Mr. Savage Landor has mixed up their beliefs regarding the Elemental Spirits with their belief in the Spirits of their Ancestors.

No one believes in the transmigration of the soul into animals, except a few Christians. These people knew better. The fact that they never ate deer or jaguar proves that these were the two original Mother Totems.

I do not think that the measurements of the head can be considered important in comparison with African Nilotic Negroes, as the practice of deforming the head is prevalent here.

They make use of an acute whistle for long-distance signalling, of which the Bororos have a regular code, the same as an African tribe at the present day. When a Boro approached Mr. Savage Landor, he deposited his bow and arrows at the foot of a tree and crossed his arms, the *left hand under the right arm*, and the right hand laid flat on the abdomen. This is a variant of some other tribes.

Mr. Landor says: "It is quite enough to visit the central plateau of Brazil to be persuaded that that continent had never been submerged under the sea! On the contrary, it must have been the oven of the world." The volcanic activity which must have taken place in that part of the world (it was not a separate continent or island, as generally believed) was quite, as I have stated elsewhere, beyond human conception, the hottest of any volcanic activity on land which had never been submerged.

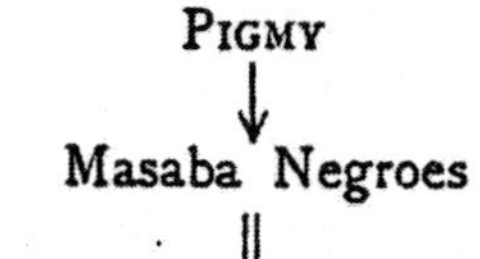

||

The Banati, The Banchi, Muyamwezi

↓

Nilotic Negroes

||

No Hero-Cult.

The Sandaiwi
The Landu
The Bongo
The Aluru
The Bambuba
The Mofu
The Bahuka
The Acholi and others

Having Hero-Cult.

The Dinka
The Shulluk
The Ja-Luo

The Abanyoro
The Madi
The Barotsi
The Laluka
The Nandi
The Nyam-Nyam
Kavirondo Tribes
The Suk-Turkans
The Nyemprias
The Wamia
The Masai Group
The Jieng
The Muhima

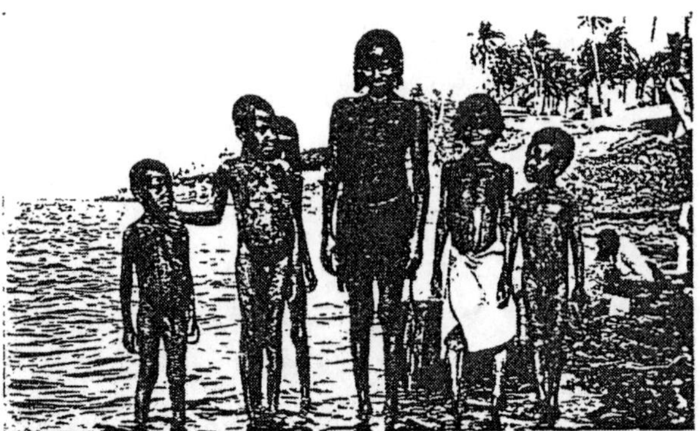

Nilotic Negroes, showing peculiar hair, long upper lip in children, and Pigmy abdomen in the children and long forearms. Males naked, females "clothed."

Woman with shaven head spinning cotton, Blue Nile.

Jurs making a grain-bin, Bahr-el-Ghazal.

PLATE XLIX.

Plate L.

Figure on the right quite a common type found portrayed amongst ancient Egyptians. Notice scars on breast and plaited hair.

CHAPTER XI

FURTHER PROOFS THAT THE NILOTIC NEGRO WAS THE FOUNDER OF ANCIENT EGYPT

Now let us turn for further proofs of my contention and see what Professor Maspero states on the earliest Egyptians, and compare this with the foregoing and with Sir H. Johnston's description ("The Uganda Protectorate").

Professor Maspero in "The Dawn of Civilization," states: "Amongst the earliest Egyptians *Maternal descent was the only one acknowledged and the child took the name of the mother alone*," and that "the Egyptians even in later times had not forgotten the ties of common origin which linked them to their forefathers, the barbarous tribes."

The marriage and other customs mentioned by Professor Maspero ("Dawn of Civilization," pp. 51–52) of the Old Egyptians are the same as those found amongst the Masai and Sūk at the present day, as well as amongst many tribes in America.

These also have similar weapons and tools. The men went naked except for a skin thrown over the shoulder. The women painted and tattooed the body and eyes, as did the Masai.

In most of the bas-reliefs of the Temples of Philæ and Kom Ombo, most of the women have their breasts scored with long incisions, which, starting from the circumference, unite in the centre around the nipple. In other cases women are seen tattooed (Maspero, "Études de Mythologie et d'Archéologie Egyptiennes," vol. i, p. 218).

Speaking of their houses, Professor Maspero states: "Their houses were even like those of the Fellahîn of to-day, low huts of wattle daubed with puddled clay, or of bricks dried in the sun. *Being oblong or square, they contain only one room. The door is the only aperture, which is shown by the ideographic signs ▭ ▯ and their variants, which from the earliest times have served to represent the idea of a house or habitation in general in the current writing.*"

There is an earlier type of ideograph for a house than this, however, viz. ⋂ ⋂, which represents the house of the agricultural Masai, as the other (*supra*) represents the house of the pastoral Masai. These pastoral Masai being the first to build houses in this form (*supra*), the above ideograph only came into being at this time; it could not have come before, although the second (*supra*) could and probably did.

Another ideographic hieroglyphic, the head-rest as a determinative to verbs expressing the idea of "bearing" or "carrying" in the text of the ancient empire, shows conclusively the great antiquity of its use. Thus we see representative objects used by the original inhabitants of Ancient Egypt, proving a record of customs, and still preserved to us, those objects critically demonstrating my claim to the "Masai group" being the founders of Ancient Egypt—the objects themselves being of the Masai group origin.

IDENTICAL.

Masai, Sūk, Turkana, and Ancient Egyptian.

Hair.—They plaited, curled, oiled, and plastered with grease, forming it into queues and erections of tails. If the hair was too short, they wore skull-caps as the Kamsai do.

The warriors wore ostrich feathers like the Masai.

Houses.—Were the same (see *suprà*).

Marriage Customs.—Were the same.

Men.—Men went about naked, except the nobles, e.g. the chief, who wore a panther skin thrown over the shoulders.

This became a symbol of the Chief High Priest as well as the chief of the tribe, the animal's tail touching the heels behind. (Note that this is what the Pigmy puts on during his sacred dances.) Some, however, were made of skins or hides.

Women.—The women painted and tattooed around the eyes like the Masai (*supra*).

They first wore a loin cloth, afterwards enlarged and lengthened until it reached below the knee, with bands over the shoulders.

Feet.—Not always covered; sometimes sandals of coarse leather, etc., were used.

Both men and women adorned themselves with ornaments, covering the neck, breast, arms, wrist, and ankles with many rows of necklaces and beads, made of strings of pierced shells interspersed with seeds, pebbles, etc. as shown in the photograph of a Masai woman (see Plate XLVII). The women of the Admiralty Isles shave their heads and have the same anatomical features as these Masai women.

Weapons.—Were the same in each case, and here again we find the origin of two more ideographic Egyptian hieroglyphics, viz. , the wooden club of the Masai and , the white mace with head of white stone (quartz) of the Masai.

Also compare the axe of Thothmes III. 

It is identical with the Masai's axe of copper hafted into a wood handle.

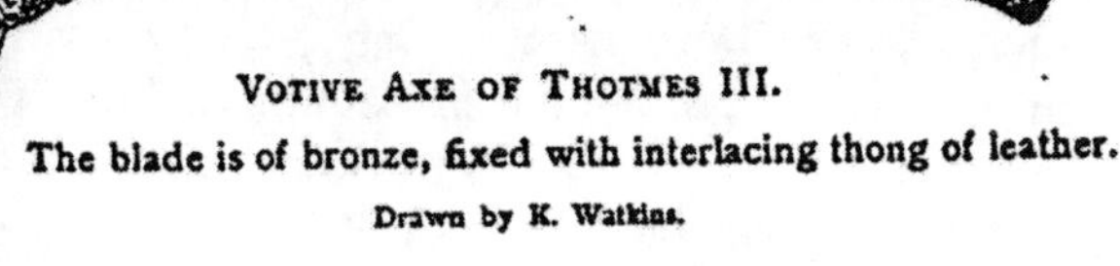

VOTIVE AXE OF THOTMES III.

The blade is of bronze, fixed with interlacing thong of leather.

Drawn by K. Watkins.

In several museums, notably at Leyden, we find Egyptian axes of stone, particularly of serpentine, both rough and polished (Chabas, "Études sur l'Antiquité historique," 2nd edition, pp. 381–82).

Clubs of wood were used, as amongst the Primitive African races. The bone of an animal served as a club. This is proved by the shape of the object held in the hand in the sign . The hieroglyphic, which is the determinative in writing of all ideas of violence or brute

WARRIOR WITH BOOMERANG AND BOW, FROM THE TOMB OF KHNUMHOTPÛ AT BENI-HASAN.

Drawn by K. Watkins.

force, comes down to us from a time when the principal weapon was the club, or a bone serving as a club (see Maspero "Notes au jour le jour," § 5, in the *Proceedings* of the Biblical Archæological Society, 1890-91, vol. xiii, pp. 310–11). (For hieroglyphics, see Maspero, "Dawn of Civilization," p. 58.) The lasso and the bola, which we still find in South America, were both used in Egypt, and are represented in hunting scenes, both in the Memphite and the Theban period.

For the games bola and lasso hunting, cf. Maspero, "Notes au jour le jour," §§ 4 and 9, in the *Proceedings* of the Biblical Archæological Society, 1890–91, vol. xii, pp. 310 and 427–29.

The boomerang is still used by many tribes of Nilotic and other Negroes.

The Nilotic Negroes worked in iron and copper, and continue to do so.

Maspero states (" Les Forgerons d'Horus en les Études de Mythologie," vol. ii, p. 313 et seq.) :—

" Metals were introduced into Egypt in very ancient times, since the class of blacksmiths is associated with the worship of Horus of Edfu, and appears in the account of the mythical wars of that God." " The earliest tools we possess in copper or bronze date from the Fourth Dynasty" (Gladstone, on " Metallic Copper, Tin, and Antimony from Ancient Egypt," in the *Proceedings* of the Biblical Archæological Society, 1891–92, pp. 223–26). Pieces of iron have been found from time to time in the masonry of the Great Pyramid (Vyse, " Pyramid of Gizeh," vol. i, pp. 275–76). M. Montelius has again and again contested the authenticity of these discoveries, and he thinks the iron was not known in Egypt till a much later period (" L'Age du Bronze en Egypt," in the *Anthropologie*, vol. i, p. 30 et seq.).

But M. Montelius was wrong; not only iron, but copper, tin, and antimony were known to these ancient Egyptians from the earliest times of their Totemic Sociology. thousands of years before the Stellar Mythos, at which period of time the Pyramids of Gizeh were built.

Bronze only appears in the First Dynasty of Babel, about 2250 B.C., whilst in Egypt laminated bronze is found in 4000 B.C.

Maspero is right in associating the blacksmiths with Horus of Edfu.

As I have stated (*supra*), the Kavirondo Nilotic Negroes work in iron, and also in copper, and amongst these people their blacksmiths are called Yothetth. There is a separate caste called " Uvino," and amongst the Gemi tribe the Blacksmiths were formed into a religious secret society, and still possess all the myths of Horus of Edfu. Horus I was the great chief in their Hero Cult, and " the Chief Artificer in Metals," i.e. he was recognized as the Chief

Hero of this clan or secret society. In Freemasonry another name is substituted, which is an innovation from the true original—Horus of Edfu—but many of their secrets even at the present time exist, not only in Africa amongst these people, but have been handed down from generation to generation by the old Turanians to the present-day Freemasons, who are quite ignorant of their origin; but these secret societies exist at the present day amongst the Copts, and an "Operative Freemason" can enter, and is recognized by the society as a "brother," the signs, symbols, and password being identical.

It was these "blacksmiths"—men who knew how to smelt iron ore and to forge the metal into weapons of offence and defence—who formed themselves into the "big clan of Blacksmiths," having Horus as their astronomical Chief, that came up from the South to the North in pre-dynastic times, and, having conquered the Masaba Negroes and the lower types of Nilotic Negroes, who were then the inhabitants of Egypt, established themselves in Egypt, making Edfu their chief city and centre. They possessed the knowledge of working in metals, brick-making, and pottery. Besides being of a superior Homo type anatomically, they could not but meet with success when warring against the inferior, because they were armed with superior weapons; troops armed with weapons of iron must be successful against those armed with weapons of flint. The Egyptians called these "followers of Horus" Mesnitu or Mesniti, which I believe was the original name for all those tribes, and which may now be applied to the Masai group. As we know, Horus was their deified God, and as Edfu became their centre, he was styled "Lord of the forge city," *the Great Master Blacksmith.* It was here that they first built a sanctuary or temple which was called "Mesnet" . The hieroglyphic here proves that these people were those belonging to the Masai ancestors. Priests were appointed to attend to the temple. The earliest legends are similar to those still found amongst the Masai group—in later times other legendary items sprang up as they advanced in their Astro-Mythology, but the great fight with Sut when he changed himself into a serpent which loudly hissed, and sought a hole for himself in the ground wherein he hid himself and lived, etc., was already established. This was the fight between Light and Darkness.

Dr. E. A. Wallis-Budge, in "Gods of the Egyptians," p. 485, states: "It is, of course, impossible to say who were the blacksmiths that swept over Egypt from South to North, or where they came from," but believes that they represent the invaders in pre-dynastic times, who made their way into Egypt from a country in the East, by way of the Red Sea, and by some road across the eastern desert, e.g. *that through the Wadi Hammamat.* They brought with them the knowledge of working in metals and of brick-making, and having conquered the indigenous peoples in the South, i.e. those round about Edfu, they made that city the centre of their civilization, and then proceeded to conquer and occupy other sites and to establish sanctuaries for their God or Gods." I contend that the evidence I bring forward is absolute proof that these people did not come from Asia or anywhere else, but from the South, and that these were the ancestors of the present Masai group, who had risen by evolution here in Africa from the lowest type of man.

Gladstone is also wrong in stating that the earliest tools in copper and tin date only from the Fourth Dynasty. Both copper and tin were worked by the early Stellar Mythos people, as proved by finds in their old buildings, and the Stellar Mythos people date some thousands of years before the Fourth Dynasty, as will be proved in this work. The Great Pyramid was built by these people (see later and in "Signs and Symbols of Primordial Man"), and the Masai tribes worked in copper still earlier. This is proved also by discoveries in South America, where we find that the art of hardening copper existed, as it did in Old Egypt. It is a lost art which I have rediscovered; I can now harden copper and make a similar metal to that found amongst the Incas and these old Egyptians.

Maspero also states that "Old Egypt was divided up

into what the Greeks have termed 'nomes,' but the most ancient native name is 'Nuit' or 'Hait,' which may be translated 'domain,'" and that "one thing alone remains stable among them in the midst of many revolutions, and prevented them from using their individuality and from coalescing into a community. This was the belief in and worship of one particular deity."

In the primary phase of the formation of the first nomes we have Totemic Sociology with Hero-Cult fully developed and a head-man elected as Chief.

Each nome represented a Tribal Totem, and the first nome of all represented Set, typified in one form as a Serpent. When Horus as Tem superseded Set as primary God, he appropriated the Zootype for Set. Another Zootype was the Crocodile. The serpent is still the sacred Totem Symbol of the Masai and the Dinka.

The First Nome was the nome named after Set or Sut, because he was the first born from their Great Mother Apt, the divinized Ta Urt, and became afterwards the Father of the Gods (see Ritual). He had not at this time "fallen," i.e. he was not yet represented as the Great Apapi.

The Second Nome was represented as, or was called, the Nome of Horus I, because he was the second born of the Great Mother, and the third, the Nome of Shu. And so on until the primary seven were formed; the representation in the primary form was by Zootype.

This was the time of the Totemic Sociology and early Stellar Mythos, but in Monumental Times, Egypt had passed out of Totemic Sociology and the Egyptians were fully evolved Stellar Mythos people.

Ta Urt, the Great Mother Earth, was the first form of Isis, the earliest Mother.

This is proved by Maspero ("Dawn of Civilization," p. 99).

Isis of Buto denoted the black vegetable mould of the valley, the distinctive soil of Egypt, annually covered and fertilized by the inundation, which brought forth their food of life—as the inundation was their water of life—and it was here that she was called *Mirit*, the Earth Goddess, or Great Mother Earth (Maspero, "Dawn of Civilization," p. 38).

The Motherhood terminated with mythology in Egypt.

They all had one common belief in the One Great Spirit, also in the Elementary Spirits and the Spirits of their Ancestors—these were propitiated by all, and still are amongst Nilotic Negroes and their descendants, in whatever part of the world we find them, and the people of the moieties into which they were divided possessed too much in common to have any differences that would lead to the "extermination" of each other.

Moreover, as there were seven different Totemic Clans, each member of his clan must keep to the land allotted to that clan, otherwise he would be an outcast, and would be driven back to his clan's territory. If he had offended against the law of Tabu he would be killed. During the time of Totemic Sociology the seven different nomes were formed in Old Egypt, each nome having its own clan and land, representing the seven primary elemental powers in Zootype form: none of these could interfere with the other without breaking the law of Tabu, and it was not until the time of the Stellar Cult that this was changed.

Maspero's description of the Oldest Egyptians ("Dawn of Civilization," p. 47) is as follows:—

"The highest type of Egyptian was tall and slender, with something that was both proud and imperious in the carriage of his head and his whole bearing.

"He had wide and full shoulders, well marked and vigorous pectoral muscles, muscular arms, a long fine hand, slightly developed hips, and sinewy legs.

"The detail of the knee joint and the muscles of the calf are strongly marked beneath the shin; the long, thin, and low arched feet are flattened out at the extremities owing to the custom of going barefoot.

"The head is rather short, the face oval, the forehead somewhat retreating.

"The eyes are wide and fully opened, the cheekbones not too marked, the nose fairly prominent and either straight or aquiline.

"The mouth is long, the lips full and lightly ridged along the outline.

"The teeth are small, even, and well set; they are remarkably sound.

"The ears are on the head.

"At birth, the skin is white (yellow), but darkens in proportion as it is exposed to the sun.

"The hair was inclined to be wavy, and even to curl into little ringlets, *but never turning into the wool of the negro.*

"The beard was scanty, thick only upon the chin.

"*The lower type* was squat, dumpy, and heavy; chest and shoulders seem to be enlarged at the expense of the pelvis and hips to such an extent as to make the want of proportion between the upper and lower parts of the body startling and ungraceful.

"The skull is long, somewhat retreating and slightly flattened on the top. The features are coarse, small frænated eyes, a short nose, flanked by widely distended nostrils, cheeks round, a square chin, lips thick, but not curling.

"These were the two principal types, whose endless modifications are to be found on ancient monuments."

The first and highest type were the Muhima and Masai, which I have termed "Masai group"; it is only necessary to compare Maspero's description with Sir H. Johnston's, or to take the present remnant of the past, still left existing here in Africa, and place him beside the oldest sculptures found, with the other evidence we have brought forward. This is a proof of our contention. These higher types still exist and allow the lower type to live amongst them, as they did in the Earliest Egyptian times.

These lowest types are remnants left of the intermarriage of the Masaba and the first and second Nilotic Negroes; types of these are still extant in Africa. These intermarried again with some of the higher types.

That this lower type constituted the original inhabitants of these parts is thus proved. As the higher types were evolved, so they gradually became the predominant race here; and, at the time they had worked out and established their Hero-Cult, we find them "forming Egypt." Whilst some of them travelled South and became the Zulu and allied nations, others travelled North to Europe and Asia, and then across to America, Australia, and other places, where we still find them in the same state as when they left the old home, having been cut off from intercommunication and having made no progress in evolution. Those that remained in the "newly formed Egypt," as their country and home, steadily continued to progress and develop in every way. The "forming of Egypt" was during their Totemic Sociology. Egypt was gradually formed into Nomes or Domains—the first, Sut; the second, Horus; third, Shu; and so on, until the seven Nomes were formed. All were depicted by a Zootype, each representing one of the seven Primary Powers, divinized. These Zootypes were the Totems of the Nomes, and all families who occupied a Nome had this Zootype as their Ancestral Totem, which they were known by. It was Totemic Sociology that kept the peace and prevented all the people of one Nome taking that land which had primarily been allotted to that particular Totem and incorporating it with their own. It could not be done otherwise, because it was against their tribal laws. The land which was allotted to A belonged to all having A as a Totem, and the land belonging to B always belonged to those having B as a Totem. A could not go and take B's land, because by so doing he would not belong to A's Totem any more, he must become a B Totem; and as he reckoned his descent from A and could not change it, he would become an outcast. Consequently, no one at this stage could become king over all, and in fact never did at that time, although we find, in the later stage, that a headman ruled over each Nome and became a prince or ruler of that Nome, often taking the divine title of the totemic Zootype.

Later, as they progressed, they re-divided these seven Nomes into twelve, and from twelve up to thirty-six in the Solar Cult, mapping out Egypt into Nomes as the heavens were mapped out in their Astro-Mythology or Stellar Mythos. Thus it was that, when this state of Totemic Sociology ended and their Stellar Mythos was fully developed, so that these seven Nomes represented the Heptanomes as seven, a further evolution took place.

Maspero in "The Dawn of Civilization" gives a good

description of how these first Nomes were formed, but fails to understand the people who formed them, or how they further developed. Personally, I cannot but think he knows, but he has "spoiled himself for ever as an authority on Egyptology" by writing the above book. Probably only he and the Society for the Promotion of Christian Knowledge know the true reason why he wrote such things in this book, as I feel confident that the greater part of the work as set forth must be opposed to his true beliefs.

Maspero's argument that "nothing, or all but nothing, has come down to us from the primitive races of Egypt, and that we cannot with any certainty attribute to them the majority of flint weapons and implements which have been discovered in various places," and that "the Egyptians continued to use stone after other nations had begun to use metal," will not bear criticism, because I have shown that these Palæolithic people were the primary inhabitants—they are still found here in Africa, and their implements as well—and if some still "continued to use stone implements to a later period" that is no argument or proof. Let him go to some parts of Ireland, the West of England, and Scotland, and he will find how primitive the people are there even now—many have never seen a train—and yet this country is a great civilized land, or is supposed to be. As regards their using and working in metals, I have proved (*supra*) that these were first worked in Africa, the present Nilotic Negro, still existing here as he did a million years ago or more, using the same methods now as he did then. Implements found in the under strata at Nagada and Abydos are undoubted proofs of a Neolithic following a Palæolithic age. Maspero must take into account that intercommunication was far different then from what it is now, and yet in our own country, how primitive remain some of the inhabitants. (Further proof of this later.)

The Egyptians always claimed to be the most ancient of mankind, and stated that their forefathers had appeared upon the banks of the Nile even before the Creator had completed His work. This, when properly interpreted by Sign Language, is the true fact, because they came here as "Inapertwa beings," i.e. before Totems and Totemic ceremonies were established. One of their names for the Great Mother was "Nit" at Sais; the first-born, when as yet there had been no birth, i.e. before the divisions took place.

De Morgan's attempt to show, by the aid of Anthropology, that the prehistoric people of Egypt were different from the historical Egyptians (whom he wishes to prove to have come from Asia) is bad logic, because if he only argued from his own great discoveries and the records of the past found, he could only form one true conclusion, which must be that these Neolithic (Stellar) people were descended from an earlier race—Palæolithic (Nilotic Negroes), remains of which have been found here; and furthermore, the objects found show that they were a progressive people. Wiedemann's argument, that because their burial customs were different, therefore they must be a different race, will not be borne out by any one who studies this question seriously.

But, on the contrary, we find that these old people had, in fact, already commenced in a primitive way to embalm their dead, or in some way to preserve them.

This is plainly shown by Fouquet in his craniological examinations. He states that there exist in the skulls of the rude stone epoch in Egypt (Totemic Sociology period) deposits of bitumen mixed with cerebral substance, and this bitumen could not have been introduced by the nasal passages, the brain not having been removed, but only through the occipital foramen after the head had been cut off, and Professor Petrie tells us that this custom of cutting off the head was common.

In the Ritual, chapter 166, it states: "Thou art Horus the son of Hathor, the flame born of a flame, to whom his head has been restored after it had been cut off. Thy head will never be taken from thee henceforth. Thy head will never be carried away;" and again, in the Ritual, chapter 43, it states, "I am a Prince, the son of a Prince: a Flame, the son of a Flame, whose head is restored to him after it hath been cut off"; and in chapter 90: "O thou who choppest off heads and cuttest throats."

Dismemberment is still practised by many tribes in South America and other parts of the world where these old Nilotic

Negroes went, also we see from Dr. A. C. Haddon's work that they performed some kind of mummifying in the Torres Straits. We found the same in America.

No doubt the earliest form was that of preserving the bones and anointing them with red ochre, this being probably the first substance employed for preserving the bones of the dead and the sacred emblems, which were exhumed periodically, scraped, and re-anointed. This is still practised amongst the Australians, the Maoris, and other native tribes, and, as may be seen from the human bones in the British Mounds at Caithness, it was the custom in these islands at the time of the later Palæolithic Age.

From these Nilotic Negroes, which are still existing in Africa at the present day, the various types of progressive evolution can be traced throughout the world, and although widely distributed from Patagonia, in South America, to Australia, New Guinea, New Hebrides, Solomon Isles, and Asia, certain types are identical, and in each case their Totems and Totemic ceremonies are the same for each type, as well as their general anatomical features, allowing for time and environment and intermarriages.

They all have Totems and Totemic ceremonies.

In the primary exodus of these Nilotic Negroes, their tradition was that they were descended from the Alcheringa, and that they were converted or transformed from Inapertwa beings into men and women, *and this primary exodus of the first or earliest type of Nilotic Negroes took place before the commencement of the "formation of Egypt."* There can be no doubt that they peopled this land, and also all the North and East of Africa; and, from implements found in North Africa, there must have been a pretty dense population, some of which crossed over to Europe, Asia, etc.

In the primitive form, i.e. a first exodus, of these Nilotic Negroes, wherever found, they are broken up into Totemic groups and everything is socialistic; there is no Council or Government except the customs and traditions of the tribes, and whenever a large number of natives are met together to perform ceremonies, there are always heads of the local groups present, who are "the authorities" of these rites and ceremonies, generally the older men, but otherwise "are not constituted chiefs." As regards these head-men there are some important points to be noted. First, in the oldest, who have kept the original form of division into "moieties," there *is only one alatinja or single individual head-man*, because these have only one great ancestor, or Totem, who gave rise to numerous other spirit individuals. The term "moiety" is meant to express the two original exogamous [1] divisions from one original Totem; later, these were again divided into what are best expressed as "classes"; these again were later on divided into "sub-classes" (the terms "gens," "phratry," "clan," etc., are misleading). In these later groups of class and sub-classes we find several "head-men" connected with one Totem. It is these head-men, who are the authorities of the Totemic ceremonies, that meet together, discuss and determine the various rites that shall be performed, and also consult and determine the punishment of any one of the members who has committed a breach of the tribal law. Thus the original authority was "more concentrated" than the later, and the individual head-man of the first group possessed by himself as much power as three or four together of the later. But there is "no form of Council meetings" beyond the discussion of the head-men who are recognized authorities in these Totemic ceremonies, and the head-man of one series of Totemic groups always belongs to one moiety, and those of the other series to the other. In a case of fighting most of these men would take the lead.

No doubt one reason the humans had for dividing into "clans" or tribes was originally self-protection.

The Pigmies and Masaba Negroes in the jungles of Africa would find that food was easily procurable, and an individual and his family could live without fear of starvation. He could also protect himself and family against wild animals found there, but when he came out in the open country and migrated from his original home, and further evolution took place to the "Nilotic Negroes" and numbers greatly increased, he found it necessary to form tribes, and band together as "Totemic classes." Each Totemic tribe

[1] Endogamy = marrying a woman in the same group.
Exogamy = the practice of seeking a wife outside the group.

took a division to itself, and that division was known to the others by the Zootype of the tribe or clan which was allotted to it. When these migrated it was very important still to keep their formation against any inhospitable environment they found; individual men would run a great risk of being exterminated very soon if they were not banded together into one Totemic tribe. An accident or starvation would be the fate of a single individual.

The strength of the community therefore had to be maintained by the practice of hospitality, friendliness, and absence of jealousy in all circumstances. They were *tribal, not individual* benefits that must be observed.

The next point is—How many different exodes were there of the Nilotic Negro?

It is impossible to say, but two at least left Egypt before the first exodus of the Stellar Mythos people. This can be proved at the present day by what we find in various parts of the world, namely, a lower type and a higher type physiologically and anatomically—lower or less developed Totemic ceremonies without mythos, or Hero-Cult, and having no mythology and no folk-lore tales, and the higher type with Hero-Cult and mythology and possessing folk-lore tales.

There is a marked difference also in the Totems and Totemic ceremonies of these two classes in whatever part of the world found. The lower type have the primitive forms of Totems and Totemic ceremonies. The higher type have further advanced in their sacred rites; they have the commencement of the Astro-Mythology, and consequently divination of the Mother Totem, along with what Dr. Haddon has termed "Hero-Cult." Therefore, there must have been two exodes of these at least. As examples, these types are easily distinguishable in the two classes of Aborigines found in New Guinea, New Hebrides, Torres Strait Islands. And in Australia the Tjingilli, Binbinga, Umbaia, Urabunna, and Dieri have no Hero-Cult or mythos, whilst the Warramunga and some others have, as will be mentioned later. In South America the Warranau Indians have Totems and Totemic ceremonies, but

Showing large distension of ears and mode of carrying children.

Showing large distension of ears.

Method of fire-making.

PLATE LI.

INDIOS BOTOCUDOS, RIO BOCE.

PLATE LIA.

INDIOS COROADOS DE NOME DES ESTRADA NOROESTE DU BRAZIL.

Showing the nude state in which they live. These are descendants of one of the Hero-Cult tribes of Nilotic Negroes. I am much indebted to the Minister of Agriculture, Brazil, and Major Albert Levy, of São Paulo, Brazil, for all photographs of Brazilian natives produced in this work, and which I understand have never been published before. I tender my sincere thanks to these gentlemen.

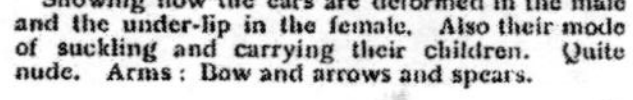

Showing how the ears are deformed in the male and the under-lip in the female. Also their mode of suckling and carrying their children. Quite nude. Arms: Bow and arrows and spears.

Showing the large abdomen in the children, which quite disappears in the adult, and a somewhat Japanese type of face in two of the children.

PLATE LIb.

INDIOS BOTOCUDOS, BRAZIL.

PLATE LIc.

These groups are photographs of the Botocudos, found in Brazil. Notice the deformed lower lip of the women and nude state. The lower photograph portrays one using the "nose flute," as found in Inner Africa, also the Tinguian in the Philippines.

no Hero-Cult and no mythology. They have long bodies, short limbs, dark skins, and are a low type of Nilotic Negro, like the Landu.

The higher type are the Arawakan and Bororo-Coroado, a higher type of Nilotic Negro with Totems and Totemic ceremonies, also with Hero-Cult and mythology.

The next higher type of these in South America is that of Monumental Type—the Nahuatlan, who are an offshoot of the Shoshoneans of North America.

Then the earliest Stellar Mythos people, the Chichimees, who were the earlier Zapotics, are followed by the Aztecs and Pipils, who are of the same type as those of the Plateau of Anahuac and Nicaragua of the present day.

The Huaxtecan were the ancestors of the Mayas, Quiches, and Pocomaus, the present Indian inhabitants of Guatemala and Yucatan. The latter have the Solar doctrines, and are of a different type anatomically, so that in America we find :—

1. The Pigmy race and type still living in the forests of Bolivia, and probably we shall find them in the more inaccessible forests of Brazil, or in some fastness in the mountain ranges. They have *no Totems or Totemic Ceremonies.* That these have existed in America for 500,000 years or more is proved by the skeleton found in the coal-bed in North America.

2. The Warranan and the people of Fuegia, the next exodus after the Pigmy, a low type of the Nilotic Negro, have Totems and Totemic ceremonies primitively developed. They have no mythos or folk-lore tales.

3. The Arawakan and the Bororo-Coroado, who are a higher type of the Nilotic Negro, anatomically and physiologically, having Totems and Totemic ceremonies fully developed or worked out by evolution with part of the mythology and folk-lore tales.

4. The Monumental types having Stellar Mythos fully evolved only.

5. The Monumental types having Solar Mythos fully evolved.

6. The various types of Indians and Aborigines found in South America. These are all easily traced to corre-

sponding ones in North America, and no doubt spread from North to South, some having their Totems and Totemic ceremonies only, others having folk-lore tales and mythos, and all originally tracing their descent from the maternal side, paternal descent being comparatively recent.

7. The Sianans, North America, and those of South America corresponding to them, use a sign and gesture language, as do the Aborigines of Australia, New Guinea, etc.

8. In all these classes in Africa, Australia, New Guinea, North and South America, (1) their "Sacred Customs," sacred signs and symbols are identical; (2) their "Marriage Laws" are the same with regard to Tabu; (3) their "Burial Customs" are the same; (4) the "Totems" correspond. The flattening of the head, as still found amongst some South American Indians, is a custom inherited from the Choctaws and Chicasas of North America, from whom these descended. At one time the early Egyptians practised this custom, and the New Britain islanders (Bismarck Archipelago) do so to this day. This practice prevails in Borneo also.

9. In Africa all these types still exist and are found at the present day, each type having certain anatomical features—Totems and Totemic ceremonies, folk-lore tales and social laws, fundamentally the same, as is the case with all those above, in whatever part of the world we find them.

Hero-Cult People of South America.

The Quimlaya have—

Abira, a creative deity	=	Horus.
Canicuba, an evil deity	=	Sut.
Dabeciba, a female deity	=	Apt.

They knock out two front teeth in the same way as the Nilotic Negroes.

The Carabs, for instance, have the same anatomical features, the same Totemic ceremonies and Totems, and the same custom of thrusting a piece of wood through the lower lip, as the Karamojo in Africa.

The tribes of Indians living on the upper reaches of the Amazon, who have that particular custom of expanding their lower lips and ears to a most ugly extent, are the same type and tribal race, and have the same Totems and Totemic ceremonies, as the present Bongo Nilotic Negroes, from whom no doubt they were an offshoot hundreds of thousands of years ago. These (both Bongo and the tribes of Indians in South America), by inserting wooden discs expand the lower lip until it has increased in size six or seven times its normal width. Anatomically and physiologically they are the same type of Homo.

The Indians round the South of the Panama district are of the same type anatomically, physiologically, and totemically as the Monfu Nilotic Negro of Africa, and have the Beetle as their most sacred Totem. A friend of mine (a doctor) was enabled to penetrate their forests and live amongst them for some time because I had given him their totemic sacred signs and told him to show them a beetle at the same time. They took great care that he should come to no harm whilst dwelling amongst them. Had he not shown these signs and the beetle, he would not have been allowed to come near any of them—in all probability he would have been killed. I shall refer again to this in greater detail in the Chapter on South America.

The Monfu.

The Monfu of Africa and the Indians of the Chaco—that enormous region of South America, removed from the Andes—have the same mythological beliefs and each have the Beetle for their Totem (a symbol of Egyptian creative power in one form).

The Badages al Kolagiri of the Nilgiris of Southern India are of the same "class" anatomically, and possess the same very extraordinary Totemic Ceremonies as a tribe in North America, one in the Polynesian Islands, and one in Inner Africa, i.e. the so-called "fire-walkers," whose ceremony is performed in all cases once a year, generally in February, and consists in spreading a layer of live charcoal 7 or 8 yards in length and 6 inches deep, over which the head priest and eight or nine chosen men walk and dance for a considerable time without even singeing a hair of their legs or feet, whilst at the same time the heat is so intense that a white man could not approach within several feet of the "fiery furnace." It has been stated that there is a tribe in South America precisely similar.

The antiquity of the descendants of the Nilotic Negro in America has been proved by the discovery of his remains in the Pleistocene era there. In a cave at Ultima Esperanza (Last Hope) in Patagonia, and a little to the North of Cape Blanco, implements of the Nilotic Negro of a Palæolithic character were discovered in Quaternary strata, whilst in the ground at some distance above were imbedded others of a Neolithic character (Stellar Cult people).

This will help to prove how long ago these people must have left Old Egypt.

The Boraro and other South American Indians use the Bull-Roarer just as the Australians and Nilotic Negroes do.

The Kobèna on the Upper Naupè are anthropophagous, also the Uitóto in the basin of the Icá, Putumayo, as regards their ceremonial rites towards the mother and towards enemies.

North of the Amazon many tribes have great drums like their African brothers, and these are beaten in a variety of ways, and may be called "telephones," because they can spread news in a rapid and astonishing manner. It is a kind of drum language.

The extermination of nearly all the aborigines of the West Indies has made a break in what was once a complete connection between the natives of the Northern and Southern parts of America.

The Cebunys of Cuba and the West Indian Caribs, and the Lucayans of the Bahamas, were some of the links between the more Northern tribes and the Caribs of the Guianas and the Arawakan group of Venezuela and the neighbouring districts.

Some of the South American Indians produce cicatrices like these Nilotic Negroes, not Hero-Cult people. Amongst the Arawak tribes the descent is always in the female line.

The Ackawois are characterized by their nudity.

The true Carib and Ackawois pierce one or more holes in their lower lip, through each of which they pass a pin or sharpened piece of wood. Their ears are often pierced and pieces of stick passed through.

Sandals of palm leaves are sometimes worn.

They tattoo their tribal marks on the corners of the mouth and on the arms.

The Botocudos have long, narrow-shaped heads; they are naked, of low type, and anthropophagous as regards their enemies.

The Ona Indians, Tierra del Fuego, are non-Hero-Cult people.

The Fuegians are about 6 feet high. Colour, dirty copper red; they are Hero-Cult people.

Their Totemic marks are important because we find the same depicted amongst the oldest Mexicans, viz. two bands across the face, one painted red, the other white, as seen in this figure.

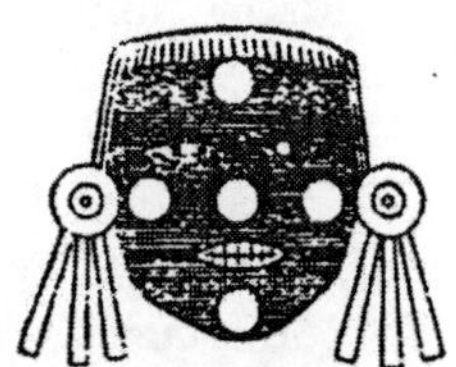

We see from this photograph that the Mexicans and Zapotecs painted their Totems across the face in earlier times, as above. This also has the Egyptian hieroglyph for daylight, splendour, etc. The five O denote the five mysteries or powers. The women of the *Inoit tribe* of the *Esquimaux* do the same. It is their Totemic sign and name, which is given them at the time of puberty, as before explained.

1. A red band reaching from ear to ear, including the upper lip.

2. A white band (chalky white) running above, parallel with and including the eyelids.

Their covering consists of a skin mantle only, thrown across the shoulders like the Masai, Sūk, and Unkana.

Their huts are made of a few boughs stuck into the ground and roughly thatched with grass and rushes.

Instruments Palæolithic.

On Plate LIc is shown the "Indios Botucudos" of Brazil playing the "nose-flute." The Tinguian of the Philippine Islands have the same instrument and customs.

These probably descended from the Bopoto of Africa, who have and play a similar "nose-flute" and have the same customs.

Among the Arawak tribes of South America the descent is always reckoned in the female line, and no intermarriage with relations on the maternal side is permitted. Amongst these tribes the curious custom of the *couvade* is universally prevalent, as it is amongst the Tibetan tribes. The body is also buried in the "thrice-bent position," and many possessions are buried with the body.

In writing of his experience in South America, Lieutenant-Colonel P. H. Fawcett states that "the Altiplanicie of Bolivia and Peru is singularly like the great plateaux of Tibet or Mongolia. The likeness extends to the customs of the Indians, and their short, sturdy appearance—and even goes farther. Piles of stones decorate the hilltops also. When an Indian woman presents her lord and master with a pledge of her affection, she is not permitted for a moment to imagine that any merit attaches to herself; on the contrary, the lord and master takes to his bed and groans in apparent pain and misery for four or five days, to show that he is really the sufferer, while the better-half carefully nurses and feeds both her lord and baby."

This is the custom also of the Indians of Central California.

The meaning of the custom of *couvade* can only be found in Egypt. It was a dramatic mode of affiliating to the father the offspring, which had previously derived its descent from the mother.

The man here impersonates the mother, because he acts as if in gestation with the child and pretends to undergo parturition. Thus he acts the part of the father and mother, both parents in one person.

It is in this sense that Sut, the first born of the Seven, is called in later times the Father of the Gods (see Ritual). In Akkad, or Babylonia, the group of Seven Males is divided into Ea, as a father with seven sons; it is represented likewise among the Zuni Indians, whose Zootype Deities are seven in number; six, with the form of God the Father as a supreme one as the seventh.

North American Indian believing in a "great spirit" which he called Wakanda or "a Wakander." He affirms that the term is applied to mythic monsters of the earth, air, and waters; according to some of the sages the ground or earth, the *mythic underworld*, the ideal upperworld, darkness, etc., were Wakanda. So, too, the fetish and ceremonial objects and decorations were "Wakanda" among different tribes. By some of the group various animals, and by others trees, besides the specially "Wakanda" cider, were regarded as Wakanda; the horse amongst the prairie tribes was Wakanda. He says the idea expressed by the term is indefinite and cannot be rendered into "spirit," much less the "Great Spirit." He finishes as follows: "Thus ends the myth crystallized into the English language by the poem 'Hiawatha'!"

But Mr. McGee did not understand Sign Language! Hence he has drowned himself in the slimy pit of ignorance. Wakanda in ideographic sign language means "Spirit"; with certain different gesture signs and intonations it means all the above. One gesture sign used with the articulate sound would mean Great Spirit; with a different gesture sign and the same sound, Great Spirit of the Underworld; with another, the Great Risen Spirit. These were and are their Zootypes representing the good and bad spirits.

Mighty spirits were supposed to dwell in certain trees by the Battas of Sumatra, who would resent any injury done to these. There was the same belief amongst the Tamils, as well as many other tribes, including many in Africa at the present time. Such mighty spirits or powers of the elements had grown up as in Egypt to become the goddesses and gods—as Hathor and Nut in the sycamore, Isis in the persea tree, and Seb in the shrubs and plants, Horus in the papyrus, or Unbu in the golden bough.

This I have proved. So much for theory without knowledge.

The mythical underworld here was the underworld before Amenta was formed by Ptah; therefore during Totemic Sociology or the Stellar Mythos. The Horse[1] here repre-

[1] No doubt this is a late adopted Zootype representation, as horses did not exist in America before Europeans brought them.

Plate LII.

Brazilians.

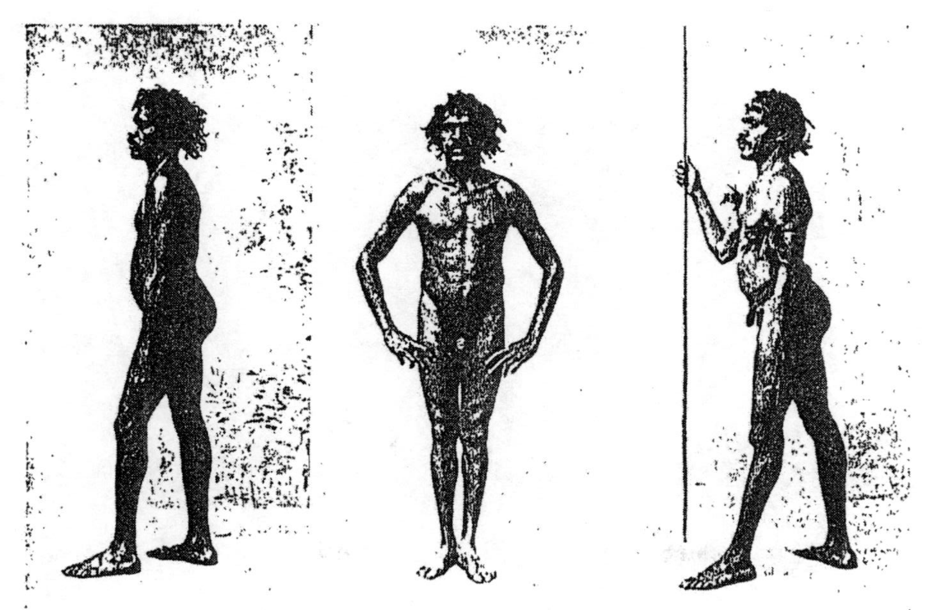

PLATE LIII.
West Australian Hero-Cult.

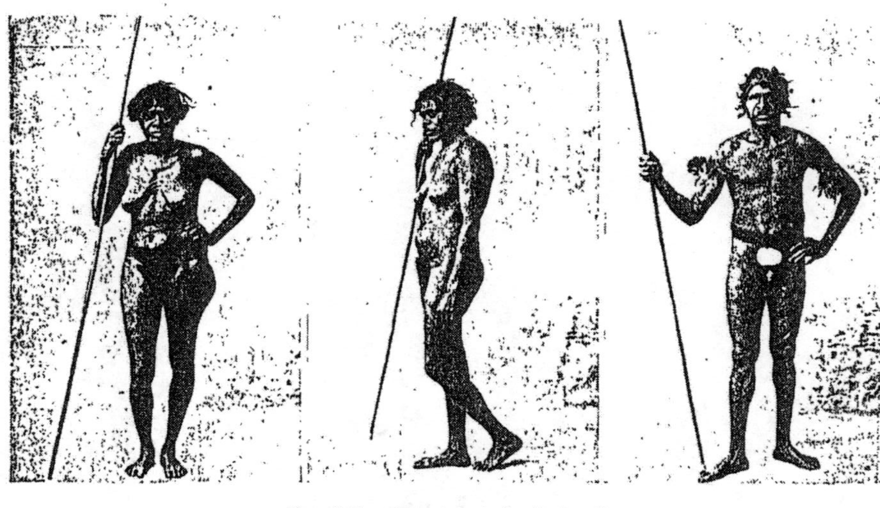

Hero-Cult natives of Australia, Coolgardie.

PLATE LIV.

Woman of Southern Cross, West Australia, showing the cicatrices made. Non Hero-Cult.
From J. B. Hardwick, Esq.

PLATE LV.

sents Anubis; with other tribes it is the Dog that is symbolical of Anubis, who was the "smeller out" or guide of the way of the underworld both in the Stellar and the Solar Mythos. He guided the departing souls through lower regions to Paradise, or to the "Great Spirit" in Paradise, situated at the Pole, represented by a house on the top of a mountain or a tree, etc.

Primitive and Palæolithic man was unable to image the superhuman powers of nature in the human likeness, although he has been credited with doing so by many, amongst others Professor Max Müller ("Science of Thought," p. 503), Herbert Spencer ("Data of Sociology," chap. xxiv, 184), Francis Bacon ("European History"), and David Hume.

But primitive man was incapable of doing it, he was too poor in possessions, and that is why myth-making man did not create the gods in his own image. The primary divinities were represented in Zootype form; for example, Sut, Sebek, and Shu, three of the earliest, were represented in the likeness of the Hippopotamus, Crocodile, and Lion, whilst Anup was imaged as the Jackal or Dog, Ptah as a Beetle, Taht as an Ibis, and Seb as the Goose.

These are the images of the superhuman, not human, powers. It is the same with the Goddesses. Old Mother Apt was imaged as the Water-Cow, Rannut as a Serpent, Hathor as a Fruit Tree, Hekat as a Frog, Serkh as a Scorpion.

Totemism was formulated with types that were the very opposite of human. That is why we find in mythology that the anthropomorphic representation was preceded by Zootypes. If primordial man had possessed the power to image human personality upon external phenomena, he would have represented the image of the male as a type of power; whereas the primal human personification is in the likeness of the female, "The Great Mother," as the primal parent in a universal type. Again, if he had been able to impose the human likeness on the Mother, Nature, the typical wet-nurse, would have been a woman. But it was not so; the woman comes last: she was preceded by the sow, the hippopotamus, the lioness, or the compound figure of the female with the head of a cat, frog, serpent, or cow on the body of the divinity, and it would of necessity have included sex. The earliest powers, however, recognized in nature are represented as being of no sex. It is said in the Akkadian hymns, "Female they are not; male they are not."

Sign or gesture language must be learnt and the mythology understood before an opinion can be formed such as Mr. McGee's. He must realize that they have no word to express by articulate sounds what he was seeking, and as he was ignorant of the Sign and Zootype language, he fell into error, as many others have done.

Another point is the intonation of the same word. Take Siamese, for instance, where you will find many words with the same phonetic sound, except the intonation; some have as many as four or five different intonations. Thus, you will have the same word representing and meaning four or five different things. Any one who did not understand the Siamese intonation would put them all down as the same.

South of the Rio Negro, in the Patagonian Plateau, wandered small groups of nomad hunters, a tall, round-headed race known as the Tehuelche, with Hero-Cult, corresponding to the Turkane of the Nilotic Negro; whilst, driven farther South, the Ona of Fuegia are a long-headed race of much lower type, having no Hero-Cult and making no pottery.

The burial customs amongst these descendants of the Nilotic Negroes here found in South America are the same as we found amongst the Ancient Egyptian Nilotic Negroes, the heads frequently having been cut off.

CHAPTER XII

NILOTIC NEGROES (*continued*)

IN "Signs and Symbols," p. 163, I have shown that the New Guinea natives have the same customs as the Nilotic Negroes. From all the information one can gather up to the present time, however, there must have been two different exodes of the New Guinea natives, as one moiety have Totems and Totemic ceremonies, but no Hero-Cult, and no mythology. The others have these, and the types, anatomically and physiologically, are of a higher and lower grade, similar to the Torres Straits natives. The New Guinea and Australian natives, many in Africa and South America, approach the camp of strangers with a green bough in their hand, as the sign of amity, as a prayer of peace and goodwill. In the "Ritual of Egypt," chap. 28, the speaker says of the God Osiris: "*His branch* is of prayer, by means of which I have made myself like him." Teru is the branch, and the same word signifies to adore, invoke, pray. It is an act of portraying, instead of saying, in sign language. They enacted the drama of mysteries of transformation in character, with masks of the Zootypes representing their requirements. "A prayer meeting asking to increase the food" to the Great Spirit—Mother Earth.

An interesting Pig ceremony is described by a writer in "The Customs of the World." He states: "All the men of the village assembled by this shrine, and there a pig was strangled by men chosen by the chief sacrificer. The body was placed in a bowl and then cut up, the purpose of the bowl being the collection of the blood, and avoidance of its flow upon the ground. This being done, the chief sacrificer took a piece of the pig's flesh and ladled some of the blood out of the bowl with a scoop of cocoanut shell. He then entered the shrine, carrying the piece of flesh and the blood-filled scoop, first putting away his bag and washing his hands, lest the ghost should reject him with disgust, and called out: 'Harmunae! Chief of war! We sacrifice to you with this pig that you may help us to smite that place; and whatsoever we shall carry away shall be your property, and we also will be yours.' Then he burnt the piece of flesh in a fire upon a stone in the shrine and poured the blood upon the fire. The fire blazed upwards and the shrine was full of the smell of burnt pig, a sign that the ghost had heard. The remaining part of the pig was afterwards eaten. The ceremony just described was for the benefit of all the people's success in approaching battle." These people, who inhabit the island of San Cristoval in the Solomon group, are here shown to be Hero-Cult people with the pig as their great Mother Totem.

"Harmunae" here represents one of these three Heroes, the God of War. He was not a ghost nor did they think him a ghost, he was a "Great Spirit," *but not a Spirit risen from one who had lived.* It is the same as one of the early stories of the Masai. They propitiated, sacrificed to their Mother Totem, and supplicated her as a great superhuman power (not supernatural as our friend states) to help them in their coming battle.

The Pig being their Mother Totem, she was solemnly eaten, and Harmunae, as her child—elementary power—was given some of the sacrificial feast. All must be consumed and no blood must be spilt; that was the most important part of the life, and so all "must be consumed." A time had been in the far-away past when the Mother of the Clan would have been eaten all alive, but now they are not anthropophagous, and it is only the symbolic Mother that is sacrificed and eaten.

If it had been a sacrifice to a ghost, food of other kinds would have been offered, but then not by all the clan; perhaps one of the chief warriors might offer some propitiation to a departed one of his own family. When all do so, this proves that the sacrifice was to one of their

Heroes—and these had never been men, but elemental Spirits divinized (as herein fully described). Horus, in one form, was the God of battle.

The writers in "The Customs of the World" have mixed up one class of Spirits with another, and do not understand the Totemic Sociology or Sign Language that they have portrayed in this work. The pity of it! How futile for the publishers to blast the trumpet "Britain leads the world"! Where? Into a pit of "thick fog and ignorance" by her present professors in this branch of science.

In New Guinea there are men said to have the power of divination, information being usually conveyed to the people by a Spirit speaking through the mouth of the Seer, who, meanwhile, is unconscious. It is the same with the Nilotic Negroes at the present time.

Omens.—The belief in omens is widely spread throughout Melanesia. Among the Mafula Mountain people the appearance of a flying-fox or firefly would be a bad omen which would cause a hunting or fishing party to turn back. But if a "Koita party" were going fishing, it would be lucky for a flying-fish to leap into the canoe. The Koita, in some of these superstitions, distinguish between the left and right as regards good and bad omens.

The photograph depicting these Totemic ceremonies on p. 90, "Customs of the World," which is not correctly explained, means that the "white face" and white paint around the masks only, with the central white wand and Serpent above, proves that it is the dance of "transformation," the change from the death of the Corpus into the Spirit. The "white ball" on the top, with the heads of the Serpent also above, represents symbolically where the Spirits go.

The natives have forgotten what these marks represent, although they still dance the old dance.

There are two classes of tradition derived from Totemism concerning the descent of the human race amongst these Nilotic Negroes in whatever part of the world they may be found, whether in the first exodus, which has no mythology nor Hero-Cult, or in a later exodus, which has some part of the mythology and Hero-Cult, and therefore folk-lore tales.

1. In one, human beings were derived from the totemic animals, birds or snakes (the Haidohs, in Queen Charlotte Sound, claim descent from the crow), and one here in New Guinea from a dog.

2. According to the other tradition, the totemic Zootype are said to have been brought forth by the human Mother. The Bakalai tribes of Africa told Du Chaillu that their women gave birth to the totemic animals—one brought forth a Snake, another a Crocodile, another a Calf, etc. (Du Chaillu's "Explorations and Adventures in Equatorial Africa," p. 308).

The Moqui Indians affirm that the people of the Snake Class are descended from a woman who gave birth to Snakes (Bourke, "Snake Dance of the Moqui," p. 177). This proves that the primary Totems were representatives of the Mother, wherever the alleged descent of the totemic animals was from human originals, which were female. The Mother being primary, it follows that the earliest human group was a Motherhood. This is still extant in the Oraon Maharis, which are the Motherhoods by name (Dalton, "Ethnology of Bengal"). There being no father known, the descent was in the female line from the Mother to the eldest daughter. These are the two women in Totemism, the two Mothers in the mythology. The Ainu of Japan say that their eldest ancestor was suckled by a Bear. The Bear here was the Mother Totem, i.e. their totemic Mother was a She-Bear, and the fact is still kept in memory when the Ainu women suckle the young bear, which is invested with a necklace and adorned with ear-rings like a woman, then killed and solemnly eaten at the annual festival. The same Totemic ceremonies are carried out by the New Guinea natives at the present day; but in this case a Sow takes the place of the Bear of the Ainu.

It is the same with the Snake Class of Africa and of Arizona, who claim descent from a woman who gave birth to Snakes. She was the Mother of that Totem, and the Snakes were her children. The same with the Girura natives, only it is a dog in this instance.

In Egyptian mythology, we have the Mother Earth in the form of the Goddess Rannut represented by a Snake as the renewer of life and vegetation. This, in the beginning, was a type of the Great Mother, because the first person recognized in Totemism was the Mother, and Totemism in Egypt was the basis of all its mythology, and all clans with the Snake for their Mother Totem took their origin in old Egypt from this.

The Indians who claim their descent from the Spirit Mother and a Grizzly Bear acknowledge, like the Ainu, that it was a She-bear, consequently a Mother Totem.

The Tugas claim descent from the She-Wolf.

The Tufans claim descent from the She-Dog.

A tribe of Indians still living in North-Western America claim descent from a Frog. The Frog here was the "Totemic Mother," or Mother Totem; but the Frog was also represented in the mythology as a type of the renewals of time, and in Lunar Cult the renewals of the Moon, or Light for æons of years. This, in Egyptian, would be descent from the Goddess Hekat, who typified the Divine Mother in the transforming Moon. The divine Cow of the Todas is an extant type of the "Great Mother" as the giver of food (in mythology). It is represented in the Egyptian, and is equivalent to Hathor, the Egyptian Venus, the Cow that protected her son with her body, primarily when the Mother was a Water-Cow.

Mr. Beaver states that the Girura natives of New Guinea claim descent from a Dog, and that they possess five Totems which are drawn on all their houses. This means that their "original" or "primary Mother Totem" was a Dog, and that all her children were originally "Dogs," who are now divided into five other Totems; the names of these he does not state. These may be termed sub-classes.

He further states that their houses are from 400 to 500 feet long and from 60 to 80 feet wide. The centre of this huge building is a kind of common hall which is used for the men only, while the walls of the structure are divided into cubicles in three or four floors, access to which is gained by a ladder. The women are not allowed to enter by the same entrance as the men.

The chief claims to be able to separate his Spirit from his body and send it on various missions.

Amongst some tribes great respect is paid to snakes, none of which are allowed to be killed. We find this also amongst some of the Nilotic Negroes (Dinkas and Masai) at the present day.

In the Theban Khonson, a statue of granite, we see the same style of wearing the beard as is practised by the Giruras of New Guinea, i.e. twisting the beard into a kind of rope so that it hangs down long in one mass. These New Guinea natives are head-hunters, but not cannibals.

The probable reason why we find various tribes remaining so distinct in character, as we do in Australia, New Guinea, and other places, is on account of the original territorial disposition of the tribes. The strictness with which each tribe was compelled formerly to keep within its own boundaries may well have prevented anything like circulation over the continent for a great length of time.

The Hippopotamus was a Mother Totem with the natives of the Zambesi. Livingstone's pilot would go without food rather than cook it in the pot which had contained any of the meat (Livingstone, "Zambesi").

Herodotus tells us that the first Mother of the Scyths was a Serpent woman. The king of Abyssinia's line of descent was traced from the Serpent—in both cases the Serpent is the Mother Totem.

The Vulture in Ashanti is the same sign of royalty as with the Egyptians. Ellis states that the Coomassie Vultures are considered birds sacred to the royal family, or else, as in Egypt, of royal and divine maternity. Any molestation of this bird was punishable with death (Ellis, "Tshi-speaking People," p. 213). It is a Mother Totem like the Vulture Neith, which was both royal and divine as the Bird of Blood—the Mother Blood of the royal family.

An Australian tribe considered themselves to have been Ducks, who at one time were changed into men. In this case the Duck would have been the Totem of the Mother as a means of tracing their descent in the female line.

The tribes of North America are divided into totemic divisions, and the Crow tribe has two interesting "signs,"

PLATE LVI.

FOUR PIGMIES FROM SADDLE PEAK RANGE AND TWO PAPUANS.

These Papuans are descendants of "Nilotic Negroes," originally of the Dinka or Shulluks. Here is a proof of my contention that the earliest Man may exist in the same land and at the same time that another Homo exists there, who was born thousands of years later and is far advanced along the scale of progressive evolution.

From *Country Life*, March 4, 1911.

tokens of friendship and brotherhood. If these are given and answered by strangers, they are safe amongst them.

1. Crossing the arms on the breast.

2. Raising the right hand to the side of the head with the index finger pointing to the great Spirit and then reversing.

It is said by the Amazulu that when old women pass away they take the form of a kind of lizard. This can only be interpreted by knowing the ideographic value in the primitive system of Sign Language in which the lizard was a Zootype. The lizard appeared at puberty (the Totem was given to her at that time), but it disappeared at the turn of life, and with the old woman went the disappearing lizard.

Miss Werner reports a specimen of primitive thought and a mode of expression in perfect survival. It happened that a native girl at Blantyre Mission was called by her mistress to come and take charge of her baby. Her reply was, "Irchafuleni is not there, she is turned into a frog." She could not come for a reason of Tabu, but said so typically, in the language of animals. She had made the transformation which first occurs when the young girl changes into a woman, and figuratively becomes a frog for a few days of seclusion.

The Cawichan Tribes say the Moon has a frog in it, and with the Selish Indians of North-West America, the Frog (or Toad) in the Moon is equivalent to our Man in the Moon. They have a tradition that the devouring Wolf, being in love with the Frog (or Toad) pursued her with great ardour, and had nearly caught her when she made a desperate leap and landed safely in the Moon, where she has remained to this day (Wilson, *Transactions* of Eth. Society, 1866, New Series, vol. iv, p. 304). This means that the Frog as a type of transformation was applied to the changing Moon as well as to the Zulu girl, Irchafuleni.

It is related in a Chinese legend that the lady, Mrs. Chang-ngo, obtained "the drug of Immortality" by stealing it from Si-Wang-Nu, the Royal Mother of the West. With this she fled to the Moon and was changed into a Frog, which is still to be seen on the surface of the Moon

(Denny's "Folk Lore of China," p. 117). In Egypt, the Mother of the West was the Goddess who received the setting Sun and reproduced its light. "The drug of Immortality" is the Solar Light. This was stolen from the Moon. Chang-ngo is equivalent to the Frog-headed Hekat who represented the resurrection. The Frog, in Egypt, was a sign of "myriads" as well as of transformation. In the Moon it would denote myriads of renewals when periodic repetition was a mode of immortality. Hekat, the frog-headed, is the original Cinderella. She makes her transformation into Sati, the Lady of Light, whose name is written with an arrow.

The Batavians, we are told, believe when twins are born that one of the pair is a crocodile. But it is a primitive mode of expressing their belief that man is born with, or as, a Soul. They knew that no crocodile was ever born twin with a human child. It is only in Egypt that we can find the explanation. One of the earliest types of the Sun as a Soul of Life in the water is a Crocodile. Thus we see the Mother who brings forth a crocodile as the Goddess Neith, depicted in human shape as the suckler of young Crocodiles hanging to her breast. Neith is the wet-nurse personified, whose child was the young Sun God. As Sebek he was imaged by the Crocodile that emerged from the waters at sunrise. Sebek was at once the child and the crocodile brought forth by the Great Mother in the Mythology, and because the Crocodile had imaged a Soul of Life in water, as a superhuman power, it became a representative type in Sign Language of the human Soul.

Amongst the Hottentots the Jackal finds the Sun in the form of a little child and takes him upon his back to carry him away. When the Sun grew hot, the Jackal shook himself and said "Get down." But the Sun stuck fast and burnt the Jackal, so that he has a long black stripe down his back to this day ("Black Reynard," p. 67).

The same tale is told of the Coyote or Prairie Dog, which takes the place of the Jackal in the mythical legends of the Red-man. The Jackal in the Egyptian representation is the guide of the Sun upon his pathway in Amenta, when he takes up the child Horus in his arms to carry him over the waters. In the Ritual, the Jackal who carried Horus, the young Sun God, has become the bearer and supporter of souls. In passing the place where the dead fall into darkness, Osiris says, "Apuat raiseth me up" (chap. 44), and when the overwhelming waters of the Deluge burst forth. he rejoices, saying "Anup is my bearer" (Rit., chap. 64), Here, as elsewhere, the mythical type extant with the earlier Africans had passed into the Eschatology of the Egyptians.

In the Book of the Dead (chap. 144) the adorations are addressed to the Great Mother Sekhet-Bast as the supreme being, to her who was uncreated by the Gods, and was worshipped as the "Only One," who existed with no one before her, the only one mightier than all the gods, who were born of her, the Great-Mother, the All-Mother when she was the "only one." This Great Mother is shown to be the only one who could bring forth both sexes. As Apt and again as Neith, the genitrix, or creatress, she is portrayed as female in nature, but also as having the virile member of the male. This was the only one who could bring forth both sexes. She was figured male in front and female in the hinder part (Birch, Egyptian Gallery). The Mother was the only one in the beginning, however various her manifestations in nature. She was the birthplace and abode. She was Earth Mother as the bringer forth and giver of food and drink, who was invoked as the provider of plenty. As the Great Mother she was depicted by a pregnant Hippopotamus. As a Crocodile she brought the water of the inundation. As Apt, the Water-Cow, Hathor, the Milch Cow, or Rerit the Sow, she was the suckler. As Rannut, she was the Serpent of renewal in the fruits of the Earth. As Mother of Life and Vegetation she was Apt in the Dom-palm, Uati in the papyrus, Hathor in the sycamore Fig, Isis in the Persea tree. In one character as the Mother of Corn she is called Sekhet or field, a title of Isis. All this preceded her human appearance, for she was the Mother divinized. This is the "Only One" who is said to have been extant from the time when as yet there was no birth (Brugsch, "Thesaurus In. Eg.," p. 637). The Mother gave birth to the child as Horus, who came by water as the fish, the shoot of the papyrus, the branch of the tree, and other forms of food and drink that were needed, all imaged by Zootypes.

Ancestral Spirits.—The Ancestral Spirits, as we have shown, were recognized and propitiated by the Pigmies in their earliest phases, and although Gerald Massey did not believe this, as we have proved (see Pigmies), it was, and is, an undoubted fact. They propitiated the spirits of those amongst them who had died and gone before, by building little spirit-houses and placing food therein. In the Nilotic Negro we find this further developed, and possibly it has reached the highest point with the Japanese at the present day. Ancestor worship as we find it in the Rit., chap. 52, was believed in by the old Egyptians, also in chap. 110, and in "Records of the Past," vol. iv, pp. 131–4, it is stated: "If thou readest the second page it will happen that if thou art in Amenta thou wilt have the power to resume the form which thou hadst upon earth." There is the key to much that is not understood, yet still practised by most African and other tribes, as well as some people of the present day—a belief of an existence hereafter, and that the souls of those whose corpus is dead can return and manifest themselves again for good or evil. This belief dates as far back as primitive man. With the Egyptians their funeral feast was a festival of rejoicing, not of mourning. When Unas makes his passage it is said, "Hail, Unas! Behold thou hast not departed dead, but as one living thou hast gone to take thy seat upon the throne of Osiris" (Budge, "Gods of Egypt," vol. i, p. 61.) From Papyrus IV at the Bulaq Museum we learn that offerings to the dead were taught as a moral precept. "Bring offerings to thy father and thy mother who rest in the valley of the tombs; for he who gives these offerings is as acceptable to the gods as if they were brought to themselves. Often visit the dead, so that what thou dost for them, thy son will do for thee."

Amongst the Brahmins the ceremony in honour of the manes is superior to the worship of the gods, and the offerings to the gods that take place before the offerings to the manes have been declared to increase their merits ("Manava-Dharma-Sastra," lib. iii, Sloka 203, also Sloka 127, 149, 207).

Great festivities were held by the Peruvians in honour of the dead in the month Aya-marca, a word which means literally "carrying the corpses in arms." These festivities were established to commemorate deceased friends and relations. They were celebrated with tears, mournful songs, plaintive music, and by visiting the tombs of the dear departed, whose provision of corn and chicha they renewed through the openings arranged on purpose from the exterior of the tomb to vessels placed near the body (Christoval de Molina, "Rites of Incas," translated by Clements R. Markham, pp. 36–50).

In Central America, as in Africa, the natives at the beginning of November hang from the branches of certain trees in the clearings of the forest, at cross-roads, in isolated nooks, cakes made of the best corn and meal they can procure. These are for the souls of the departed to partake of, as their name, "hanal pixan," i.e. "the food for souls," clearly indicates.

The Mexicans and many others offered human sacrifices to propitiate the souls of the departed.

The Behls among the hill tribes of India at the present day offer food and provisions for the Spirit. The North American Indians pay annual visits to the place of the dead and make feasts to feed the Spirits of the departed, so do the Amazulu (see Callaway, "Amazulu," p. 175) and also the Fijians (Fisons, "Kamilaro and Kernai," p. 253). The Chinook Indians declare that the dead wake at night, and the Algonkins bring food to the grave for the nourishment of the shade, which remains with the body after death, whilst the Iroquois say that unless rites were performed the spirits would return and trouble their relatives and friends. The Chinese Taoists called the Yu-Lan-Ui or "association for feeding the dead" collect supplies for those who have been paupers and have no relations on earth to offer sacrifices for them. It is here we have the origin of Ghosts and Spiritualism. The mediums were the first persons to demonstrate the facts of Spirit existence and Spirit intercourse, and those in Egypt were all trained experts, as well as in many other countries. These mediums were the highest of the Priesthood, the Divine Seers, and possessed more power than any one else, and it is so with the Inner African and other tribes of the present day; they are revered, not because they are Priests or Kings of a high standing, but because of their being intercessors with the superhuman powers on behalf of mortals, and the Spirits of the dead are dreaded or adored according to the sayings of the medium, for their good or their ill. Many of these mediums were and are women. In Egypt the woman was held to be the superior medium as seer and diviner. Duff Macdonald, vol. i, p. 61, says of the Yao people: "Their craving for clear manifestations of the deity is satisfied through the prophetess." It was by Magic that these powers and Spirits were to be controlled, as well as by propitiation and offerings to the dead.

Fetishism and Magic.—Following Totemism the next stage in the evolution of these Nilotic Negroes was so-called Fetishism—Magic, which may be considered a further development of their Sign Language, and should be defined as a reverent regard for Amulets, Talismans, Mascots, Charms, etc., and although Max Müller ("Nat. Relig.") called it a religion, it is not so, and never was an organized religious cult.

To these Negroes this "Fetish" or "Charm" represents a visible symbol of Magical Power to influence the elemental or ancestral Spirits; for example, in Africa, Apt, the Great First Mother, was propitiated as the Mother of Protection, and the protection at first was signified by types of permanence and power that were at first natural and then artificial. The power of Apt was symbolized by the Hippopotamus, and a tooth of the animal would symbolize its strength. The tooth, as a fetish or charm of magical power, is a common primitive type amongst all these Negroes, thus:—

Great Mother Totem's teeth.

Lions' teeth are worn by the Congo blacks.
Crocodiles' teeth are worn by the Malagasy.
Dogs' teeth are worn by the Sandwich Islanders.
Tigers' teeth are worn by the Land Dyaks.
Boars' teeth are worn by the Kukis.
Hogs' teeth are worn by the New Guinea people.
Sharks' teeth are worn by the Maori.
Bears' teeth are worn by the Esquimaux.

From old Egypt we discover that these amulets or charms became fetishistic because they represented some protecting power that was looked to for superhuman aid, and that this power belonged to one of two classes of Spirits or superhuman beings.

The first were the elemental powers divinized, the second the spirits of human ancestors—called Ancestral Spirits. In the Ritual of Egypt they are called the Gods and the Glorified.

Gerald Massey states in "Ancient Egypt" that in the earlier exodus of the Nilotic Negroes, i.e. the earliest totemic people, there were no Gods or Goddesses. The powers of the elements were not even divinized. For example, like the Arunti of Australia, they were at the stage of mythical representation. There were the superhuman powers in Totemism which preceded the Gods and Goddesses in Mythology. Instead of the Gods and Goddesses they had the Mythical Ancestors, who were Emus, Lizards, Snakes, etc., as totemic representations of the elementary forces in the primordial Alcheringa who were incorporated or made flesh on earth in both men and animals. At a later stage we can identify their elemental powers and trace them by name and nature to the Egyptian divinities as Gods and Goddesses. Egypt supplies the key. The human descent from the elementary powers is indicated by hundreds of traditions; but their elementary powers were prehuman, superhuman, and non-human, and, although called Spirits, they must not be confounded with the Spirits of the human.

These elemental powers were the origins of the Gods—givers of food and drink—and must not be confounded with the propitiation and worship of the Ancestral Spirits, which were spirits of human origin and have been frequently confounded with them. When the native powers are represented as human (as in the folk-lore tales) they assume a misleading phase, the primitive thought is charged with puerilities of the most recent fashion. It is these elementary souls that have been mixed up by the Hindus, Buddhists, and Greeks, and mistaken for the human soul, that has led them to think of transmigration, when they were non-human.

As soon as these early people discovered that the terrors of the elements were non-sentient and unintelligent, their chief objects of propitiation were food and drink and air, the elements of life. These were the powers born of the Old Mother as elemental forces that preceded the Gods or divinized powers. In Egypt we can trace the transformation of an elemental power into a God—in the deity Shu—who, as an elemental force, represented the wind or air. Darkness and dark clouds were blown away by Shu. This elemental force was represented by a panting Lion couched upon the horizon or mountain top, as the lifter up of darkness or the sky. This then represented an elemental power. Afterwards Shu was given his Star and he became the Red God and attained the rank of a Stellar deity as one of the seven "Heroes" who obtained their souls from the Stars of Heaven, and thus we have the Warrior God Shu, who was one of the Heroes or one of the powers in the astronomical character. There were seven of these elemental powers or animistic souls who were life-givers in the elements of food, water, and breath, not as begetters or creators, but as transformers from one phase of life to another, finally including the transformation of the superhuman power into the human product. Dr. Haddon's Heroes represent one or the other of these powers. They are early distinguishable. Dr. Budge ("Gods of the Egyptians") does not distinguish between the Great Mother Earth, Ta-Urt (who bore no children and was primary), and Apt, the Great Mother divinized (who bore children). He confuses the two and is not sure which form was first.

The Egyptians sometimes depicted her as having the head of a woman or a man, a vulture and a lioness. She is provided with a phallus, a pair of wings, and the claws of a lioness or lion, and sometimes in later times is addressed as Sekhet-Bark-Ra, a fact which accounts for the presence of the phallus and the male head on a woman's body. It proves that Mut, another name for her, was symbolized as possessing both the male and female attributes of reproduction; hence, when she brought forth her first-born, Set, he is called the Bull of his Mother, i.e. she fecundated herself by the male attribute, and as she bore seven males, each of these is called the Bull of his Mother. All are mythical.

As a summary I say :—

1. That all aboriginal native tribes, with Totems and Totemism, in whatsoever part of the world found at the present time, are from an exodus from Egypt, and descended from the Nilotic Negro after at least two exodes.

2. That they all possess Totems and totemic ceremonies. The later possess Hero-Cult and folk-lore tales.

3. That their Totems and totemic ceremonies are identical with each tribe found in whatever part of the world, when the anatomical and physiological conditions are similar.

4. That all tribes can be shown to have descended from a still extant type of Nilotic Negro in Africa, allowing for the difference caused by intermarriage with the primary Negroid women in some cases, the effect of environment in others, and the time that has elapsed since they left Africa.

5. That each tribe can be identified by "the Mother Totem" as to the time they left Egypt.

6. That the meaning of their rites and ceremonies, in whatsoever part of the world they exist at the present day, can all be traced to old Egypt, and nowhere else can the interpretation be found. It is only through the Ritual of ancient Egypt that you can read the symbology of these ceremonies. The old Egyptian Mythology is in the language of the Zootypes.

7. That modifications are found of the true types :—

First, when the earliest Nilotic Negro followed, drove out, and exterminated the little red man, and took his women as wives or slaves.

Second, when the second exodus of the Nilotic Negro, men who had advanced in anatomical evolution, etc., followed the earlier type, and, taking their women, intermarried at times.

8. That all these tribes spread over the whole world, following the Negroid race.

9. That another race (monumental) followed this second exodus who had fully worked out their Stellar Mythos and carried it throughout the world, as is proved by their burial

15

customs and remains of old ruins—with signs and symbols sculptured on them, still extant—classed as Turanian.

10. That these Nilotic Negroes did not follow the Pigmy or Negroid ancestors to the most inaccessible parts of the forests and mountains, and we find these little people still existing in these places, as primitive as they were a million years ago.

11. That the later exodus did not exterminate all those tribes which had only advanced as far as the totemic stage, and whose Mythical Mother totems had not yet become divinized, but drove them away in isolated groups. Thus we find them, at the present time, still existing in their primitive mode of life, and, having been cut off from intercommunication with the rest of the world, they could only marry and intermarry and continue to practise their old sacred ceremonies, without developing by evolution from the state they were in when they originally left Africa.

Consequently we know that there are two classes of tradition derived from Totemism, concerning the descent of the human race.

1. According to the one, human beings were derived from the Totem Animals or Birds or Snakes, etc.

2. According to the other, the totemic Zootype is said to have been brought forth by the human mother.

In the earliest exodus of these Nilotic Negroes there is no mythology mixed with their Totemism, no folk-lore tales, only "Mother Totems," "The Two Women," and the descent is always in the female line. In the later exodus mythology has commenced to be established by evolution, the two women are changed to the two mothers, who have been divinized, and this is kept in memory by folk-lore tales, etc.

CHAPTER XIII

TOTEM AND TOTEMISM

It will be well to examine and explain the meaning of Totem and Totemism, and to define some of these religious and sacred ceremonies, proving obviously their origin and connection with ancient Egypt, as against Professor Frazer's theories and expressed views on this subject, and those others who follow, in thought, his "school." It is necessary to draw attention also to the work of Professors Spencer and Gillen, containing their description of the Natives of Australia, their version of the customs found there, and the interpretation of the same, and later to Dr. A. C. Haddon's valuable work, "Reports of the Cambridge Anthropological Expedition to Torres Straits"; works of which one cannot speak too highly, as they prove the authors to be accurate observers and seekers after truth. They have given the native interpretations but have gone no farther, not understanding the true interpretation, and thus wisely recording only the facts found.

One should, however, consider how many thousands of years it took for the human race to develop the first ideas and practice of the use of the tribal Totem, to the final development of their mythology, when the characters became divinized, connecting this with Astral mythology (a further development of their Hero Cult, i.e. when they had formed their Stellar Mythos), because it is these tribes, whose ancestors left Egypt at the time of the beginning of the Stellar Mythos, who have legends and myths of having *descended* from the Celestial Mount, or the summit of the Mount, which was an image of the Pole. The races of man who descended from the Mount were people of the Pole, whose starting-point in reckoning time was from one of the stations of the Polar Star, determinable by its type as the Tree, or Turtle, or some other image representing a first point of departure. These traditions, found in various countries, show that they were born when no Sun or Moon as yet had come into existence, i.e. that they were pre-Solar and pre-Lunar in the reckoning of time, and in their legendary lore they try to tell us from which of the seven stations they descended, as a time-gauge in the prehistoric reckoning of the beginnings, which can be worked out fairly accurately by astronomers who can understand their myths and recognize their Zootype totem astronomically.

When as yet they had no names nor any art of tattooing the totemic figures on the flesh of their own bodies, the brothers and sisters had to demonstrate who they were, and to which group they belonged, by acting the character of the Zootype in the best way they could, by dancing, crying, or calling, and posturing like the animals of their Totems. This is the explanation from Howitt's "On Some Australian Beliefs," where he states that "The supreme Spirit Tharamulum, who taught the Murray tribes whatever arts they knew, and instituted the ceremonies of Initiation for young-men making, is said to have ordered the names of animals to be assumed by men." Before the names could be assumed, however, the animals were adopted for Totems, and the earliest names were more or less the cries and calls of the living Totems. The mother would be known by them, imitating the cries of the totemic animal, to which the children responded in the same prehuman language. For instance, in the case of those that have the Pig (Sow) for a Totem, as in New Guinea, the mother would call her children as a Sow and the children would try to repeat the same sound in response. Frederick Bonney, in his notes on the customs of the River Darling aborigines, observed that the children are named after animals, birds, and reptiles, and the name is a word in their language, meaning the movements or habit of one of them. Thus we find that the Totem, say, is an animal, representing first a figure; then a sound was added, ending in the formation of a name, which at first was the voice of the animal; thus we see that man was preceded by the animals, birds, and reptiles, who were the utterers of preverbal sounds that were repeated and continued by him for his cries and calls, his interjections and exclamations, which were afterwards worked up and developed as the constituents of later words in human speech. These were "things" first and these "things" were named by sounds first; then when he had not yet a name for the thing he used "Gesture Signs," and that is the reason the old Egyptian language is written in hieroglyphics or ideographic types. It may be called the language of animals, the oldest in the world—the language of the old Africans.

Behind the hieroglyphics are the living types, many of which are continued as Egyptian, having the same significance in Egypt as they had in other parts of inner Africa, and they still say the same things in the language of words that they said as zootypes. But nowhere else except in old Egypt (Africa) can you find the origins and explanations.

Neither the "Bushman" nor true "Masaba Negro"—the types which are the connecting links, or intermediate types in development, between the Pigmy and Nilotic Negro, which we find in Africa, and which have no Totems or totemic ceremonies—is found outside Africa, and, furthermore, the "true Negro," who is a development of the Masaba Negro in West Africa, is not found outside Africa, either in North or South America, Asia, or elsewhere, except those who have recently been imported into the West Indies, North America, etc.

Now, the reasons for this and the reason why the Nilotic Negro did not develop by evolution into the "true Negro" outside Africa can easily be explained, and these facts afford critical proof of my contention.

The Pigmy was the "first Homo," who multiplied and spread all over the world, but the one great progressive centre of evolution and knowledge was his home, in and around the Nile Lakes. It was here that his father and mother taught him all they knew, gained more knowledge, and grew in stature and mind. The further developed Bushman made an exodus and travelled *South, not North*, but was stopped by the sea surrounding the coast of South

Africa, where we find him still. The Masaba Negro developed here from the central home, and his followers in evolution (Nilotic Negro) travelled North-East and North-West, spread and multiplied. These types of Nilotic Negro, which differ from the true Negro, and are a class distinct, are still found here in their old home, the Nile Valley. As these Nilotic Negroes spread and multiplied, exodes took place following the little Pigmy North, East, and West, and so the two types, "Bushman" and "Masaba Negro," were cut off from going North, and would have been driven back South if they had tried to push their way North of the progressive Nilotic Negro.

Those Masaba Negroes who spread more to the South and West of "the home" through environment became the "true Negro," but those that went North and then spread East developed into the Nilotic Negro, the environment they found in different parts being conducive to the formation of that type of Homo. Some alteration would take place no doubt after existing thousands of years in a different climate and country, etc. But these changes would not, in any other countries than Africa, be conducive to the formation of the true Negro type. In each country, as these advanced and lived, there would be some tendency to alteration of the features of the original type. No doubt, as he advanced and exterminated the Pigmy, he took the women, and the second exodus of Nilotic Negroes would intermarry with those of the first. The offspring would be a modification of both the original and the later type. Natural selection should, I think, be taken into account in tracing the evolution of human races, as with other species.

Totemic Sociology was the first formation of Society into which the human group was first divided, or discreted, from the gregarious horde that mixed together in animal promiscuity.

The Totem in its religious phase was the sign not only of motherhood or brotherhood, but also of the Goddess or the God. It was an image of the superhuman power that was invoked, for water, for food, and health; as, for instance, the water-cow was a symbolical Zootype for the mother-earth who gave them water and food, etc.

Sign Language.—Totemism and Mythology were not only modes of representation, they were also the primitive means of preserving human experiences in the remotest past, of which there could be no written record. They were the records of prehistoric times. In this Sign Language (which was earlier than Totemism) *man acted his wants and wishes*, ceremonial rites were established as the means of memorizing facts in sign language; in these the knowledge was acted, the ritual was exhibited and kept in ever living memory by continual repetition, and the mysteries, totemic or religious, were founded on the basis of action. "The Inapertwa beings in the Alcheringa" who preceded men and women were "Homo," pre-Totemic, when all had one sign language, with few verbal words, and lived in a state of general promiscuity.

The earliest law of covenant, or tabu, was based upon the transformation that occurred at the time when the girl "was cut open" and became a woman ready for connubium. The girl in her initiation joins the ranks of Motherhood. She has attained her opening period. The tooth is knocked out to visualize the opening.

Therefore the Sign Language of Totemism was in existence long before two groups of people were distinguished from each other by two different signs or zootypes. Sign Language is far older than any form of Totemic Sociology. The signs now known as totemic were extant; they had served their uses, and were continued for other purposes. The very first thing to regulate in primitive marriages was the time at which the pubescent girl was marriageable. This was determined primarily by nature and secondly by the preparatory rite. For further proof see "Signs and Symbols of Primordial Man," p. 47, *et seq.*

I take it that the marital or sexual relations were first promiscuous; then there was a division of the gregarious into two communities in which the primal promiscuity was regulated by group marriages, with the totality divided into two halves, and subdivided afterwards by the Totems into totemic groups. These were extended more and more until they reached the Chinese "hundred families."

In the first phase the Arunta have traditions of a time

when a man always married a sister of his own Totem ; this, as tribal, followed the marriage of the brother and sister of the blood in natural endogamy as found in African Totemism. The Royal Families of the Incas, like those of Egypt, married their sisters; a way of keeping the blood-motherhood pure. This endogamy was primal and existed for a long period. The Kalangs of Java and the Goajiros of Colombia, South America, still practise this endogamy,

In the next phase there were only two divisions, A and B. as in the primary, but all the women of A division were common to all the men of B division, and the children born all belonged to A's Totem. The mother's Totem and the descent was thus always recognized in the female line.

The women of A were tabooed to the men of A, except during some special ceremonies. All the women of B, division were common to all the men of A division, and the children which were born all belonged to B's Totem, the women of B being tabooed to the men of B with some special exceptions.

The further development was the division into four, when the children reckoned their descent from the male side, which was afterwards followed by a division into eight, when the children's descent was not taken from the father or mother Totem, but from a different one—one of the subdivisions.

Group marriages were the universal custom at first, and a modified form of individual marriage followed.

The Wonkgongaru, Tjingilli, Binbinga, Umboia, Urabunna, and Dieri are only divided into two exogamous moieties and reckon by maternal descent. These were the first exodus after the Pigmy. They have no-Hero Cult and no myths. The Mara, the Warramunga and some others, with division into four, now reckon their descent from the male side, although the names of the original two moieties still persist. These have myths. Amongst the great Arunta tribes, the tribes are either divided into four or eight intermarrying groups, and the children are of a different Totem from that of the father or mother. For instance, if a Panunga man marries a Purula woman, the children are *Appungerta*; if a Purula man marries a Punanga woman,

PLATE LVII.

Natives in the Bismarck Archipelago have the same dress and customs as these.

the children are *Kumara*, these being subdivisions of the original.

Dr. A. C. Haddon, F.R.S., in his introduction to "Customs of the World," states that "the persistence of custom is not due to any mystic property of 'the East,' but is merely the result of the permanence of geological conditions and the suitability of the customs of that mode of life, for instance, the herders of domesticated cattle who practise a little agriculture—such as the Zulus and similar tribes of South Africa on the one hand and the Ancient Germans as described by Tacitus on the other. There is similarity between these people so widely separated in space and time—there cannot have been any cultured contact—and again the sense of solidarity is common to mankind; nowhere is it stronger than amongst savages, and not a few customs are concerned with strengthening this sense of solidarity. Ornaments, clothing, and various mutilations of the body are outward and visible signs of 'the consciousness of kind.' In these and other matters the individual has to conform to the usages of his group—nonconformity is almost unthinkable." It is certainly not any mystic property of the East; it is the customs of inner Africa, or old Egypt, that have been handed down from generation to generation. Take for instance "the herding of domesticated cattle with a little agriculture." The Madi and Masai were the first people who practised this. The Zulus are an offshoot of these who spread South, and so carried the same customs with them down to South Africa. The ancient Germans and earlier people throughout Europe who practised these customs were an exodus to the North of these same people; hence the same customs, and if Dr. Haddon will travel to South America he will find that the herding of cattle and agriculture are the same there, even to the forms and designs of building their houses—all originating from Africa. Like their totemic ceremonies, Totems, Religious Conceptions, Ornaments, Signs and Symbols and Language, their mutilation of the body, each exodus had its own Totem and Totemic ceremonies, and these the individual carried throughout the world with him, and so each individual has to conform to the usage of his group. Nonconformity would not only be "unthinkable," but would spell punishment and, in many cases, death. These customs have been carried on from generation to generation by those who left old Egypt, with very little alteration where the tribes have been isolated and cut off from intercommunication. Our Professors of the present day have not even attained to Anthroposophism, and certainly are not so wise in these matters as the old Egyptian Theopneustics. But Dr. Haddon is right when he states "an unbroken continuity can be traced between most of our customs and those of our barbarian forefathers, who in their turn received them from their savage ancestors."

Plate LVII is a photograph of some natives of Umtata, Africa, performing the "Abakweta" dance in totemic costume, precisely similar to the natives in the Bismarck Archipelago. These have the same customs and totemic dresses (see C. W., p. 25).

I am much indebted to my friend Dr. Lewis Lewis for sending this to me direct from Umtata.

Information from Bulletin No. 2, Northern Tribes, by Spencer and Gillen.

In Australia of the Northern Tribes the following only circumcise :—

1. Worgait.
2. Warrai.
3. Mandot.
4. Djanan.
5. Nullakin.

The following tribes both circumcise and sub-incise :—

1. Mudburra.
2. Yungman.
3. Nungarai.
4. Mara.

Amongst the Melville Island natives and Port Essington natives descent is counted in the female line; thus, for example, a Stingarree man marries a Crocodile woman and the children are Crocodile.

In all these tribes cicatrices are made on the bodies of both men and women, and amongst the Melville Island natives they are wonderfully well developed, forming very definite lines of V-shaped marks on the chest, abdomen, back, and arms (see Bulletin of the Northern Tribes).

These cicatrices made to pattern, or otherwise, are found in all primary Nilotic Negroes, in the earlier stages before they mastered the art of tattoo. Amongst these primitive people (Nilotic Negroes) the tribal badges sometimes represent their Totem.

CHAPTER XIV

HEIDELBERG AND NEANDERTHAL TYPES

THE Heidelberg and Neanderthal types were early Nilotic Negroes, probably amongst the first who migrated North from old Egypt. Their implements were those of the early Nilotic Negro, and not those of the Pigmy race. As these men went North, they would, by reason of their constant warring against wild animals and the Pigmy, develop into a hardier, fiercer, and more brutal type. As they only married their own women, or the Pigmy women, they would not rise in the scale of evolution. We must take environment also into consideration.

They probably inhabited most of Europe and Asia, and existed for many thousands of years after the next exodus left Egypt. This type of man would exist for a long period by brute force alone, warring against savage animals, etc., until driven into mountain fastnesses, where he could probably hold his own against the superior race which left Egypt after him. Many centuries would have passed before all were exterminated, and it is for this reason that his remains have been found in higher strata than those of a much superior race. No transformation took place from these in the evolution of the human race to a higher type.

As evolution did not take place outside Africa, at this epoch, all the first Nilotic Negroes would become extinct for they would not have been allowed to take the women of the superior race.

Even at the present day very few "black men" marry white women; in that case, as a rule, the progeny are superior to the "black man." From this, it may be deduced that the "black man" of himself does not breed a higher type, save in a few isolated cases, when the "result" invariably dies out.

But when we come to the Stellar Mythos people, we find that the result of the marriages and intermarriages taking place is in each case a modification of each type, and very often a great advancement in the progress of evolution is the result.

These pictures of living examples (Plates LVIII–LXI may assist my readers to understand the argument and the evolution of the human race.

1. The Masaba Negro (see Plate XXXVIII) is the next in development from the Pigmy; already there is a distinct development of the beetling brow and massive jaw.

2. As a comparison of the next development from these take a man of the Tjingilli tribe, a living type of the Australians, as representing a partly Neanderthal type (Plate LVIII).

These Australians are descendants of the first exodus of the Nilotic Negroes who have Totemic ceremonies but are not "Hero Cult" people. It is easy to see from anatomical comparison the relations which these bear to each other.

These are the lowest type that were evolved from the Masaba Negro, and they left Africa next to the Pigmies. Probably Australia is the only remaining continent outside Africa where they can be found, and the lowest type in South America that the author has, up to the present, been able to discover is the next higher type in evolution.

The Pigmies outside Africa always remain Pigmies, although environment has produced some effect on their features and the hair, etc., in some places.

The first exodus of the Nilotic Negro remains immutable outside Africa and the second exodus people remain the same.

In the last exodus of the Nilotic Negro a great advancement is, in many cases, to be observed, the reason being that the type mixed to a certain extent with the early Stellar Mythos people.

In Africa itself must evolution be looked for, and there

only can it be found, in the many and varied types of Pigmy, from a very low type to a greatly developed Pigmy, in the Masaba Negro to the greatly developed Nilotic Negro, and thence to the higher type of Masai and Muhima.

On carefully examining the anatomy and cults of these Nilotic Negroes, it is possible to trace the type from which sprang the various Stellar Mythos people; another important factor in this discrimination consists of the various stages of advancement in Totemic Sociology. There is the very tall race whose remains we find scattered over Europe and other places, and the splendidly developed man with as high cranial capacity as the average present race. Their originals are still to be found here in Africa, but nowhere else. But we frequently come across "throws back," as, for instance, we frequently see a very tall man or woman with spindle legs out of all proportion to the body, a "throw back" from the tall races of the Nilotic Negro.

When we take into consideration the origin and evolution of the human race, as set forth in these pages—and this is the only key with which to unlock the mystery—it is by no means surprising that any particular race, which existed here thousands of years ago, should have become extinct. As one race came up from Egypt after another, the new-comer would be superior to the old inhabitant, being further and more highly developed in the scale of evolution. In cases where the former were not allowed to take the women of the latter, it would be solely a question of time before the type became extinct, save only when the aborigines could escape from the invader and live as a tribe or nation by themselves.

Thus, we find the Nilotic Negro disappearing from Europe before the onslaught of the Stellar Mythos people; nevertheless, these Nilotic Negroes still exist and continue to live in Australia, New Guinea, New Hebrides, and some parts of South America, whither the Stellar Mythos people did not penetrate.

At the present day they differ but little from their prototypes who left Old Egypt a million years or more ago; having little or no intercommunication, and being cut off from all others, they would still continue to exist, much in the same way as when they left Africa. The type would, in all probability, improve up to the highest point of the latest invasion, when they were allowed to marry and intermarry, but nevertheless the two types are quite separate and easily distinguishable at the present day, both anatomically and by their Totemic ceremonies.

As in Africa, so all over the world, we discover here and there a remnant of the older type, that has not mixed with a more recent development, and been exterminated by more highly developed surrounding tribes. In South America, the various tribes have separated from each other, and still retain the same characteristic features, both anatomically and totemically, as their old ancestors who came out of Egypt, where the primary types may still be found living. But we never find the connecting link, i.e. the Masaba Negro, in America.

Professor Frazer, in his recent large work, has come to the conclusion that Totemic ceremonies, Totemic rites, etc., originated separately and independently in various parts of the world amongst various tribes; this, surely, is opposed to all the facts of the case and cannot be accepted by science. He also supplies a map which gives some idea as to where these totemic tribes still exist. But Totemism and Totemic rites exist all over the world; and amongst all tribes, past and present, they are the same, have the same meaning, and could only have sprung from one original, and that must have been Egypt and the Nile Valley. That the various tribes have a different zootype for their Mother Totem only shows and proves at what time each left Egypt, and the different fauna of the country—the ceremonial rites are the same in each group that possesses the same anatomical and physiological condition, and if one of these tribes from one country visited a tribe in another country possessing the same Totems and Totemic rites, he would be admitted as a brother amongst them at once. This has been proved, and there can be no sort of argument, logical or otherwise, that, because we now find isolated groups in different parts of the world, these evolved all their ideas, etc., from their own surroundings separately. Many tribes in North and South America have the same Totems and

Totemic rites as we find in New Guinea, New Hebrides, and Central Africa at the present time; all these could not have evolved the same ideas, the same ceremonies, using the same signs and symbols and many Egyptian words, performing the same sacred rites, without one common origin. They also manufacture the same kind of implements and pottery. That they are now only found in some parts of the world is explained by the evolution of man; and the evolution of the human race, always working towards a higher state, can and does explain and give the key to the explanation why exodes from the centre (Egypt) were constantly taking place, from the first original Pigmy Homo to the time of the highly cultured priests of Egypt, who possessed the greatest astronomical and eschatological knowledge. As each exodus had a higher standard in every way, those that went before would be driven farther and farther away, cut off and broken into tribes and clans. So the original totemic people are still found to-day, in those islands or isolated places where intercommunication has been suspended, or to which there is very little access. The white man in the last few hundred years has had much to do with cutting up the tribes in America, which has been a rapidly progressive country, and also in New Zealand. In Australia and Africa there has been, during the past hundred years, a vast "breaking-up" and dispersal into groups of the original inhabitants.

However much my critics may disagree with me, there is no other solution of the question if all known facts are taken into consideration. No other theory can be advanced that will not lead the author of the same into a deep pit from which he cannot be extricated without an Egyptian ladder, and, taking all the evidence brought forward in this work, the conclusion arrived at must be final, whatever may be said to the contrary. Certainly it may be years before Anthropologists and Ethnologists acknowledge the contents of this book to be the true solution of the whole question, but a future generation will, I am convinced, affirm that it is so.

I do not expect the present generation, imbued as it still is, to a very great extent, with preconceived ideas and dogmas to be able for many years to grasp and follow all

Showing large abdomen of the Pigmy in childhood, which disappears in later life.

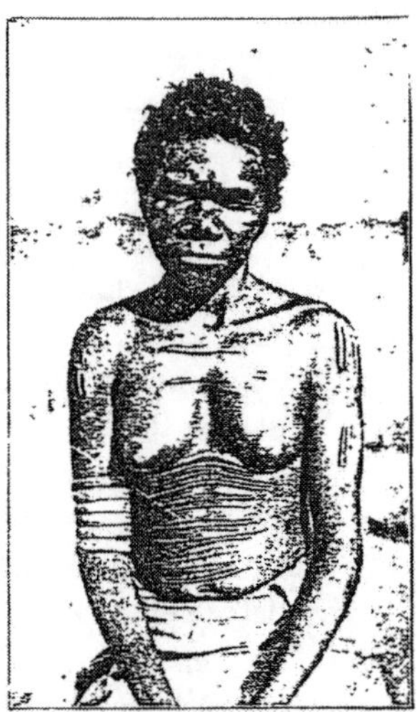

Front face, showing the early characteristic lips of the Nilotic Negro which I call Pigmoid.

Side view of first Exodus, showing prominent supraorbital ridge, curly hair, etc.

Full face of first Exodus, showing Pigmy nose; "beetle-brows," cicatrization marks, curly hair. Non Hero-Cult. The next type above Masaba Negro.

PLATE LVIII.

AUSTRALIAN TYPES OF DESCENDANTS OF THE NILOTIC NEGRO

A Dinka, White Nile.

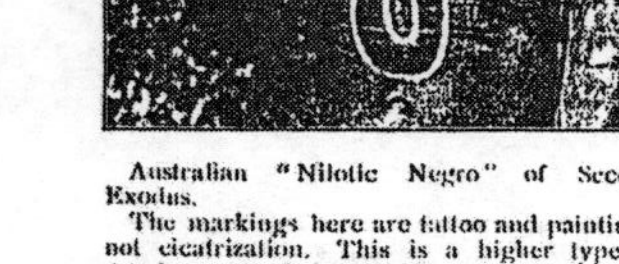

Australian "Nilotic Negro" of Second Exodus.

The markings here are tattoo and paintings, not cicatrization. This is a higher type of development of the human. They have Hero Cult and myths.

PLATE LIX.

HERO CULT PEOPLE.

Fidjelu woman, Lado Enclave.

Australian descendant of type of Second Exodus of Nilotic Negro.

Tattoo marks, but little cicatrization shown. A finer type and further developed human. They have myths.

PLATE LX.

HERO CULT PEOPLE.

Plate LXI.

Type of First Exodus of Nilotic Negro.

Note Pigmy nose, lips, also the incisor tooth knocked out, and the hair tied up. Note also cicatrization—no tattoo. Non Hero-Cult.

the facts and arguments that I have brought forward, but there is no doubt that the future will prove I am right.

In his recent writings Dr. Frazer states, ". . . a difference between Magic and Religion, more than that, I believe that the evolution of thought magic—as representing a lower intellectual status than Religion—has probably everywhere preceded Religion." I am surprised that Dr. Frazer, whose works and writings are taken as a standard of truth, and anthropological, ethnological, and scientific knowledge, should hold this view, which I am sure is unsound. In the evolution of man, magic is later, in comparison with religion, by thousands of years. *Magic does not occur at all amongst the primary Negroes* and non-Totemic peoples, but is a development which was first practised at the time of the Nilotic Negroes, when "the elementary powers were born of the first Great Earth Mother." It is not practised by the Pigmy or the Masaba Negro. These are primary, and of a much lower mental status than the Nilotic Negro, as may be proved by their lower brain capacity, their anatomical and physiological features, and the difference in development of articulate language, as well as by other physiological conditions as stated (*supra*).

I have shown in the preceding pages that *Religion had already been developed in the Pigmy*, and further in this work I have explained the meaning and origin of Magic.

If Dr. Frazer will read this, he will perhaps modify the opinions he has expressed in recent works, and thus prevent the dissemination of false ideas due to ignorance of the fundamental principles underlying Totemism, Sign Language, Mythology, and the Ritual of Egypt.

No objection can possibly be raised to any man, or woman, holding such private opinions as they choose, but here is a very great objection to men, occupying the positions of teachers or Professors, disseminating exploded, antiquated, and wrong ideas which scientific research and material evidence have proved to be false. It is, also, quite impossible to agree with Dr. Brenton when he too advances the Autochthonous theory with regard to the prehistoric people of America. He states that it is impossible that any affinity exists between the Old World and the

New; that Mongols are the roundest-headed of people, and that Americans are in nearest contact therewith; that the Eskimos are a long-headed people. This argument is of no value whatsoever—the people are all of different exodes from Old Egypt, where I have shown the long-headed and the round-headed exist and were even evolved. He has to acknowledge that the so-called "Mongolian eye" exists, and I have shown where it came from, and from which division of the Nilotic Negro. There are many very distinct types in America, and the reason of the origin of each can be traced by the various migrant waves of humanity from Old Egypt at different times. Again, his theory would involve the evolution of the Indians here from the Pigmy, without the two anatomical and physiological links, which are missing here, but are found in Africa only, namely the Bushman and Masaba Negro, and the evolution of these Indians from the Mayas, Incas, Aztecs, etc., with arts, architecture, and religious doctrines. The designs, patterns, and games, the hieroglyphics, language, signs, and symbols are in many cases the same as those we have in Egypt. I think it unnecessary to discuss the point further; it is too absurd, in my opinion, for any thoughtful man to entertain it for one moment, considering all the evidence that we now have knowledge of. His *knowledge* on these subjects is of the same value as his knowledge of the Maya language, and worth just as much—nil.

Thus we see that, during the Totemic period, the human race made great strides up the ladder of evolution. The advance from the lowest type of Nilotic Negro, with his primitive Totems and Totemic ceremonies, to the highest type of that race, with fully developed Totemic ceremonies and Hero Cult, must have occupied many hundreds of thousands of years. We have only to examine the fundamental difference at the present day between the Masaba Negro, to the south-east of Uganda, and his neighbour, the Masai, to grasp some idea of the time that must have elapsed before the human race rose to the higher level in anatomical and physiological advancement alone; not to mention the cycles that must have gone by before further advancement was achieved in articulate words and religious conceptions;

these latter must have been formed solely from observation of the elemental powers and the general laws of nature, for there were no books or writings at this early period to act as a guide.

Then, compare the other Nilotic Negro tribes found at the present day, and follow their offspring throughout the world; the remains of many are still extant. How many hundreds of thousands of years must the process of evolution alone have involved? Seven or eight hundred thousand years at least would be required to work this change and development in human anatomy and physiology from the first Pigmy to the Masai alone.

That we do not find the osteo-anatomy of the very early human remains is not surprising. Indeed, it would be more surprising if we did. Take the burial customs of all these primitive races; they only bury in earth, a few feet below the surface, and entire disintegration must have taken place within a few thousand years at the most. Their wooden weapons, clubs, bows, and other primitive arms, would all disappear in the course of a few years. Their stone implements would remain, and these we find; but how primitive were the first!

Probably thousands of years elapsed before a human first discovered that by chipping flints he could fabricate a scraper or knife. We should never recognize his first implements, even if we saw them, for there would be little or nothing by which to distinguish them. When, however, we come to that step in evolution, the Nilotic Negro, even the instruments used by the first exodus of this race can easily be distinguished from those of the Pigmy.

So, not only anatomically and physiologically have we progressed in the phase of evolution, but also in religious ideas and ceremonies, and in the fashioning of instruments, all of which are now easily distinguishable from the fundamental characteristics of the Pigmy race, the first and only one that left Old Egypt and spread throughout the world before the advent of the Nilotic Negro.

We have, therefore, now traced the origin and evolution of—

(*a*) The First Man who went out and peopled the world.

(*b*) The SECOND or connecting race of men—which never left Africa.

(*c*) The THIRD race, which followed the Pigmy all over the world.

As groups they may be distinguished by—

(i.) Their implements and the fact of being pre- or non-Totemic people.

(ii.) Their implements and the fact of being Totemic people.

These latter may be divided into—

I. Totemic people with no Hero Cult, lower in the scale of evolution than

II. Totemic people with Hero Cult.

Again, all these different tribes may be classified under one heading—Palæolithic people.

It is now recognized as a scientific fact that mankind is specifically of one family, and that he must have had one original "Cradle-land" from which the peopling of the earth was brought about by migrations. The "Cradle-land" was Old Egypt, and the evidence shows that, first of all, the world "was peopled by the little red man," and that the Nilotic Negroes followed him. How did primary man migrate from this home into various parts of the world, some of these places being islands? Geology has been able to show the existence of earlier continents, or of a prearrangement of continents; and also that the earth's crust has undergone great changes. Complete land communication existed where now none is. An Asia-African and a Euro-African continent probably existed, and the extension of Australia and New Guinea formerly was probable. America was connected with Europe on the one side, and with Asia on the other, in the earliest times; other continents may have existed in the Pacific and Atlantic, but of these we have no positive proof. Thus these early people were free to move in all directions over Africa, Europe, Asia, America, and Australia.

The above, therefore, must be a conclusive answer to all hitherto preconceived and erroneous ideas—that the human race had its origin in Asia, or in some mythical region—fertilized in the brains of those who persistently seek the cradle of the human race away from its true home.

Such proofs and facts as I have brought to light cannot be ignored by any, save those who possess a blind belief in the Biblical tale—for which there is no corroboration critically, and which, derived as it was from the Ritual of Ancient Egypt, has been much perverted—or those, it may be, who, obsessed with one idea, wander away from "Home and its true affinity," and resolutely refuse to be influenced by fact or reason.

Thus we have traced Homo from his birth to the formation of his Totemic Sociology and of the early "Nomes" in Old Egypt. He built houses, tamed wild animals, engaged in agricultural pursuits, worked in metals, built boats, and began the formation of his picture-writings in the Hieroglyphic Language; migrations went out all over the world during this time of evolution, and the descendants of many of these still exist in isolated groups, just as we find remnants of the earliest types in Africa and nowhere else in the world. We must now follow the evolution of man to a still higher type, and show how here in Old Egypt he must be classed as belonging to the first stage of Stellar Mythos.

Thus the terms Aurignacian, Solutrean, and Magdalenian used by our present Anthropologists and Ethnologists are wrong, and absolutely misleading if applied to the age, or period, at which the human remains found in places under these terms lived, *and only lived, at the time that those particular strata were formed.* Yet, after all the above facts are known, we find the great University of London arranging for a course of Lectures to be given by a foreigner on the "Art of the Palæolithic Period"!

I here give a cutting from the *Times*, February 7, 1913.

"PALÆOLITHIC ART.

"The Abbé Breuil, Professor at the Institute of Human Palæontology, Paris, gave recently a lecture on 'The Art of the Palæolithic Period.'

"The Lecturer, who spoke in French, said that the history of palæolithic art began with the human populations which immigrated into Europe at the beginning of the upper palæolithic period. We did not know whence they came, but we saw them suddenly substituted for the men of the Neanderthal type, from whom they differed as much as we ourselves. The new civilization which came, as it seemed, *towards the end of the last glacial period, cut flint, principally*, in long blades, very different from the thick and short splinters of the earlier period. It worked already in ivory, bone, and reindeer horn, of which it was rapidly to perfect the industry. As did its predecessors, it sought its subsistence exclusively in the hunting of mammoths, the woolly rhinoceros, horses, bison, and reindeer, which inhabited Central and Western Europe. The camps of these people, sometimes in the open air, were sheltered for choice at the base of projected rocks or in the entrance to caves. Their vestiges had accumulated there during thousands of years in the form of kitchen and industrial débris, mingled with the ashes of extinct hearths. Search in these mounds revealed various stages in the evolution of this civilization. Very often new discoveries or fashion introduced types of instruments which had been modified subsequently before disappearing, and which admitted of the subdivision into numerous shorter periods of the three great principal phases of this age—the Aurignacian, the Solutrean, and the Magdalenian."

I state emphatically that the "Art of the Palæolithic Period" proves by which race these implements were fashioned—the Pigmy—Early Nilotic Negro—Late Nilotic Negro—and then the Neolithic people—Stellar Mythos people, as I have shown in my Lecture at the Royal Societies' Club and also in this work. The present teaching I have proved to be absolutely wrong and misleading.

It is sufficient to mention that "the present type of man" has been found in the Pliocene strata, and that the Pigmy still lives in New Guinea, to show how erroneous and futile it is to continue teaching under the above terms the twaddle that emanates from "our most potent, learned, and grave signori."

CHAPTER XV

TOTEMIC PEOPLE

Dr. A. C. Haddon, F.R.S., having conferred upon me the honour of asking me to read his great work and to write my views upon it, I have the very greatest pleasure in complying with his request, and I trust that those who read it will not think any criticism I have made is too caustic, because such criticism is not set forth in any unfriendly spirit of opposition, but rather, I would say, in a cordial and brotherly way, the more to emphasize my opinions for whatever they may be worth.

Many years' study of this subject, personal visits and observations amongst numerous native tribes, and a knowledge of the Ritual and old hieroglyphic language and glyphs, convince me that my ideas are correct; but their acceptance first by scientific men and afterwards by the general public would involve them in the unlearning of so much that I do not look for their acceptance during my lifetime; nevertheless I am confident that future generations will, in the light of new discoveries made, and from the acquirement of a more intimate knowledge of the Ritual of Ancient Egypt, and the study of the Zootype language, acknowledge them to have been the true solution.

Dr. Haddon towards the end of his volume vi, "Reports of the Cambridge Anthropological Expedition to Torres Straits," states as follows: "We did not discover in Torres Straits anything like an All-Father or Supreme Being." He acknowledges, however, that the natives give mythical beings as their ancestors. You ask a native about his ancestors, if they were married, or had parents. He looks at you and replies, Certainly not! they came in their own manner, without mother or father—"Nole e nole Kosker Kak e nole babu Kak e Tabara tonar!" as though you had committed an act of desecration in asking such a question. Dr. Haddon then tries to make a great distinction between these tales of mythical beings and folk-lore tales of what he called "The Culture Heroes." No doubt these natives were surprised that so learned a man did not understand that their forefathers or "foremothers" were the Inapertwa beings (the same as the Australians), men and women, not yet made men and women, i.e. their ancestors who were not Totemic. The cutting and opening of the female and the circumcision and sub-incision of the male, had not yet been brought into existence. There were no marriages, tribal and otherwise; so far these had not been established, but when the Torres Straits people became men and women, it was an indication that Totemism and the cutting and opening had been introduced, and that tribal marriages took place; initiation into men and women had been accomplished. That he did not discover anything like an "All-Father" is not surprising; indeed, it would have been very astonishing if he had. These natives, as all others in their Totemic stage, acknowledge the "mother," not "father." The women were bearers to the tribe; a child knew his mother because of the Totem given to her on arriving at puberty. If she were a Lizard, then her children would all be Lizards; if a Crocodile, then her children would all be Crocodiles; but the father was unknown and remained unknown up to a much later date. Until the Solar Cult there was no "All Father"—it was all "The Great Mother" and Mother tribal—both represented by Zootypes, and there is often some difficulty in distinguishing between the Great Mother, who was Mythical, and the tribal Mother, who was human, each sometimes being represented by the same Zootype. For instance, the Frog and the Crocodile or Lizard, represented the Great Mother in the waters of space, and the same totemic Zootype was given to the pubescent girl on reaching her maturity.

Had he been initiated into their most sacred rites, and able to read and understand their sign language, gestures, or the meaning of their signs and symbols—in other words,

PLATE LXII.

THE MANUKA OF WHAKATANE.

A Maori talisman of Life and Health. It was regarded as the essence and semblance or personality of health, of life, and of spiritual prestige.

Plate LXIII.

Wooden Idol from Santo Domingo.

Imbert Collection, about one-fourth natural size.

to talk the Zootype language—Dr. Haddon would have understood what these mythical beings and the "Culture Heroes" of the folk-lore tales were. With but little variation they are found amongst the Nilotic Negroes and their descendants.

Here is Dr. Haddon's description: "Waiat depicts a wooden figure, without eyes or ears, the arms cut off at the elbow and the legs cut short at the knee, no sexual parts indicated. He was head or chief of the tai, but *tai* was not made in the islet of Widul, when Waiat always remained in his square house in the Kwood. A mythical figure of no sex."

This represents Horus as one of the primary Heroes, the Blind Horus, or the elder Horus, or Horus I of the Egyptians. "Being situated on an island and always remaining in his square house and the chief of the Tai"—in the old sign language it would mean "the island in the waters of space." "His house in the Kwood" is his sacred place, afterwards represented as the sacred city of Paradise, situated in Polaris. The chief of the Tai would be the chief of the Pole Stars – $6 + 1$ (himself) $= 7$. All were situated in the Celestial North, representing elemental powers. The Mexican and Central American Indians have precisely the same representation, as well as some tribes in inner Africa. Also the Maori of New Zealand (see Plates LXII and LXIII).

The Chief of the Tai shows also that these came out of Egypt at the time of their Totemic Sociology. Tai is an Egyptian word (Tai) — "localité entre Migdol et Adida" (Marcette).

The story of the six blind brothers is explained only through mythology; it is one of the earliest stories of the Precession of the Pole Stars. The same tale is found in Africa and America and represents six elementary powers.

Kwoiam and his four brothers are the same as Horus and his four brothers. This was the earliest representation, afterwards they were represented as the four children of Horus. The totemic dance, when one party wears white feathers as head-dress, and the other black plumes of the cassowary, represents in its primitive form the fight between Sut and Horus, the contest between light and darkness. One moiety of these people therefore represents Horus and the other moiety Sut, and they dance the tale of their ancestors in symbolic form, telling in Sign Language what occurred.

The story of Kari the Dancer can only be read like the others, and explained by the Mythology of Egypt. It is part of the story of Tabu (see *supra*) the bamboo rattle padatrong = the Bull-roarer, and is used to drive away the women from the sacred ground of initiate ceremonies and also to inform them that the great evil one is about. In Egypt it was also used as a magical instrument by the women to call the young men.

Kwoiam, who killed his mother for cursing him and then went and killed a number of people, is read only through the Egyptian Mythology. These and the boy Upi and Sida are all mythical personages and are interpreted easily by mythology, but not in any other way. Markep and Sarkep, in his comic tales, are types in Egyptian Mythology of the Sun and Moon.

Spirit children derived from elemental powers of air are described in the Ritual as the younglings of Shu (the God of breathing force). The Sun was looked upon at one stage as the elementary source of a soul and was represented by a Phallus. Thence also rose a belief that the sun could impregnate young women. There is the same idea in the young Arunta women, fleeing from the embraces of the wind, for fear of being impregnated with the elemental spirit child.

The records of cannibalism in Dr. A. C. Haddon's folk-tales in the story of Nageq and Geigi, and Kaper Kaper, were mythical, and therefore there is no question of morality or cannibalism to be considered; just as, when the Egyptians in the time of Unas speak of the deities devouring souls, *this is no proof of their being cannibals at the time.* Mythology commenced in the most savage, or crudely primitive, state in the most ancient Egypt, but the Egyptians who continued to repeat the mythos did not remain savages.

Irado (p. 51) represents the Egyptian "primary Great Mother": Ta-Urt or Mother Earth. The Egyptian

Ta-Urt or "Great Mother Earth" had no children, neither has Irado, but she supplied food as she does here, and was a bringer forth of life, before she was divinized. It is necessary to distinguish the primary form, Irado = Ta Urt, of no sex, mother of no children, who only supplied food and life, because in the next phase of the mythology we find that the Great Mother "Apt" brought forth children and was given a constellation. She was divinized as both male and female, as Apt, the goddess in human guise with the head of a hippopotamus; this was the next stage of development in evolution (see *supra*).

Ankern and Terer.—This shows the origin of the commencement of Stellar Lunar mythology.

"*Terer who lived with his Mother; he had no father.*"—This was Horus as the Calf born of the Cow, and a prehuman type when the fatherhood was not yet individualized.

Ankern here represented the Mother Moon, an early form of Hathor. In this case Har-Ur, or Horus the Elder, the son who does not know his own father, is the same as in the Maori tale of Kokako; the boy is called a bastard. Also in the tale of Peho the child is a bastard. This was the mythical son who was born but not begotten. As Egyptian, Horus was the child of light that was born of Hathor in the Moon, when the Moon was the mother of the child, and the Father source of light was as yet unidentified. It was before Amenta was formed that the Moon god had not discovered that Isis was the wife of Osiris in the underworld, where they met every twenty-eight days for three days, the result once a year being Horus the child. Here the Moon goes down into the underworld and disappears, and once a year appears with a son; these people did not know that the moon met the sun in the underworld. The appearance of Terci after meeting with Zogo (a primitive fight with Sut), and the disappearance of both by diving into the sea, means "transformation"—phases of the Moon. The boy was called "bastard" because his father was unknown, but in the Mythology at this time he was the "fecundator of his Mother."

The story of Mutuk (p. 89, vol. v) is a primitive form

of, and is the same as, the later Biblical one, Jonah and the whale, which is part of the Mythology of Egypt. The Hebrew writers have omitted the name, and effaced the prototypical figures because they did not understand the mythology; they have foisted on the world revelations they themselves cannot explain or understand. In this legend the belly of the fish is identical with the belly of Sheol, the uterus of the underworld, portrayed by the Great Fish. The Great Fish in the form of a crocodile (in this case a Shark) was one of the types of the ancient mythical Great Mother, who brought forth Sebek-Horus from the Nu, as her young Crocodile, just as she brought forth her first Sut, her young Hippopotamus. This is also portrayed in the Mexican Codices and on the walls of the remains of ancient cities in Central America, as we have shown by photographs, etc., in "Signs and Symbols of Primordial Man"; both of these figures are depicted in that book, i.e. the young Crocodile in the act of being born from the belly of her great Mother (p. 116, fig. 50, 2nd edition), also Horus being cast up on land from the Great Fish (p. 105, fig. 42). This folk-lore tale here set forth is one of the primary forms of this part of the Mythology of Egypt. On p. 344, vol. v, figs. 64 and 65, there are primitive figures of the Stellar Lunar Mythology, Mudu Kap. Dr. Haddon states that the people of Maburg had a Kap or dance when the star called Kek first appeared just before sunrise, but he does not understand the meaning of this. It was the helical rising of Sothos (here called Kek and still so-called by the Masai), Keskes, or Kos, or Kek, and a great many Totemic dances and ceremonies were connected with this star. It is the same as the Zapotec Pelle-Nij and the Mexican Citlalpol.

Let us turn to Egypt for an answer. It was the opener of the year in the first place, and Sothos was the place whence all the White Spirits came forth after travelling through the underworld (before Amenta was formed in the Solar Mythos). It was the star of Annunciation, and heralded the birth of Horus, the Morning Star of the Egyptian year. In Lunar Mythology it was the Star of Hathor and her infant son, Horus. And above all, to these the oldest observers, it was the herald of the inundation, telling them when the Nile would come down and fertilize the lands, and food would be plentiful as the result of a good inundation.

Thus amongst all these Totemic people, as well as those of the old Stellar Mythos, in whatever part of the world you find them, these people have most important and sacred rites and dances, all identically the same, for the purpose of keeping in living memory all the primitive forms of mythology connected with the above star, Egyptian Sothos, here Kek. And it is only in Egypt that you can find the solution to the question as to the meaning of these sacred rites and dances, etc., which Dr. Haddon could not understand because these originated in Old Egypt, where the importance of the appearance was observed, and indications were understood and carried out from them.

The language of these Totemic natives, as well as of the Pigmies who are not Totemic, is a monosyllabic Sign Language, illustrated by Signs and Symbols, the primary ideographs of written language, which these people have, but not as a written language in the common acceptance of the term, which was developed by the old Egyptians from these Ideographs, and which we call Hieroglyphics.

Tabu—the Snake Totem.—This is an Egyptian name for the same. The name is on the sarcophagus of Seti (Bon. II,

A. 30), and has a snake for determination. In the Calendar of Esneh there is a feast on the 14th day of Thoth in honour of Tutu, "the son of Neith," and the text gives the important determinative of a serpent. Tutu is variously written, but the evidence here given is sufficient to prove identity with the English pronunciation of Torres' Straits natives' phonetic.

It is the same as the Totem of the Dinka and Masai "Nilotic Negroes." In the totemic stage it represented the Earth Goddess Rannut, renewer of life and vegetation, and in a later phase we find it representing the God Tem. You must be careful not to confound the good Snakes with the bad, as both types are represented. The Great Apap in one form was represented by a huge Snake with coils winding around the mountain of light (moon), obscuring the light; this was a symbol of the greatest type of evil, whereas the Asps worn on the crowns of the Egyptian Kings and Queens—as well as by the Mayas and Peruvians—was emblematic of the invincible power of royalty; because it was a symbol of the Sun or Solar fire.

The interpretation of Dr. Haddon, "The miraculous birth of a stone from a virgin"—the Moon as father—is the primitive equivalent to Hathor and the young child Horus of the Egyptians.

"The two stones erected Adi" are the representations of the North and South Poles.

"The story of the six blind brothers" is also an early story, which we find amongst the Mexican and Central American Indians and other peoples as well, and must be referred to the Precession of the Pole Stars, the same as the origin of Hammond Rock. A drowned man becomes a rock, his wives drown themselves and change into rocks—the Precession of the Pole Stars—as they change, i.e. as each Pole Star changed in precession, the race sinks down into the water of space and is drowned, to re-enter life again after the others have served their turn.

His "Naga" was "the instructor of Totemic ceremonies" here, i.e. he was the transformer of the Inapertwa beings into men and women, and instructed them in the ceremonies or initiations.

Land Tenure.—Dr. Haddon states: "They have a primitive way of acquiring land personally," but he states he had failed to find and decipher how and why this is. Here again we must return to Old Egypt to find the marks and meaning of the same, which was ᒥᒥᖴᖴ, and this was not altered until after the Turanian people, so-called, which

I term "Stellar Mythos people," had carried the same throughout the world.

He states that the chief clans were three and two minor ones (whatever that may mean), probably he means that they were sub-classes. The three chief were:—

1. Sam = Cassowary (Bird).
2. Kodel = Crocodile.
3. Tabu = Snake.

These were the big Auged or big Totem. In other words they were the representative Zootypes of the Mother of the original tribe, i.e. their original mother was a Sam, Kodel, or Tabu, and they claim their descent from "this Mother," and were therefore Cassowaries, Crocodiles, or Snakes. The two "minor ones" may represent a subdivision of these tribal Mother Totems. I cannot say for certain from the evidence here given, but it is quite evident from studying this work that there are two different types of the Nilotic Negro here, a primary exodus, having Totems and Totemic ceremonies with no mythology, and therefore no folk-lore tales, and a later exodus having part of the mythology, folk-lore tales, and greater developed Totemic ceremonies. Therefore in the latter case one of these tribal Mother Totems may also represent "the Great Mythical Mother," Kodel, or Tabu, or Sam, for instance. In the one case the Great Tribal Mother was human, in the other she was not, but was a mythical personage.

In Totemism she is the human woman, or the two women, Mother and daughter. In Mythology she is the Mother or two Mothers, not human and never had been human.

In the subdivision, a Totem would be given to the girl, or Mother of the Clan, at pubescence, when she would become bearer for the tribe, and all her children would be named after the Zootype given to her, and would be known as her children. Therefore if she was made a Shark all her children would be Sharks.

The Great Tribal Mother Totem was the original one, and held sacred; if a Kodel man killed the zootype representative of this Totemic Mother he would be killed by the other Crocodile sons. The Cassowary here represents, as does the Australian Emu, a type of the Earth Mother. It was a bird of earth, like the Goose in Egypt, and as layer of eggs is the representative of the Mother of Food.

Dr. Haddon's expedition to Torres Straits establishes the fact that at least two exodes left Egypt during the time of the Totemic Sociology of the Nilotic Negroes, both here and in the Pacific Islands generally. The first exodus of these Nilotic Negroes possessed Totemic ceremonies much less developed than the second; as Dr. Haddon puts it—"a Hero Cult with masked performers and elaborate dances spread from the mainland of New Guinea to the adjacent islands; part of this movement seems to have been associated with a funeral ritual that emphasized a life after death. The new cult possessed these two elements of strength, individualism, and the assurance of immortality, corresponding to those found in Fiji, etc." (p. 45, vol. vi, "Rep. Cambridge Anthropological Expedition to Torres Straits"). This Hero Cult found here corresponds with the Egyptian. The representative is Bomari or Malu or Malo with his two brothers, here called Siagea and Kula. These advanced ideas and knowledge had been worked out in Old Egypt, and undoubtedly the second exodus of these Nilotic Negroes carried them throughout most parts of the world. Being still Nilotic Negroes, although of a higher type, they would be received with open arms, coming from the old Mother Country with the latest up-to-date knowledge, and thus would find no opposition in the way of "great battles and war." Also *it was no alteration of their previous beliefs*, it was *an addition by more elaborate ceremonies*, which would appeal to the nature of these men. When these ceremonies in their fully developed form were brought from New Guinea or Australia does not matter, and at the present time it is difficult to say, on account of the different geographical formation of the world. Then tradition tells one that this knowledge came from the West, and that is sufficient for my argument, the West being here Egypt.

Dr. Haddon gives several mythical tales which correspond to Kalulu and the Dzimwi in the African folk-lore tales. Kalulu is the type of the good power and a Zootype of a new or young moon. The Dzimwi is the evil power, like Apap, the swallower of the water of light, who is very cunning but is always outwitted by Kalulu. The killing or swallowing of Kalulu's mother by the Dzimwi corresponds to the devouring of the Lunar light by the Dragon of Darkness or Eclipse. The Moon myths are far older than the Solar, and the Stellar are much older than the Lunar, "The Mother of the young child of light."

Gerald Massey states that the earliest slayer of the dragon was Lunar, but we have it earlier than this. It was undoubtedly in the Stellar Mythos from evidence we have collected, here represented by the dragon of Darkness—Apap = the night, and The Light, = day, represented by a type of Horus I.

Here she is killed by the Dzimwi. Then Kalulu comes with a barbed arrow with which he pierces the Dzimwi through the heart. It is the fight between Light and Darkness and between Horus and Sut in the primitive form. Amongst the Andaman Islanders the Devil is a big Toad; amongst the Iroquois or Huron, a gigantic Frog, who also drinks up all the waters of the world. We find this also amongst the Indians of the Central States and in South America, as witnessed in the pictures and the various Codices, and on mural paintings found in the remains of the ancient cities in these parts (see Mexican antiquities, Bureau of Ethnology, published by the Smithsonian Institute, Washington). The aborigines around Lake Tyers say that all the waters were drunk up by a monstrous Frog. In Sumatra it was the black Cat which typified the Power withholding the rain. Amongst various tribes in different countries the Apap monster or Set was represented by a Zootype entirely black or red. This in Egypt was the Black Boar of Set. In the folk-lore of various races the human soul takes the form of an Egyptian Zootype, and to understand these tales it is necessary to know the "Talk of the Zootypes."

Dr. Haddon and his companions need not have been shocked by their morals, as, all being mythical, morals could not enter into the subject of discussion.

Kibu, "the island home for Spirits," is the same as the Egyptian Khui Land (the word Kibu is equal to, or is the same as Khui—"Egypt," i; see *supra*).

The Ari, or Companions.—This is an Egyptian word and name for the people of the 17th nome of Upper Egypt and proves that they had advanced to Totemic Sociology. These people have the "Footprints," as have the Australians and all other tribes of this totemic exodus, and these refer to Horus (see Ritual).

Amongst these natives a man must not name his mother-in-law or a wife her father-in-law. This same Tabu is found amongst all Totemic people in all parts of the world, the Nilotic Negroes in Africa, in Australia, and North and South America. (What a loss to human happiness that this does not still universally exist!)

"The time of performing certain ceremonies is fixed by the appearance of certain stars." This we also find amongst the natives in Africa (Makalanga and others), Mexico and Central States of America, and in South America. Also, "after death being changed into white people" is only the symbol and sign language for becoming spirits in the next world, and a belief therefore that they have a life hereafter and a belief in a spirit world. There is a tribe in Central Africa who dance this Totemic ceremony on the banks of a river, when one moiety paint themselves white and cross the river, calling upon the others to follow. This represents the spirits calling upon their brothers to cross the water of death to rejoin them in the new spiritual world.

"Their Spirits go to the West." This would correspond to "the Westerners" in the Egyptian. Their funeral song, "To the West, to the West," as given and translated by Maspero, fully proves this.

The song was a very old one in Ireland before Russell's time, to whom the origin was wrongly attributed, and the people of America copied this from Russell later, but it was originally Egyptian, as Maspero proved by his translation from the Hieroglyphics. The manes entered Amenta, or the Underworld, in the West, passed through many trials and dangers, and rose or came out at the East as a glorified Spirit. The meaning of the symbology "The Keber of the Pop le op," p. 135, vol. vi, is not the rising and setting of the Sun, but the enactment dramatically of Birth, Death, and Spiritual Life.

The index finger pointing downwards indicates that he

descended, i.e. from the West he entered the Underworld. Pointing upwards, it indicates that the White Spirit had passed through the Underworld from West to East to the star Sothos.

Mr. Bruce should not delude himself by thinking that in the "Spirit dance" the women and children have the implicit belief that this is really a ghost (p. 141, vol. vi). They know quite well that it is only a "cinematograph picture," or a representation or portrayal of their beliefs, a part of their Totemic ceremony. All these are only dramas of the Mythology of Egypt. The place where they rose was called Hematit —a place near the eastern horizon only mentioned once in the Ritual (78).

But at the time of this Totemic people's exodus from Egypt Amenta had not been formed. But they had an Underworld which was the same all through Totemism and the Stellar Mythos—a place deep down underneath the earth, which their manes entered in the West, when they had to travel through a desert of darkness, where there was no life but all was darkness and desolation (inhabited by a huge monster, Apap). After travelling through this, they arose again in the East in spirit form, always depicted by painting their bodies, or part of them, white, and were conducted to the circumpolar paradise by a Zootype representing Anubis.

The Markai.—Dr. Haddon interprets this as a spirit, or ghost, of the recently departed. The Puro men of West Africa have the word, as Mr. Fitzgerald Marriott gives it, Mama Koome, signifying the same, i.e. spirit that may return. It is the Egyptian , Maa-Kheru; and in the Ritual of ancient Egypt it is stated: "a spirit who is Maa Kheru: every door in the underworld opened itself and every hostile power, animate and inanimate, was made to remove itself from its path." These were the spirits or ghosts whom these poor natives looked upon to come back and do them good, and whom they propitiated not to do them harm, a primitive form of ancestral worship. His Mamoose (Mamus), meaning Red Man or Red Hair—Mamoose boy at Erub, is a type of Sut, who was always depicted with red hair. This same type is found amongst the primitive races in all countries; even amongst ourselves, the tale is still extant in some counties.

The same primitive mode of preserving skulls, etc., is practised here that we find in the remains of the earliest races in all parts of the world, even in these Islands, as is witnessed by the ossified remains of our forefathers found in barrows in Scotland, Ireland, and England, viz. scraping and cleaning the bones and then painting them with red ochre as a preservative; also the same mode of decapitating and removing the brains through the foramen ovale. This was an early Egyptian custom, as is proved by the researches of Fouquet. Therefore it originated during Totemic Sociology. The dead bodies of their ancestors were desiccated and kept in their huts for the purpose of Spirit communion, and oracles were supposed to be given from their skulls; probably this was the first reason, at least, for making and preserving the mummy.

The *Bull-roarer* is used here as amongst all natives of the totemic class wherever found, and the explanation is given in the Egyptian alone.

The *Mudab*, as shown on p. 347, is a very important find, as it is the most primitive formation depicted symbolically of the Egyptian Spirit of Amsu as risen Horus I, and of Ptah-Seker-Auser of a later Cult (Solar). It is depicted here as the resurrection of the Spirit. "The association of the sexual act with agricultural fertility, and the story of Sida" is the primitive "Phallic Festival," wrongly called Phallic Worship; it was not a worship, but a festival of seedtime and harvest, and fructifying, increasing, or generating; and in this primitive community it is acted in their totemic dances, etc., in order to ask for food and show that it was such that they required in plenty. Our maypole dance is a remnant of the same.

Dr. W. H. Rivers in this work (vol. v) states that a good many instances were found in which a man with a given Totem married a woman with the same Totem, the most common cases being marriages between individuals when the Totem was "Kodal" = Crocodile.

This might have been the Great Mythical Mother, or a tribal Totem alone, but it might also have been a Totem given to a girl at the time of puberty, therefore this would be the Zootype representing both, i.e. "The Great Mythical Mother" and also the tribal Totem. Dr. Rivers does not distinguish between the type of "The Great Mythical Mother" and the tribal Totem, which might have the same symbolic Zootype, and therefore the same ideographic name, and yet have no connection.[1] The one is mythical, the Great Mother, the other "Mother who has been given the Totem" is not; she is human, although both are depicted by the same Zootype symbolically, therefore it is more than possible that there is some mistake here in his interpretation of what was told him by the natives, or if not, we have here the earliest form of Totemism, i.e. where brother married sister (see *supra*).

He also tries to trace the genealogies of these people through the *male* line! He might as well try to degenerate back again into a Pigmy—both equally impossible. *The descent was always through the female* amongst all the earliest Totemic people, wherever found. He acknowledges that he cannot get back beyond the fourth generation, *and then he commences with mythical personages*!

If Dr. Rivers had inquired of these people whether they were descended from the Kodal, or if the human mother had brought forth "Kodal children," we should have learnt something. Dr. Haddon states that if a "Kodal man" killed a Crocodile, the other Kodal men would kill him, and if a Umai man killed a Dog his fellow clansmen would fight him, and no Sam man would kill a Cassowary. If he did, his Cassowaries would kill him. Therefore here we have three "Mother Totems" from which these descended, viz., Kodal = Crocodile, Umai = Dog, Sam = Cassowary.

One moiety of these people have the Turtle for their Great Mother Totem, just as some tribes in North and South America and Africa, and their Totemic ceremonies

[1] Without a thorough understanding of the Mythology and Zootype language of Old Egypt, there is great difficulty in distinguishing the one from the other.

correspond in each case. Another moiety have the Crocodile for their Great Mother Totem, which corresponds in every particular to tribes in Africa and America, New Guinea, and other places. The magic of these people corresponds, and is the same, with little variation, in all totemic tribes, and the meaning is only explained by the mythology.

The records obtained and set forth in these volumes are incomplete in one respect, i.e. in the explanation and interpretation of the same, but the whole forms valuable data which, when deciphered with the correct key, adds additional proof that Africa was the original home of all —the starting-place of "original man," because all their Totemic ceremonies, mythical ancestors, and "Hero Cults" must be traced back to Africa—nowhere else in the world can they be found. Moreover, if the writers had known the mythology of Ancient Egypt and African folk-lore tales, which followed as an outcome of the mythology, they would have been able to state fairly accurately the different ages of development and the time of exodus of each tribe from their old home; but until one learns the language of the Zootypes, the Sign and gesture language, and the meaning of their sacred Signs and Symbols, it is useless to expect any definite advance either in Anthropology or Ethnology. Those who do not learn this will always be wandering away to the dark shore of Asia or mythical Atlantis or some other such region. The natives found here are in the Totemic state of Sociology, and whilst these natives have stood still as regards progress in knowledge and bodily development, and possibly some may have retrograded, having been cut off from external communication with the outer world, the old Egyptians at this time still continued to advance, from the time these people left, up to the Third or Fourth Dynasty. The traditions of these people are that they came from the West, and there are certainly two different types of the Nilotic Negro, a lower and a higher, each having particular characteristics, corresponding to the Dinkas and Niam-Niam in Africa, the inferior type having the Totemic ceremonies and symbols of a more primitive time, and no mythology or folk-lore tales, the superior having the folk-lore tales developed from the mythology, and so-called Hero Cult. Totemism and mythology did not spring from various sources, as Professor Frazer would have us believe. They are one as a system, one as a mode of expression, one as a mould of thought, and all the great primordial types are universal and originated in Egyptian Mythology, the oldest in the world, and were not an explanation of natural phenomena, but a representation of their earliest primitive times. The masks used in these native ceremonies are not made in human likeness, but made to represent some elemental forces and powers of nature, which were superhuman primitive ideographs, not worshipped, but adopted for their representative value.

The clear and lucid manner in which Dr. Haddon has recorded his observations is a great gain to all scientists, and although he and his companions were not conversant with the mythology and the symbolism, and therefore not able to set forth the true meaning, yet the accurate observations, the close study of these people, and the careful manner in which he has recorded them, render it easy for one who can understand and speak the Zootype language to interpret the same. These people are fast altering their old faith, and so for the future very little of the truth of their origins, etc., will be available; on this account, Dr. Haddon's work will stand as a "beacon of light," one made from personal examination and observation on the spot, before all their traditions had become distorted or lost, and thus will add a great and important chapter in the study of the evolution and development of the human race. The beauty of his work consists in the plain, straightforward, and lucid manner in which he has recorded his observations and the folk-lore tales, as told to him, without any of those embellishments we so often find, or those imaginary ideas in trying to interpret the true meaning which he obviously felt might be only theoretical. Such students of nature are rare and should be placed on the highest pedestal science has to give.

CHAPTER XVI

THE TRIBES OF BORNEO AND THE TODAS

MESSRS. HOSE AND McDOUGAL state in their "Payasi Tribes of Borneo" that they do not believe that the customs of these tribes of Borneo can be regarded as institutions surviving from a fully developed system of Totemism now fallen into decay. But on reading the book carefully, there is only one conclusion that we can deduce from it, and that is this. At least two different tribes were here before the influx of the Chinese and the Southern Mongol influence, one with Hero Cult, and the other without. There is quite sufficient evidence that Totemism or Totemic Sociology was fully developed amongst the later tribes; that the Crocodile was the Mother Totem of the one tribe and the Pig or Sow of the other.

These gentlemen have confused the "animistic spirits" with the "Spirits of the Ancestors."

The Punan are nomad hunters, short in stature, with long bodies and short legs, of pale yellow colour and small eyes. The head is sub-brachycephalic.

They cultivate no crops and have no domestic animals.

Their dwellings are merely rude, low shelters of palm leaves, supported on sticks to form a sloping roof which keeps off the rain. Some live in caves of the limestone mountains.

A Punan community consists generally of some twenty or thirty adult men, women, and children. One of the older men is recognized as leader or chief.

The Crocodile is the Great Mother Totem.

They believe in omens and pray to "Bali Penyalong," who is their principal object of trust. They have great faith in "charms," and believe in elementary spirits and in the spirits of their ancestors.

They live in forests and cannot count beyond three (ja-dua-telo). Larger numbers are "many."

The Indonesian and Tari tribes include several classes or tribes; all these have Hero Cult, and some the commencement of the Stellar Mythos. Dr. Haddon recognizes five main groups, which might be divided into the Indonesian stock and are dolichocephalic, and the Proto-Malayan stock, which are brachycephalic.

But there is a great deal of mixture at the present day, and up to the present time no positive evidence of the primitive race that existed on this island; probably they have been destroyed. There is sufficient evidence that the Punan are descendants of the Nilotic Negro, but the indigenous population has been so mixed with emigrants from the mainland that it is difficult to say to which branch of the Nilotic Negroes the originals belong. The fact that there are no ceremonies to initiate youths into tribal mysteries points to the conclusion that the decay of Totemic Sociology has taken place, and that the beginning of the transition stage to Stellar Cult was brought here by the Indonesian element, which has affected the primary tribes found here.

More especially is this shown by the sacrifice of "animals," instead of the "human mother," as a sacrificial rite, and their interpretation of Orion as the figure of a man, "Lafang" (like the old Stellar people of Egypt); his left arm is thought to be wanting, i.e. it has not yet freed itself (Amsu of the Egyptians), and the story of "Usai" (p. 142) can only be interpreted by the early Egyptian Shu.

The image of the "Hornbill" used at ceremonies represents the Egyptian Great Mother Apt, or Ta-Urt.

All the class of Spirits called "Toh" are Spirits of the Ancestors, i.e. the living spirits of the "dead heads." This is proved by their being associated with the dried human heads that hang in most of the houses. It is from the fear that these spirits (of the dead) may act malevolently towards them that propitiation is offered to them. Messrs. Hose and McDougal state that "the Toh play a considerable part in regulating conduct, for they are the powers that bring misfortune upon a whole house or village, when any member of it ignores 'Tabus,' or otherwise breaks customs, without performing the propitiatory rites demanded by the occasion. Thus on them, rather than on the gods (elementary), are founded the effective sanctions of prohibitive rules of conduct. For the propitiation of offended Toh, fowls, eggs, and the blood of fowls and young pigs are used, the explanations and apologies being offered generally by the Chief, or some other influential person, while the blood is sprinkled on the culprit or other source of offence." They further state: "The belief and practice of the Kenyaks and the Klemantans in regard to spirits of this class are very similar to those of the Kayans. They designate them by the same general name Toh."

This is another proof that the Toh are considered "ancestral spirits," because if "elementary spirits" they would be quite indifferent to observing "Tabu" and tribal customs, whereas ancestral spirits would be angry if the old traditions were neglected or not observed.

The Kayan's attitude towards a Crocodile when he says "a Crocodile may become a man like himself," shows that he is only speaking in Totemic language. The Crocodile was originally the original Mother Totem, and all her children would be young Crocodiles.

Laki Ncho and Bali Flaki are the two messengers, like the two Hawks amongst the Australians. Bali Sungeri is the name given to a being, thought of as embodied in a huge serpent, or dragon, living at the bottom of the river. He is supposed to cause the violent swirls and uprushes of water that appear on the surface in times of flood. He is regarded with fear, and is held responsible for the upsetting of boats and drownings in the river. Here we have the same great evil serpent as we find amongst the Australians and other natives, the original myth is found connected with the Nile (see "Signs and Symbols," 2nd edition), and represents the Great Apap of the Egyptians.

Bali-Atap = Horus, is associated with the wooden image of a hawk.

Do Tenangan, identified with Pa Silong and Bali

Penlong, the supreme gods of the Klemantans and Kenyaks, corresponds to the Egyptian Mother Apt.

Bali Penyalong or Balingo, God of Thunder, corresponds to Shu.

Messrs. Hose and McDougal state that "the Kayans believe in the reincarnation of the soul," although they state "this belief is not clearly harmonized with the belief in the life of another world." These gentlemen have confounded the "Totemic soul" with the Human Soul, or Spirit. These people do not believe in reincarnation of the *human soul*, and no natives ever did, as we have shown—they only believe in "Totemic Souls or Spirits," which were never human and never could be.

The beliefs and traditions of the various tribes in regard to the other world are like the old Egyptian, and the rough map depicted in their book might be compared with the Humboldt fragment found in Mexico, and the underworld of the Egyptians.

The mode of burial "in jars" practised by the Murats is the same as we find amongst some tribes in South America and other countries, and as we have found was a custom of the early Egyptians.

The distension and elongation of the ears and the bands decorating the Iban women, the women's corsets of brass-bound hoops, borrowed from the Malohs Pcat (195), are precisely similar to what we find amongst many Nilotic Negroes in Africa at the present day, as is also the adornment of the edge of the ear with many brass rings or beads (see Masai women).

Furthermore, Messrs. Hose and McDougal state: "We have here numerous cases in which a whole community refuses to kill or eat an animal which is believed to protect them by omens and warnings and in other ways, and in which the animal is worshipped with prayer and sacrifice (e.g. the hawk among various tribes). We have at least one instance of a community claiming to be related to a friendly species (Long Patas and the Crocodile) and having as usual an extravagant myth to account for the belief. We have the domestic animal that is sacrificially slain—its blood being sprinkled on the worshippers and its flesh eaten by them—and is never slain without religious rites (pig of the Kenyahs and Kayans). We have the animal, that must not be killed, tattooed on the skin of the men (the dog) or its skin worn by fully grown men only (the tiger-cat), or images of it made of clay or carved in wood and set up before the house (the Hawk and Crocodile). We have also the animal that is claimed as a relative, imitated in popular dances (the Dok monkey of the Kayans). The belief that the souls of men assume the form of some animal that must not be killed or eaten (deer and the Arctogale among the Klemantans), the observance by individuals of a very strict avoidance of contact with any part of an animal that must not be killed or eaten in any case (horned cattle among many Kenyahs and Kayans)."

All the above is conclusive evidence that we have the old Egyptian Hero Cult here, the elementary spirits divinized, Set, Horus, and Shu, with the Egyptian Mother Apt.

The sacrificial pig represents, as it does with the New Guinea people, Rannut; the pig is sacrificed in place of the Mother of the Tribe and eaten as a sacrificial meal.

The Dog (Anubis), the Tiger-cat, and Deer, and "the image made in clay or carved in wood," and the customs and ceremonies observed, prove that these were former Totems of the tribes, and "the belief that the souls of men assume the form of some animal" proves this, because these were Totemic souls and not Ancestral souls, which they clearly differentiate, but which are not understood by the writers of the book. Messrs. Hose and McDougal's conjecture that the "Indonesian element" originated from a common source with the early Greek and Roman myths is, in my opinion, correct. Some of these were Hero Cult people in the transition stage to Stellar Mythos and came out of Egypt. Whilst some travelled North-west through Europe, others went North and East. The earliest Mongol families spread over Asia, North and South, and as no doubt Borneo was at one time part of the mainland, we find them here. They intermarried with the native tribes already in Borneo and imposed upon them the higher cults, which in time caused the original natives to abandon some of their Totemic ceremonies and rites.

The word "Tenangan" is derived from the Egyptian, and signifies "the water-cow of Apt," in Apta. Here it represents "earth as the Great Mother and giver of water."

In Africa the first Great Mother of all was constellated as the female Hippopotamus Apt, or *as a Crocodile* (here found). In Greece she was imaged as a female Bear. Set, her son, was represented by a Jackal, which here became a Dog through change of fauna. The whole of the customs, beliefs, and rites of these people can easily be read by and through the Egyptian myths, and are proofs that we have here some of the descendants of the Nilotic Negroes without Hero Cult (the Punans), and others with Hero Cult (Kenyahs and Kayans), and also the transition stage to the Stellar Mythos brought on by the Indonesians intermixed with each other.

The Todas.

Dr. Rivers, who visited the Todas and wrote a book called "The Todas," has therein set forth very interesting remarks on these people. He states that the Todas are divided into five clans; that every clan has a god (a Totem); that two gods stood out pre-eminent amongst the rest, one a male, the other a female. The female was here called Teikirzi, who now rules over the Todas.

He states that "they were father and daughter" (that is not so, they are mother and son); that they also have a number of gods.

"On," he says, was the son of Pithi ("Pithi" and "On" = the same), who created all the buffaloes, and after the death of his son Piu he went to Amnödr and left the Todas to be presided over by the goddess Teikirzi.

All the dead go to the West to enter Amnödr, and "On" now presides over it.

The earliest god was Pithi, who was born, i.e. came. On was his son by his wife Teikirzi ("he was the bull of his mother"). Later, death came in the person of Piiv, who was the son of "On." He also states that they have "River Gods," Teipakl and Tarkhwar (god and consort goddess). They all came from the summit of the Hills and

Mount, where there are circles called "pun" (to which they are rather indifferent).

There is also Koraten or Kuzkaro, who was the son of Teikirzi (? another name for "On").

There is another named Kwoten who murdered Parden (see Masai). There is no doubt that they also propitiated the Spirit of their ancestors, as well as elemental spirits, but Dr. Rivers has done what a great many other men have done, i.e. he has rather confused the *Ancestral Spirits* with *divinized elemental Spirits*, and then does not find himself clear on the point, and cannot believe that there are elemental spirits because they do not here represent thunder, lightning, or other elemental forces. How does he know this? As a matter of fact they do here represent these elemental forces, but not as elemental forces, as Dr. Rivers represents or thinks. These forces *became divinized at the time of the latest phase of Totemic Sociology.* The Buffalo here is a sacred representation of the Egyptian Cow—as a very early form of Totem for Hathor, the Cow depicting her as a Zootype for Hathor.

The Palal is a representation of one of the Priests or Seers, but inasmuch as the Priests of the Temples of all the goddesses in Egypt were subordinate to the High Priest or High Priestess, he does not occupy the high station as if it was a temple. Here he obviously personates the divine Son and is dispenser of blessings to the world for the divine Motherhood that was represented by the Cow, as her Zootype—the giver of food. Typical of Hathor, the Egyptian Venus, the Cow that protected her Son with her body, primarily when the mother was a water Cow. Teikirzi, "the ruler of the world," here represents the Egyptian Great Mother Apt, who brought forth first Set, second Horus, third Shu, as the three Khamite Heroes. "On," here, who represents the ruler of the world of the dead, is Set, the first-born of Apt, who became the King of Darkness in one form after his fall, when Horus superseded him.

Here he is also represented as a "Creator" of the Buffalo. Dr. Rivers states that he became the ruler of Amnödr, where he lives, i.e. the Egyptian "Underworld."

The four most important "Gods," as he calls them, were the "Old Mother Apt" and her three first-born, who were the three Khamite Heroes—the bulls of their mother who became Creators in Totemic Sociology, and afterwards others were added. I quite agree with Dr. Rivers that there has been much mixture here—both Stellar and Lunar. I would suggest that it is a religion that has been corrupted, "rather than degenerated." The Todas were originally Hero-Cult Nilotic Negroes; they still offer supplications and sacrifices to the "Gods" or "Heroes" for help and protection—as well as to Ancestral Spirits or Spirits of the departed. The "circles on the hills" which are found here and made of stones, are remains of the first Stellar Cult people, who evidently lived here at the time that Set was primary God; how many, or what became of them, it is not possible to say, but the two circles are quite a characteristic formation of early Stellar Cult people. Dr. Rivers can find no trace of Totemism amongst them. Yet the Totem is here—the Cow or Buffalo—which was the very earliest Totem of Hathor in Egypt, a later name for the Great Mother Apt.

The Women tattoo, and the fact that the Priest is allowed marital relations with women is an old custom of the Egyptians, which has been carried on to much later times in many countries.

Dr. Rivers states that there is a near affinity to the Hindu deities, and he might add also to the Stellar Cult—the reason being that both these Cults have corrupted the Todas, having been in contact with them at some time, probably the Stellar, very long ago; evidently a small moiety came to this part of the world, as the Todas were not destroyed by them.

There is in Africa (South) a tribe of Nilotic Negroes, at the present day, who have the same traditions and practices as the Todas—Hero Cult people, with the Cow or Buffalo as their Totem.

PART IV

STELLAR MYTHOS PEOPLE

PLATE LXIV.

Neolithic "Hut" and remains of ancient town opposite Nodden, Dartmoor.

CHAPTER XVII

STELLAR MYTHOS PEOPLE

Characteristics.—Osteo-anatomy, present type of man, with supra-orbital ridges a little more in evidence in the earlier exodes of these people than in the later, but in the earlier they are much less prominent than in the Nilotic Negroes:—

Facial appearance of a Mongoloid or Turanian type, broad across the zygomatic process. Eyes almond-shaped.

Hair.—As a rule straight. Transverse section round.

Implements.—Chipped on both sides and polished, in flint and hard stone. Recent excavations and research in Egypt have yielded the most perfect examples of the flint-knapper's art known. Flint-tools and weapons, more beautiful than the finest Europe, Asia, or America can show, have been found. Also worked and used iron, copper, bronze, gold, silver, etc. Pottery beautifully made, decorated with iconographic figures.

Buildings.—Built with large polygonal stones and monoliths; used a very fine cement made of the same material; iconographic, corners bonded, could not build arches, always lintels.

Language.—Spoke an agglutinated language. Invented the art of writing in hieroglyphic ideographic Symbols, and commenced to form alphabet.

Worked out their Astro-Mythology and part of the Eschatology (if not all). Represented God and His attributes by Zootypes, i.e. pre-human forms.

Burial.—Buried in the thrice-bent position, either sitting or lying, with characteristic amulets and implements interred with the body. For the first 52,000 years the face was placed facing south, for the next 250,000 years the face was towards the north.

In the next stage of the evolution of man, and the next exodus from Egypt—after the Nilotic Negroes, who carried with them the so-called Hero Cult—were those whom I have termed the first or early Stellar Mythos people.

Here in the Nile Valley man had been progressing and developing his body, mind, and religious conceptions, all now gradually assuming a higher type, until, in his spiritual notions or thought formation, we find that the transformation of an elementary power into an attribute of God has taken place. To the ancestors of the "Masai group" we must turn for further developments of all the Stellar Mythos people. I have shown how these people came up from the South and conquered the original inhabitants—Pigmy, Masaba and Non-Hero-Cult Nilotic Negroes—and "commenced to form Egypt by dividing it into Nomes"—a state of "Totemic Sociology" which lasted, according to Manetho, for over 16,000 years. *This is the only record left of any date* of the human race *before the Stellar Mythos period.* So from the birth of the first Pigmy until the advent of the Nilotic Negroes, who possessed Hero-Cult, and who first settled to form the "Nomes" of Old Egypt, we have no record of the time in years that must have passed. This can only be gauged by approximately calculating how many thousands of years it took to produce the difference, anatomically and physiologically, between the first Homo and the Nilotic Negro with Hero Cult; probably a period of 1,000,000 years or more.

The evidence of Manetho is most conclusive in the statement he makes *as regards this period of time,* and also as to the approximate date at which the Great Pyramids were built. He states that "the Pyramids were built at the *end of the reign of the Gods and the Heroes, who reigned over* 16,000 *years.*" That is to say, *the end of the reign of the Gods and the Heroes was at the end of the Totemic Sociology.* Therefore, it must have been during the time of the Stellar Mythos, which was the next evolution in progressive development, that the Pyramids were built. Moreover, it is stated that the Great Pyramids were built by the followers of Horus, who were the Stellar Cult people.

An inscription at Dendera states that "In the great foundation in Dendera were found old writings, on decayed rolls of skin of kids [parchment], of the time of the followers of Horus. These writings were found between the walls of brick of the south part, in the reign of Pepi."

This is borne out by the evidence of the Ritual, as well as the evidence, external and internal, of the Pyramids themselves. That these were not built during Lunar and Solar Mythos is proved by the orientation, which is north, not east. The supposed builder, according to most Egyptologists, was Cheops, or Kufu, who, Herodotus distinctly states, reigned during the Osirian Cult—which was the latest phase in their evolution of Solar Mythos before the downfall of their Empire. Therefore, the Pyramids were built many thousands of years before Cheops or Kufu. This evidence, added to what I have already written in "Signs and Symbols of Primordial Man," makes quite conclusive both the great antiquity of these wonderful monuments and the past greatness and learning of that period—mathematics, astronomy, and eschatology—to which the old Egyptians had attained. In this no nation since, up to the present day, has surpassed them.

The progression and regression of the exodes and their descendants, owing to glacial compulsion, would affect the peripheral nations or tribes and leave Old Egypt untouched. Hence there is reason to believe that in the latter a very high type of man was evolved. The incalculable ages through which these central people lived undisturbed, and persistently observed and noted, are sufficient explanation of their scientific progress even without resort to the explanation of inspiration.

It has all come about probably through the perfection of the creative thought that set rolling an evolution, both material and spiritual, *ab initio*.

In considering the sentence "the followers of Horus," you must read it in the same sense as "the followers of Christ."

I use the term "Stellar Mythos People" because these were the first, or earliest people, who *kept time*, and they reckoned time by the observation of the Precession of the seven Pole Stars, Ursa Minor, as witnessed by the Ritual of Ancient Egypt and by Herodotus. These people did not reckon the "year" as we do. They had one great year of 25,827 years, that being the time it takes for the Pole Stars to perform one cycle of precession, also the period of one revolution of the sun around its centre. This they divided again into smaller periods of 3,689 years (approximately), that being the time of the duration and changing of each one of the Seven Pole Stars. They divided the year into twelve parts, with two different measurements of time, the one civil and the other astronomical. How long Stellar Mythos Cult lasted it is not quite possible to say, but one must come to the conclusion, from evidence found, that it lasted for about 300,000 years, because we have found a record of their observation of the precession of the Pole Stars *for ten great Periods*. As this was written *during the Stellar Mythos*, 258,270 years must have elapsed before the record was made. The history of the creation of the world, version B, states: "They covered up my eye after them with bushes twice for ten periods."

A period being one Great Year, i.e. 25,827 years, ten periods would be 258,270 years. This gives the duration of time observed by the old wise men of Egypt, during which, up to that date, they had recorded the precession of the seven Pole Stars (Ursa Minor), and this record has been handed down to us in the Ritual. The knowledge of this was known to those who left Old Egypt and spread over the world, as witnessed by the fact that it is also found in the Central American States (see later).

The "covering up" must be translated by Sign Language, and means that for two periods the seventh star Polaris, "Herakles," did not belong to, or represent, Horus—as primary God—until after two periods of precession, i.e. 51,654 years. Then Horus was given the primary place as God of the Pole Star North, and Set was deposed and became "The Hidden of the Khas." Thus we have a record of time left to prove how long Set reigned as primary God, viz. 51,654 years, during which period all the Stellar Mythos people were buried with their face to the south and their temples were round in form at first, with steps leading up from the four cardinal points. I shall give further proof that they worked out the stars of the Southern Hemisphere first, and associated them with Set, the first-born of the Old Mother Apt, as their first or primary God (El Shaddai of the Phœnicians). Set was called Father of all the Gods.

Nowhere except in Egypt would the climate allow of this uninterrupted observation. In other countries the clouds, rain, monsoons, and atmospherical conditions would render it impossible.

During the period of the Stellar Cult, man made greater progress in evolution here in Egypt in every form than at any other period. The first exodus of these may be classed as Neolithic people, man having now passed the Palæolithic stage.

We find that all these Stellar Mythos people spoke an agglutinated language. The earliest probably only wrote a few signs, symbols, and ideographs, but the later exodes, who had further developed their faculties, wrote fully the hieroglyphic language. The original signs and symbols —ideographs—had multiplied, and certain sounds were used to denote objective things, which were ideographic.

Prefixes and suffixes were subsequently added to them. From the beginning the whole was progressive learning, from the primary formation of a few hieroglyphics until, with gradual additions to and combinations of these, they perfected their alphabet.

Herodotus (Eut. IV) states that "all agree that the Egyptians first defined the measure of the year, which they divided into twelve parts; in this way, they (the Egyptian Priests) affirm the *stars to have been their guides.*" Further, he states: "They were *the first also who erected altars, shrines, and temples; and none before them ever engraved the figures of animals on stone*: the truth of which they sufficiently authenticate."

They worked in iron, bronze, silver, and gold, as well as in flints, etc. Their first and primitive instruments are easily distinguishable from those of the Palæolithic people by reason of their beautiful finish and polish.

A progressive advance of culture had been developed here in the Nile Valley; they cultivated the soil, tamed wild animals and bred them, understood the art of spinning, weaving, pottery making, working in leather, iron, copper, tin, and gold, etc., and had worked out by observation the time of revolution of the Sun, Moon, and Planets, the Great Bear, and other constellations.

This relatively advanced culture had been progressive in the Nile Valley, whilst in Asia, Europe, and other countries, the earlier primitive man and primary Nilotic Negroes were still struggling for existence with the mammoth and other ferocious animals. When the intense cold of the glacial epochs occurred, those who had gone to the North were driven back South again; they would thus meet the early Stellar Mythos people coming up from the South in ever increasing numbers, and with more and more cultured development—and war would be waged. We know from remains found on the hills, in this and other countries, that these first Stellar Mythos people formed the first villages or "towns," and protected themselves against wild animals and the fierce inferior tribes found as they progressed along towards the North, gradually exterminating them as they interfered with their progress.

Probably some of the first of these Neolithic people were the descendants of the Madi class, who dig deep trenches round their villages, throwing up the earth on the inner sides into a parapet, and planting a stout stockade of poles along the top. This Madi class were agriculturists, keeping goats, sheep, and cattle, corresponding in every particular, as far as the evidence left can prove, to the first old Neolithic people found in this and other countries.

Good examples of their old remains still exist in this country; perhaps one of the best may be seen on Dartmoor, Devonshire. It is situated near the head of the River Lyd, and opposite Nodden. Two very well-preserved "huts" are here (one photograph is shown here taken by my niece, Miss E. Churchward). One has two almost perfect jambs, and lintel; the other has evidently been disturbed by elemental forces (Plate LXIV).

The deep-dug trenches and high walls composed of

enormous boulders, surrounding a large encampment which must have existed here, and the formation of the "lanes" leading into the "town," with the two circles in the centre, are still sufficiently preserved for any one to reconstruct the whole.

There are many others found on Dartmoor and in Cornwall.

The first exodus were comparatively primitive, but as time passed on they developed to a higher degree, mentally as well as physically; their advance in works of art became great. We find remains and monumental ruins of a later exodus of these people existing in almost every country, as well as in South and other parts of Africa, showing and proving their progressive evolution in language, works of art, buildings, and religious conceptions. In China and Japan, in Java, the Central States of America and South America, in Easter Island, Pitcairn Island, Tahiti, the Marquesas, Tonga or Friendly Isles, Lele and Pnape of the Caroline Islands, and the Marianas or Ladrones, we find traces of Stellar Mythos people. The Easter Island statue at the right of the entrance to the British Museum, brought here forty years ago by H.M.S. *Topaze*, which is fairly well sculptured, represents "Horus and Set," the twin Gods in Stellar Mythos, but it is evident that at the date of the sculpturing Horus was in the ascendant, because Set is represented at the back as a much smaller figure.

In various parts of Europe, Asia, Africa, and America we find the evidence of a past great race, all buried in the thrice-bent position, facing north or south, with signs and symbols common to all, and with the old Ritual of Ancient Egypt as their religious beliefs and doctrines. These are all facts written in stone, still extant, and open to all who can, or will, learn to read them. But we find no evidence that they ever went to Australia, Tasmania, the extreme part of North and South America, nor to New Guinea and the New Hebrides Islands.

One of the proofs of this lies in the finding of skeletons of these people, all buried in the thrice-bent position, facing *south or north*, the oldest facing *south*, with amulet.

and instruments, etc., in the tomb or buried with them. This is *quite characteristic.*

There is a proof of their progressive evolution in Egypt still extant, viz. remains found in the tomb of Nagada and other places in Egypt. Skeletons were found buried there in the thrice-bent position with the face to the north—in cists—or on the side, the same as those found at Harlyn Bay, Cornwall, and in Yucatan.[1] The writings found at Nagada in Egypt are of the same kind and character as those found in Cretan, Ægean, Breton, and various other ruined monuments throughout the world (see later).

We call the first exodus of these the "old Neolithic Stellar people," who may be classed—as far as their osteo-anatomy is concerned—as a "modern type of man." When did the first exodus of these take place? As before stated, Professor Ragazzoni, between 1860 and 1880, discovered human remains in the *Pliocene strata* in Italy, and Professor Sergi, one of our highest authorities, states that these remains "were undoubtedly of the modern type of man." Many call into question their authenticity, but Professor Sergi is far too observant a man to be deceived, and his little book, "The Mediterranean Race," should be read by all who take an interest in this subject. By his writings and photographs he clearly proves the African origin of the various types of skulls found in Europe. Implements of Neolithic people have also been found in the Pliocene strata, which were formed not less than 600,000 years ago. *Therefore the type of modern man* had been evolved here in Old Egypt at that period, and had travelled as far as Italy at least.

That the remains of pre-Neanderthal man (jaw found at Heidelberg, 1907) and the Neanderthal man type, both Nilotic Negroes of the first exodus, were found in a more recent formation, viz. Pleistocene, does not affect the question in the least, as all our friends appear to think. They both probably existed here in Europe at the same time, and I have already explained and given the reason for this. The analogy of my friend, Professor Keith, with

[1] See "Signs and Symbols of Primordial Man."

regard to the Australians would not "hold water." He, like Professor Sollas, has never taken into account that the Pigmy is a much lower type in evolution in every way than the Australians, and yet they still exist in the same country as the present white man. How greater, therefore, would it certainly be at that long distant period, when the exodes coming up from Egypt would have to travel on foot, and make good their way in their progress north, surrounded on all sides by the more barbarous clans. They could not extinguish them in a continent like Europe for perhaps more than 100,000 years.

As regards the religious conceptions of these Stellar people, their earliest division of heaven was into two, south and north, which they depicted by two circles, and all these so-called temples were south and north. The remains of these two divisions are found in many countries as two circles of stones, especially in Cornwall and Devon, North Scotland, on the hills in India and in America, formed by twelve stones or monoliths for each.[1] It is the oldest form of temple built by man.

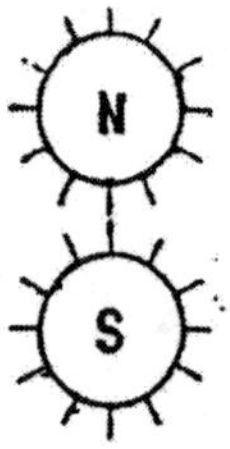

N

S

For knowledge in the development of religious doctrines we must turn to the old Stellar Mythos people, if we wish to follow the true evolution of man, and we have the Ritual of Ancient Egypt as a guide to decipher that which we find otherwise.

The Nilotic Negroes, having divinized the Elemental Powers and having observed the two fixed Pole Stars which never set, named them the two Eyes. One—the South—they assigned to Set, the first-born of the Great Mother Apt. The other—North—was assigned to Horus, the second-born of Apt. Thus to these two brothers were assigned the two Pole Stars on high, often represented by the two Eyes, and depicted in their ceremonies by the two Poles, North and South. These are sometimes referred to as "the twin brothers."

As these people migrated north, so the South Pole Star

[1] I do not mean that these temples were composed of twelve monoliths only. These circles were built of stones as other temples, only there were twelve monoliths surrounding them, or inside in the form of a circle.

would gradually disappear and the North Pole Star would rise higher and higher in the heavens, and as this North Star was associated with Horus, so he would gradually be placed in the ascendant and made Primary God, appropriating all the attributes they had created for Set, with more added, and Set was then consigned to be the God of Darkness.

To each one of the Pole Stars (Ursa Minor) was assigned a zootype representing a god, or one of the powers or attributes of Horus, and these were called "The Seven Glorious Ones," or "The Khuti."

These Pole Stars, Ursa Minor, were always to be seen, whereas the Great Bear or Ursa Major was not; it partially dropped below the horizon, out of sight of the Egyptian observers.

The next phase was when Shu lifted up the heavens and represented them as a Triangle, and although many have supposed that the old Egyptians meant by this that "the lifting up of the heavens" was literal, they did not mean any such thing. They knew better. The term is used in sign language to indicate that, as they came north, the North Pole Star rose in the heavens from the horizon.

This was the primary Trinity—Horus, Set, and Shu; later the four quarters were filled in to form heaven as a Square, first by two pyramid triangles, and later, into a square of four quarters, as will be shown in the Solar Cult.

Thus we see amongst these Stellar Mythos people a progressive advancement in the portrayal of the heavens in three different phases.

We find that the transformation of an elementary power into a God had taken place. This can be traced back to Egypt and nowhere else on this earth. An example has been given in the deity Shu, who was first an elemental force of wind, or air, or breath. The power was animistic or elemental. Shu was then given a star, became the Red God, and attained the rank of a Stellar Deity, as one of the seven Heroes who obtained their souls from the stars of heaven. The three principal elementary powers which preceded the existence of the Gods, or powers divinized, were Set, Horus, and Shu. These are Dr. Haddon's "Heroes" in his Hero Cult, in one form or another.

These were the beneficent powers that were divinized as male deities in the Egyptian Mythology, and were the three "Kamite Heroes." They were propitiated and recognized as the bringers of food, drink, and the breath of air, as the elements of life, by the Hero-Cult Nilotic Negroes of the later exodus, in whatever part of the world they are found. These three powers, which represented the elements of water, air, and earth, were elementary powers or animistic souls, who were "the life-givers," *not as begetters or creators, but as transformers of the super-human powers with the human product.* There were seven of these powers altogether, *not derived from the ancestral spirits, once human, and no ancestral spirits were ever derived from them.* In these we find the commencement of the Stellar Mythos. *Six of the seven were pre-human, the seventh (Horus) was imaged as human.* Two lists of the the seven are given in the Ritual (Chapter 17–99–107): (1) An-ar-ef "The Great," (2) Kat-Kat, (3) Burning Bull, who liveth in the fire, (4) The Red-eyed One in the House of Gauze, (5) Fiery Face which turneth backwards, (6) *Dark Face in its hour,* (7) Seer in the night.

Light	Horus
Darkness	Set
Breathing power	Shu
Water	Nu or Hapi
Earth	Seb or Tuamutef
Fire	Khabsenuf
Blood	Child Horus

—each one having a star in the heavens assigned to him, they were now called "The Seven Glorious Ones," or "The Khuti."

These correspond to the two categories of the elemental powers and "The Glorious Ones" or "The Khuti." *The leader of that divine company is An-ar-ef, the Great by name.* This, then, was the beginning of their Stellar Mythos, a result of progressive evolution in their Mythology. *An-ar-ef is identified with Horus I,* or the Blind Horus.

The title here identifies the human element as the sightless mortal, i.e. Horus, who was incarnated in the flesh at the head of the seven to become the first in status, he who had been the latest development. In this chapter of the Ritual the seven now become astronomical, with their stations fixed in heaven. In the astronomical mythology the elementary powers were raised to the position of rulers on high; Manetho describes the beginning of this as "the Gods," the primary class of rulers, whose reign was divided into seven sections, or in a heaven of seven divisions; that is, the celestial Heptanomis. Egypt also being divided into seven divisions or Nomes, we have already seen and shown how the first three were formed under their Totemic Sociology, or Hero-Cult (Nilotic Negroes). Some of these can be distinguished in the ancient heavens as figures of the constellations which became their Totems. The Egyptians have preserved a portrait of Apt, the Great Mother, in a fourfold figure as the bringer forth of the four fundamental elements of earth, air, water, and food. As a representative of earth she is a Hippopotamus, as representative of water she is a Crocodile, and as representative of air she is a Lioness, the human Mother being imaged by the pendant breasts and procreant uterus. Thus we see that she was the bringer forth of first these four elemental forces and next three others, seven in all. Six of these seven powers were represented by a zootype, the seventh was later given the human image. Evidence for a soul of life in blood was furnished by the incarnation, the elemental power of which was divinized in Horus I, the eternal child, and later Atum, the man. The earliest type of the man, even as male power, was the bull, the bull of his Mother, who was a Cow or Hippopotamus. The Great Mother was imaged as a Cow, a Serpent, a Crocodile, or other Zootype, ages before she was represented as a woman. The same with the powers that were born of her as male—six were portrayed as Zootypes before the seventh in the likeness of a man or child. These powers were divinized as the primordial Gods—but *were not human. The Egyptians had no God which was derived from a man.* Herodotus states that they told him no God had ever become a man;

Atum was a God in the human likeness, but he did not himself become a man, and no human ancestor ever became a deity. The Spirits of human ancestors always remained "The Glorified."

Neither did these old priests believe in the transmigration of human souls into animals, as Professor Frazer and others have stated. It is only these "wise and clever men" who have been born recently that have brought forward this doctrine, because these "learned men" have confused the elementary spirits or souls which were never human, and never could be human, with the Ancestral Spirits or Souls, which had been human, and were now "Glorified Spirits," immutable and eternal. The old Egyptian priests never confused them; they were not merely Anthroposophists and knew better, but the present wiseacres have done so, being ignorant of the true gnosis, nor do they understand the representation of these by Zootypes and Sign Language.

The Elder Horus is the divine child and Atum the perfect man, but only as an anthropomorphic type. These two classes of Spirits are separately distinguished in the Hall of Righteousness when the Osiris pleads that he had made "oblations to the Gods and funeral offerings to the departed," Ritual, Chapter 125, and in Chapter 126: "The oblations are presented to the Gods and the sacrificial meal to the glorified." In the Ritual, Chapter 17, the seven have now become astronomical, with their stations fixed in heaven These were then the first seven Elemental Powers born of the Great Mother Earth, Ta Urt. Then the Ancient Mother who had been a Cow on earth was elevated to the sphere of the Cow in heaven, Apt, and the seven Elementary Powers were now her children, known as the children of the Thigh when that constellation was hers. These seven became the seven wise masters, the seven Ali or Elohim, etc. Here they acquired souls as Gods in the Astro-Mythology. In the Stellar Mythos they were the seven Khus, with Horus on the Mount as the eighth. They are the seven Taasu with Taht in the Lunar Mythos, the seven Knammu with Ptah in the Solar Mythos, and in the Eschatology they are the seven souls of Ra, the Holy Spirit, and the seven souls glorified with Horus as the eighth in the resurrection of Amenta.

The order in development, then, was—

1. The Elemental forces or Animistic nature powers;

2. Ancestral Spirits, and finally

3. The one Great God over all, who was imaged phenomenally in the Kamite Trinity of Asar-Isis in Matter, Horus in Soul, Ra in Spirit, which three were blended—the One Great God.

It is somewhat difficult to trace from the Ritual alone how far these Stellar people had advanced in the Eschatology, because it has been so obliterated by priests of later Cults, and one name often substituted for another; but the evidence I shall bring before my readers will, I think, be a critical proof that these people had worked out everything. This opinion is absolutely opposed to all published works that I have ever seen, even those of my late dear friend, Gerald Massey (from whose great work "Ancient Egypt" I have freely quoted), and during the last twenty years of his life we both worked together on these subjects, but he did not grasp the total evolution that had taken place so early. *It is only from fresh evidence found since his death that I have formed this opinion*, viz. that during the immense time that the Stellar Mythos was in existence the old wise men of Egypt had worked out the greater part of the Eschatology, which later priests carried on when they discovered "this Stellar Time-gauge" to be incorrect. They then changed to Lunar time, and afterwards to the Solar time, and *perfected* the Eschatology, which was the final point they reached before the downfall of the Egyptian Empire. Then their religious Cult was brought on by the Copts as the Earliest Christian doctrines. The evidence in the Ritual is distinctive and conclusive that the Stellar Mythos must have existed for immense ages of time, and the monumental evidence still extant is even more conclusive. In the Ritual we find the birth of Horus the Child—the man-God Horus—God of the North and South, East and West, who lived, was crucified, and rose again as Amsu (Horus in Spirit).

That these people worked out the Southern Hemisphere first there can be no doubt, nor that Set, the first-born of Old Mother Apt, as Set-Anup, was elected as their primary God of the South Pole Star, or Southern Heavens, with various attributes; and he remained as primary God for over 51,000 years, as may be seen from records left (the El Shaddai of the Phœnicians). As these people came North, and the Northern Hemisphere was watched and the time recorded, by observation of the precession of the Pole Stars (Ursa Minor), the South Pole Star not rising again, and the North still keeping its place in the ascendant, Set was gradually deposed, and his younger brother placed as primary God of the Pole Star North, carrying all the attributes hitherto assigned to Set, and many more being gradually added. From records on the monuments still extant it is evident that for a time these two brothers were co-equal, but finally, after working out the time by observation of the Pole Star North, and this star still being in the ascendant, the South Pole Star having gone down and not risen again, Horus became the primary God—God of Light, etc., and Set was deposed and regarded as the King of Darkness.

That the seven Glorious Ones were the stars of the Little Bear is distinctly proved in the Ritual. These are spoken of as the stars which never set (ähmiu seku), whilst the stars which set, the Great Bear, were called ähmiu uretu.

When Set (or his star) went down and disappeared below the horizon, he gradually sank to the lowest parts of the Abyss, which was afterwards converted into Amenta, in one form. Here his character was changed, "as the opener of roads or ways in the Astronomy," and this form was continued in the Eschatology as "Ap-Uat or the Jackal"—"conductor of souls"; he thus became the "King of the Darkness."

There is monumental evidence of this.

The Pyramid of Set, next to the Great Pyramid of Ghizeh, has all the above records, and more, written on stone in Sign Language. (For further proof, see "Arcana of Freemasonry" and "Signs and Symbols.")

In Ritual, Chapter 130, is mentioned "The slaughtering block of the Armed God" Sept, called (Unas, 282), and Septu Ābu, "armed with horns." Renouf translates it as "armed with horns, that is, rays of light." This is a very important passage in the Ritual, because in the remains of the oldest temples of the Stellar Mythos people, this "slaughtering-block" has been found—namely, amongst the remains of the old temples in Tunis, in Africa, and a fine specimen now in the Museum of the City of Mexico, found in the oldest temple in that country.

In pictures he is represented *as a Hawk,* armed with bows and arrows, and there is one picture, in which "he is in the form of Bes, destroyer of the Menti."

Here we have a proof in the Ritual as well as monumental evidence that Set or Septu (as here written) was the primary God, and was superseded by the Hawk-Horus. Another proof that they had worked out their astro-mythology in the South first may be found in the Ritual, Chapter 164, when eight Gods offered Sekhet-Bast words of adoration in the language of the Nahsi (Masai), which proves that part of the Ritual at least was composed in the language of the Negroes of Southern origin. Sekhet was a Stellar Goddess, and thousands of years before Ptah; the Put-Cycle of Nine did not then exist. Again in Ritual, Chapter 143, the Great Mother is identified as *Apt* of Nubia.

The King of Egypt as the *Suten* dates from *Set.*

The Osiris says to the powers, "Grant ye that I may have the command of the water, even as the mighty Set had the command of his enemies on the day of disaster to the earth. May I prevail over the long-armed ones in their (four) corners, even as that glorious and ready god prevailed over them" (Renouf, chap. 60).

This subject matter is very ancient, and refers to the time when Set was "The God, and to the inundations of the Nile, also to the time when Set had not lost his place in glory.

In the same chapter, Osiris has superseded Set as Lord of the Flood, so we can see how the old priests substituted one for the other; more especially is this demonstrated by the fact that after Set was depicted as the Lord of the Flood—in the earliest Stellar Cult—we have Horus portrayed as the Lord of the Flood in the later Stellar and Lunar Cults, and it was thousands of years after this that Osiris was given the title, during the latest phase of the Osirian Cult.

Ritual Chapter 125 is very corrupt, and, unless the key be known, it is difficult to decipher; Sir Page le Renouf himself has been unable to translate or decipher the real meaning, because it comprises the very earliest Stellar Mythos records which the priests have tried to convert into the Solar Cult, substituting the name of Ra in some texts.

The name of Set is mentioned in others even later, but in the oldest, Ba, we have the reading with determination, and Bekenrenef has = Apep or Set. In some texts it is "I am Tmu"; these words are not in Ba, but they occur in all other copies, and Renouf adds that the omission of the divine name which stops the crocodile is an evident fault—but it is not; the original was Set, and as a proof we have in Ritual Chapter 32, "Tam-Set drives away the Crocodile of the South."

Set and Thoth were the inseparable characters and signs of the commencement of a Great Year or cycle of time, and so they remained to the latest times. In all the old monuments where the Ideograph for the name of Set has not been obliterated, we find it. Perhaps the best examples remain, and are still extant, on some monuments in Central and South America, where the name has not been erased, and Horus or Osiris's name substituted, as we find so frequently in Egypt, this being the work of the priests in subsequent Cults. The reason for this is obvious, because Stellar Cult commenced with these people in Egypt, and Set was made the first primary God in their reckoning of time. From the record left (*supra*) we find that these people had watched and marked down two cycles of the precession of the Pole Stars (Ursa Minor) before Horus was made primary God, i.e. while Set still reigned as such and before he became "Hidden of the Khas." I wish to draw the attention of Drs. Edward Seler, E. Förestmann, Paul Schellhas, Carl Sapper, and E. P. Dieseldorff to the above fact, because in all their papers on Mexican and Maya Chronology "they have gone astray," by not understanding the gnosis of what they find. The Ideographs for Set and Thoth in Mexican, Zapotec and the Maya have all the same meaning, and I shall emphasize this again in the Mexican chapter of this work. The difference they find between the Ideograph for the Chronology in Mexican and Zapotec on the one hand, and the Maya on the other, is the difference between Stellar Cult and Solar Cult. Mexicans and Zapotecs had Stellar Cult and reckoned by Stellar time. The Mayas had Solar Mythos, and reckoned by Solar time.

In some of the chapters of the Ritual there are more of the old Stellar Myths obliterated than in others—Chapters 17 and 125, in particular, and the misunderstanding of this has misled translators of these texts. Even Sir Page le Renouf is not correct when he speaks of "the two Eyes." "Horus," according to the Pyramid Texts, has two eyes; a Light one and a Dark one. But "the Eye of Horus" is most frequently spoken of in the singular number. It is certainly meant for the Sun. This is not so. Horus as God of the Pole Star, in Stellar Mythos, is meant, and the Light Eye is meant for and attributed to the Pole Star God North. The Dark Eye (Set) South—it was part of the old Stellar Mythos—in the first part of Chapter 125 is undoubtedly Stellar, and, as Sir Page le Renouf justly states, is much older than the second and third parts, which have grown out of the first; in other words, the Ritual shows and proves most distinctly and truly the changing from the Stellar to the Lunar and Solar, and how, brought on from one cult to another, they overlapped and tried to obliterate the old with the new. The forty-two gods and the negative confession, i.e. the number of certain sins in the second part, is a higher number than in the first, and is Solar.

The writing of the Ritual was not commenced for a long time after Stellar Mythos had been in existence. The old priests transmitted everything they knew orally, and it was only known amongst themselves.

That Horus in the Stellar Mythos suffered dismemberment, as did Osiris in the Solar, is proved by the legend of Necheu. The Limbs of Horus had been thrown into water, and when, at the prayer of Isis, Sabek threw his net, he brought up two fishes into which the arms of Horus had been turned. The Apis tablet gives the name of the place (Zeitoche, 1882, p. 22) as Ta-kerk-en-Har.

The "Two Fishes" is the name of Mer of the second northern nome. In another part of the Ritual his head has been cut off and given to him.

In Ritual, Chapter 13, it states: "I know the mystery of Necheu: Horus and that which his Mother did for him, when she herself uttered the cry, 'Let Sabek, the Lord of the Marshes, be brought to us.' He cast the net for them and found them, and his Mother made them fast in their places."

"Sabek, the Lord of the Marshes, said, 'I sought and I found the traces of them under my finger on the strand I netted them in a powerful net, as the net proved to be. And Ra said, 'Verily those are fishes in the hands of Sabek and he hath found the two arms of Horus for him, which had become fishes'—and the hands of Horus were brought to him and displayed before his face, on the feast of the 15th day of the month when the fishes were produced."

This story is preserved in the names of several localities, but we see from the text how the priests of the Osirian Cult tried to obliterate the Stellar, and merge the one into the other.

Hence we have the names of the Solar Cult as Mother and Father, i.e. Isis and Osiris, but in tracing back the details we possess and which are still extant, we find that —in the Stellar Mythos—Horus was conceived by the Goddess Sekhet and the Goddess Sebek-Nit gave him birth, but in the above legend the priests of the Solar Cult have substituted the name Isis. In the Stellar Cult his name as a child was P-ub-Ta, Lord of the World, God-Son of Horus, and the hieroglyphic representation is a child crowned with a triple reed crown and goat's horn (see p. 503, Bunsen, Dict., No. 101).

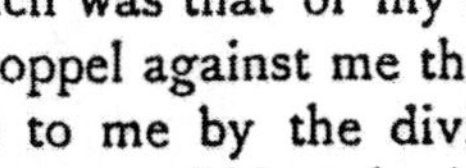

Therefore there is quite sufficient evidence in the Ritual that the tale of our present Cult must date back to the Stellar Mythos. In Ritual, Chapter 30, we have these words: "Heart mine which is that of my mother; whole heart mine which was that of my coming upon Earth. Let there be no estoppel against me through evidence, let no hindrance be made to me by the divine circle; fall thou not against me in presence of him who is at the Balance. Let not those Ministrants who deal with a man according to the course of his life give a bad odour to my name. And, lo, though he be buried in the deep, deep grave and bowed down to the region of annihilation, he is glorified, then—Lo! how great art Thou—The Triumphant One." This shows and proves it was the Stellar Mythos. It was "The Mother" and not the Father—the Father had not yet come into being—The Glorious Ones or "the divine circle" were those attributes of Horus represented symbolically by the stars of Ursa Minor. This text is very old, and there is one copy on a Scarab in the British Museum where "the divine circle" is written [hieroglyphs]. As Sebak-em-saf is written, it shows that [hieroglyph] (a wall of enclosure) is ideographic of the whole word, and this sign in the hieratic, when placed upright [hieroglyph], has given rise to the [hieroglyph], which takes its place in the earlier texts. This enclosure was the circumpolar Paradise or Heaven, and, as I have shown, it was Sekhet the Goddess who conceived and Sebek-Nit who gave birth to Horus in the Stellar Cult. The four brothers or four children of Horus are here referred to "as being on the side Lord of Horus," i.e. with him in his youth or earthly career. These were four out of the divine circle of twelve which was established in this Cult. At the same time, we see in other parts of this text how the priests of later cults tried to obliterate the original and to substitute the Solar for the original Cult. "To the course of his life" the determinative ☉ shows that [hieroglyph] is here to be taken in the sense of the duration of *human life*, and the pronominal suffixes [hieroglyph] or [hieroglyph] show whose life is spoken of.

The four brothers or children of Horus were the representation of the Evangelists in the Christian Cult out of the twelve Apostles.

We have here distinct evidence that the Lord came on this earth, and that the whole of the Christian doctrines which are the latest evolution of religion were really the first. Probably the event took place over a million years ago; certainly not 1921 years ago. That is a tradition of which we have no history; of the above older date we have written history still extant on stones and papyri, so that all we have to do is to alter dates.

I find that the story was so ancient in Egypt that in the time of Amen-Hetep it was humanly applied to his child and to his consort Mut-em-Ua, as the divine woman, the mother who, like Neith, was ever-virgin.

In the picture (Plate LXV) is shown the story of the Annunciation, the miraculous conception (or incarnation). The birth and adoration had already been engraved in stone and represented in four consecutive scenes upon the innermost walls of the holy of holies in the Temple of Luxor built by Amen-Hetep about 1750 B.C.

The first scene on the left hand shows the God Taht, as divine word or logos, in the act of hailing the virgin queen and announcing to her that she is to give birth to the coming son, i.e. to bring forth the royal Repa in the character of Horus, the divine heir.

In the second scene the god Kneph, in conjunction with Hathor, gives life to her, i.e. the Holy Ghost or spirit that causes conception, Kneph being the spirit both by name and nature. Impregnation and conception are depicted in the virgin's fuller form.

Next the Mother is seated on the midwife's chair and the child is supported in the hands of one of the nurses.

The fourth scene is that of the adoration. Here the child is enthroned, receiving homage from the gods and gifts of men. Behind the deity who represents the Holy Spirit, on the right three men are kneeling offering gifts with the right hand and life with the left. The child thus announced, incarnated, born and worshipped was the Pharaonic representation of the old Stellar Cult Egyptians. (This was copied by Sharpe from the temple of Luxor.) Thus we see how the divine drama was represented humanly by the royal lady, Mut-em-Ua, who personated the mother of God with her child in the Osirian Cult, which, as I have now proved, was brought on from the Stellar Cult.

It is amongst the traditions of the Masai Group—the first Stellar Mythos people—that the man-God first came on earth. How many hundreds of thousands of years ago? And how much has been perverted and changed by ignorant priests and discoverers amongst themselves ever since? How many nations have risen and been destroyed since? Yet through all the past ages there has been sufficient evidence left to the student who can "read the writings on the wall" to decipher and learn the truth. The whole tale and history of the world and man is there, to read, learn, and profit by, if one so wishes, a guide for the future and "the danger signals" for a nation to avoid, if a nation wishes to become great or to retain its supremacy. May the contents of this book ever preserve this nation from becoming socialistic and degenerating into a non-religious nation, for as surely as it does, it will be destroyed, as others have been in times past. It will no longer exist, but will be erased from the face of this earth as a nation, and others will take its place. (For further proof see "Arcana of Freemasonry.")

Dr. Eduard Seler, in "Mexican and Central American Antiquities," published by the Smithsonian Institution, Bureau of American Ethnology, Bulletin 28, mixes up these

The Annunciation, Conception, Birth, and Adoration of the Child.

PLATE LXV.

Zapotec, Mexican, and Maya names, and attempts to translate them into years and days!

He states that Kan, Mulac, Ix, and Canac stand for and mean years, which begin with the days Beeu, Eznab, Akbal, and Lamat, or in the Mexican characters Acatl, Tecpatl, Calli, and Tochtli, but he acknowledges that the years correspond to the north, west, south, and east. But, as I have proved, these are only the different names in Toltec, Zapotec, and Mexican for the same four Brothers or Children of Horus of the Egyptians. As in Egypt, they represented here the four quarters of heaven and the four supports thereof, and the four seasons of the year and four colours. Amongst the Zapotecs they were also termed Cocijo, or Pitao, the translation of which means the Great Ones, the Gods, the Holders of Time. The Zapotecs were more ancient than the Mexicans, and the latter more ancient than the Mayas. Dr. Seler's *Five Cardinal Points* represent the first five "Glorious Ones," the divinized Elemental Powers of Apt, seven in all (see *supra*). In Mexican we often find these represented symbolically as five houses.

He states that amongst the Aztecs they believe that the direction of the realm of the dead is to the north (p. 30). This is not correct; the direction of the realm of the dead was here, as in Old Egypt, to the west, but after they had entered the underworld in the west and passed through to the east, from there they were taken to the Paradise at the north—if they were good spirits, but not otherwise. From the Mount of the East—Orion—in the Stellar Cult, they were taken by attributes in Zootype form to Paradise. In the Solar Cult it was the Mount at the exit of Amenta, where they entered a boat to be taken across the starry firmament to Paradise situated at the North, after having changed boats in the West (see Ritual).

The Zapotecs and Mexicans were Stellar Cult people, the Mayas were Solar.

Dr. Seler's supposed translations and opinions as set forth in his papers contained in the above book are about as erroneous as one could conceive, as I have clearly demonstrated and proved in this book and in "Signs and Symbols."

Mr. Percy Newberry's paper, "Notes on Some Egyptian Nome Ensigns and their Historical Signification," in "*Ancient Egypt*," Part I, p. 5, 1914, is only interesting as showing how much injury can be done to the cause and progress in the knowledge of Egyptology by "guessing," on the part of men who are supposed to be authorities on this subject, and the publication of their opinions in such a paper as "*Ancient Egypt*," thus disseminating to the uninitiated false ideas. It is quite useless for Egyptian students to attempt the decipherment of these old Ideograms without the knowledge of Totemism, Totemic Sociology, and the old Mythology of Ancient Egypt; by "guessing" they only "educate" the public to believe that which is not true.

Let me put Mr. Newberry right about these "Nome Ensigns," or, as the correct term should be, "The Totems of the Nomes."

The author of this paper is not correct in attributing the titulary name of Horus to the first kings, as he states on p. 5. The Set name was attached to the primary, and the Horus name later, i.e. when Horus had superseded Set; but for a time the two were equal in the eyes of the old Egyptians. I have already written fully on this.

The Totems, or as he calls them Ensigns, originally were the Zootypes of the Motherhood in one form, which now in Totemic Sociology here had become the Totems for the Nomes, and these were carried on into Dynastic times. *They do not represent "different Cults,"* as Mr. Newberry believes, and his ideas "that the chieftains of the Set Nome had overthrown the Horus chieftains and seized the throne," is about the most futile rubbish that he or anyone else could ever pen. During Totemic Sociology there could be no overthrowing of one Nome by another for reasons of Tabu (as explained *supra*), and this lasted for 16,000 years according to Manetho; then this Totemic Sociology was by evolution converted into the Stellar Mythos Cult, and no "overthrowing" ever took place then in Egypt in the Nomes—only others were added to the primary seven. Mr. Newberry has confused and confounded the celestial with the terrestrial battles, and made the "gods" human, which they never had been, nor ever could be. His remarkable group of compound ensigns, in which behind the distinguishing sign of the Nome there is a figure of a Bull—"in the Old Kingdom there were four of these ensigns and later a fifth appears"—these are the Totems of the first five Nomes that were formed in Egypt—there were seven in all—with the determining sign of the Bull: the seven Elementary Powers divinized, children brought forth by Old Mother Apt, and called the "Bulls of their Mother," hence the determinative. I give a drawing of these (by Mrs. K. Watkins). No. 1 represents Set. (This is a Set animal and not a calf, as Mr. Newberry supposes.)

No. 1. No. 2. Var. 2 No. 3. No. 4. No. 5.

This was Apt's first-born, and the first Nome that was formed, was dedicated to Set; at that time he was one of the Heroes. A later variant of this No. 1 reads "The Lord Set." Tb-ntr certainly means the Divine Calf, but this name, although generally associated with Horus, was first a name given to Set before his "fall." In Sign Language it means the Calf of his Mother—the Great Cow Apt. Later the name was attached to and associated with Horus. He is spoken of in the Ritual as the Divine Calf on the one horizon and the Bull on the other—one of his dual types. This is plainly depicted in

No. 2 and variant. This is Horus of the Double Horizon. When the second Nome was founded it was dedicated to Horus, then the chief Hero of the Masai group.

Mr. Newberry reads this *as the Cult of the wild "Bull of the Desert"*! It is Horus of the Double Horizon, here graphically portrayed in the Sign Language. The first figure here depicted is a single horizon, i.e. the Eastern, and the second figure, or, as he calls it, a variant, portrays the double horizon—the Western. He is the Calf in the East; he has gained double strength or double power in the West, where he is the Bull of his Mother in the Mythology. All seven children of Apt were called the Bulls of their Mother.

I think the author of this paper knows very little of the old Egyptians, to suppose that they would ever connect the *Bull with the Desert*. They were not simply Anthroposophists; in all their actions, signs, and symbols they studied from nature and nature's laws, and would know that no bull could live in the Desert, and they never connected him with it—they were not so ignorant as that.

No. 4.—The third Nome formed was dedicated to Shu (here shown as No. 4) depicted by a feather—which was an Ideogram for his name in one form. As regards "the ostrich feather being a Libyan Sign"—it was used by the Libyans certainly, but ages before the Libyans it was an Ideogram for Shu; it also represented truth, justice, etc., and had the phonetic value of Maat. The God Set has his crown composed of two of these feathers only (see the God Set), but primarily it was the "La-Gou-Ma" of the Pigmy, and its origin must be taken from the Central African district.

No. 3.—This should have been placed in rotation as No. 4, because it is the Totem for the fourth Nome formed. Mr. Newberry asks what was the Cult of Kam. There was no Cult of Kam. Kam was the name of the black land of Egypt, and the oldest name for Egypt, here depicted by a crocodile's tail.

It was one of the names for Set or Tuamutef, and also another name for Hapi. This was the Totem and Nome of Tuamutef.

No. 5.—Mr. Newberry states: "Whatever O represents does not concern us here, but the interesting fact is that the sign serves to differentiate the Nome ensign from the other ensigns of the Bull group"!

Does it not concern us? It is the Ideogram or Sign and Symbol for Elephantina (Bunsen, Dict.). It was the fifth Nome that was formed, and we therefore know which of the "gods" it was dedicated to from the Ritual (see *supra*), and his idea that "it serves to differentiate it from the Bull group" is quite ununderstandable, because "*the*

Bull" is here as the determinative. If his argument was correct, then both No. 3 and No. 5 have nothing to do with this Bull group. I am of opinion that Mr. Newberry does not know the meaning of the Ideogram O, and has therefore written the "rubbish" we find in this paper; what he has written is of no value, but the Ideograms portrayed are of great value, as proving my contention that these Nomes were formed at the earliest period of time of Egypt, and were dedicated to the seven Elemental Powers, which were divinized during the Hero-Cult. Their Totems and names are given and brought on to Dynastic times. His mixture and suggestion of different Cults of the Nomes can have no meaning or value to those who understand the Totemic Sociology and Mythology and Astro-Mythology of Ancient Egypt, except as a proof that these Nomes existed, and that a Totem or ensign was attached to each one.

What is very interesting, however, in this paper is the portrayal on p. 7 of the origin of the old Egyptian Ideogram, originating from the Masai group—the Shield and Arrows, which have, however, no connection with "Cults":

Herodotus describes the inhabitants of the cultivated portion of Egypt as the best informed and most learned of mankind.

Clement of Alexandria states that the Egyptians in his time had forty-two sacred books. The last six of these books treated of the art of medicine, which had taken root in Egypt in the darkest ages of antiquity, and boasted royal authors, from Athotis down to Nechepso.

He surmises that the original form of the old constitution must have been a really elective monarchy, as from a passage in Synesius (Opp. p. 94) we find that "the priests stood next the candidates for the Throne, then came a circle of warriors, and last of all the people. The priest declared the name of the candidate, who must have been one of them, or have been initiated into their highest rank; every soldier's vote counted for one, a prophet's for a hundred, priests of subordinate rank for twenty." The crown became hereditary with Menes, and the right of succession was extended during the Second Dynasty to the female line (in the so-called third century of the Empire). Whether it was hereditary before Menes it is impossible to state.

Eratosthenes, according to Strabo (Strabo, XVI C. 4, p. 767), elucidates the historical connection between the native tribes of South Africa, Asia, India, and the Egyptians. "The four principal races of South Africa have not only a well-regulated monarchical constitution, but also stately temples and royal palaces; the beams in the houses are arranged like those of the Egyptians." Speaking of Bab-el-mandeb, in Arabia, he says: "Here must have stood the pillars of Sesostris, inscribed with hieroglyphics."

That Egypt was once the most densely populated country of ancient Empires may be gathered from both Herodotus and Pliny, who state that at the time of Amisis Egypt contained 20,000 cities and fed 8,000,000 people, besides the populations of Italy and other countries. That they were the most learned nation in science and art is also borne out by Herodotus. "As Egypt was then famous for the sciences and arts, the Greeks, who were beginning to emerge from barbarism, travelled thither to obtain knowledge, which they might afterwards communicate to their countrymen."

CHAPTER XVIII

STELLAR MYTHOS PEOPLE (*continued*) AND PROOFS OF WIDESPREAD DISTRIBUTION

THE foregoing demonstrates conclusively that it was the Stellar Mythos people who existed in Egypt and Italy at least 600,000 years ago, as proved by the skeletons found in the Pliocene Strata (of the present type of man), and by the amulets and implements found (see *supra*). They also worked out the whole of the doctrines of the man-God being born here on earth—the man-God who, having lived, was crucified and rose again. The proofs shown are such that there is no answer, nor could there be any answer to disprove it. The whole tale was introduced in the Lunar and Solar Cults, and finally in the Christian doctrines. These later Cults have copied and embellished the same, obliterating as much of the Stellar as possible. The Yezidis, who inhabit the mountains around Mosul in Asia Minor, and number about 20,000 at the present day, still practise this Cult, as do also the old Chinese, although at the present time there is much contamination and mixture of other Cults.

Some learned men who have visited these Yezidi have described them as "devil worshippers." This was recently done by one who visited them, and published in the *Illustrated London News* (1912) photographs of the entrance of the temple depicting the Great Serpent. *It is acknowledged that these Yezidi deny it*, but these learned men say "statements have been made that there is ample evidence to justify the epithet 'devil worshippers.'" We have, however, here a living book, the alphabet of which they do not understand, to prove how ignorant they are of the past history of Egypt, of its Cults, and of the past history of the world.

The Serpent is the *Symbol* of the *God Tem* (another form of Horus). Therefore, it is not the *Zootype Symbol* that the Yezidi worship, but the god it represents. So, much for our present learned men's "devil worshippers."

I shall now trace some of these people throughout the world, giving critical proof of their common origin. From the higher type of the Nilotic Negro, the "Masai group" were evolved, the highest type of these Negroes, a fine race of men who made much progress towards civilization. They still have in their texts the same Cosmic Myths, moral and religious tales, religious beliefs and traditions that we find in the old Egyptian papyri of the Stellar Mythos, which proves what progress they had made to reach a higher state, as is depicted by their physical and mental characteristics, their social and military institutions and their weapons, some of which symbolize the Primary Ideographic Hieroglyphics. On many of the ancient tombs and mural sculptures, the faces and forms of the oldest Egyptians that we find are undoubtedly similar to these Masai of the present day, who are now fast dying out; moreover, the representation on the sculptured frescoes found at Palanque and other Central American ruins are of undoubted likeness to the Masai group. Here in Egypt, from their totemic ceremonies, were evolved their Mythos and Astro-Mythology, by observation of the celestial bodies and elementary powers; and progress continued until they had completely worked out their Stellar Mythos and divinized these powers. They had, through many thousands of years, advanced and improved in stature, had risen to a high state of knowledge and wisdom by observation and learning, and from their original monosyllabic language, which was ideographic and symbolic, had commenced and formed a further development of speech — the agglutinated language — by adding prefixes and suffixes to the ideographic symbol, and later using linear additions to the original hieroglyphic; they had, in fact, learned the commencement of an alphabet. Indeed, nowhere else can this be traced except in Old Egypt, i.e. the original ideograph used in the oldest hieroglyphic writings, followed in the same country by a further development of the same hieroglyphic by prefixes and suffixes, thus developing the old monosyllabic language into an agglutinated language; this occurred and co-existed with the development of Astro-Mythology and their perfected Stellar Mythos. It was at this time that an important exodus took place which travelled to far-distant countries, the proof whereof still remains amongst the various nations that have been great and have fallen; only ruined cities, relics of their past glories, are left to their children and descendants to read the lesson of God's divine laws of evolution. The great lesson of the rise and fall of all nations—and none was so great as that of Egypt—is left to us as a heritage to read, learn, and profit by. It is written in stone that has not perished, or else in the degeneration of their children, who were once great, but are now either no more or only remnants, fallen into decay or oblivion. The remains of the old Stellar Mythos people are still found at the present time in various parts of the world, and this will prove how great were the migrations of these people, and for how many thousands of years they were in existence. We have the Osteo remains of these people here in England, at Harlyn Bay, in Cornwall, and probably other places. The early people of Tibet and some of the North, Central, and South American Indians were of this exodus, as is proved by their anatomical features and their religion, which they call "Bonba," but which is the old Stellar Mythos of the Egyptians.

Mr. C. Beaumont wrote an article published in the *Times*, June 27, 1911, which is extremely interesting, but he has fallen into the same pit as all the rest who place the origin of man in the mythical Lemurian Continent, a pit from which they cannot extricate themselves. Many people reading the article would be led to the belief that Peru was a much older country in civilization than the rest of South America, viz. Brazil and the Argentine, but this was not so. Mr. Beaumont has not taken into consideration the old ruins of cities found in Brazil, pottery and implements discovered there, which were all of Stellar Cult origin. He also somewhat mixes up the Central American States, i.e. he makes no difference in point of time between the temples of Stellar Mythos and those of Solar Mythos found in these States. Stellar Mythos was diffused throughout part of North America and Central and South America, at least as far south as the south of Chili, and its implements have been found in some parts of Patagonia, but not Fuego. This is prior, by thousands of years, to the time when Solar Mythos came to Yucatan — the Solar starting-point in America—brought there direct by the Egyptians, who landed in Yucatan, spread a little north and a little west, south down to Peru, but not, as present discoveries teach, entering Venezuela, Brazil, or Bolivia, or Argentine. It never reached the west of the Central American States or North America. The evidence found so far is positive proof of this, to all who can read the signs, symbols, and glyphs on these old ruins, as well as the remnants of the older races, as confirmed by their anatomical features and continuity of language, customs, buildings, pottery, and instruments. The natives of North and South America are substantially the same in race and character. The Mexicans, Peruvians, Bolivians, and indigenes of British Columbia at the present time are of a type which shows a common stock. But the South American races are rapidly dying out as a result of the advance of the European settlers, who treat them with inhuman cruelty, and of various diseases introduced amongst them by the white man. It is a great pity that these whites cannot see and recognize that here they have an early civilized people, now degenerate, but who, under humane conditions, would supply the labour market.

But the future of the American native races will be of short duration; they will vanish and become people of the past.

The lower class of South American Indians are still in a half state of savagery, but they are strong, healthy peoples, and are all totemic with the exception of those hereafter mentioned.

Those of a higher type are the remains of the civilization of the old Stellar Mythos people, or the Incas, who were Solar people, and may be classed at the present day as partly civilized. These have been so ill-used and trodden down for so many generations that they resemble the Egyptian natives and take very little interest in endeavouring to improve themselves and their surroundings. South America, like Egypt, has suffered so many years of misrule and oppression that the old Stellar and Solar people have lapsed into decay, but we are pleased to see that some of the South American Governments are now beginning to improve their position, just as we have done in Egypt. The Arucanian and Aymara Indians here are the remnants left of the old Stellar people. These are precisely of the same type and class, and wear the kind of head-dress, etc., that we find depicted on the monuments of Central and South American States, as found, for instance, on the pottery from a tomb at Zaachilla. They were probably the same exodus from Egypt as the early Bitiya of Asia, and all speak an agglutinated language. The Arucanians, like the Masai, are divided into two groups, one entirely nomad. They were distinguished by their indomitable courage, which defied both the Incas and the Spanish. The Arucanian-speakers south of the river Maule were almost entirely nomad, but the northern branch, extending from the river Maule to Atacama, were at most semi-nomad, as they practised agriculture and lived in permanent huts.

The Atacamas built their houses like the pastoral Masai, almost in every particular except that they used stones for the walls. They buried their dead in the thrice-bent position. Similar buildings are found along the Peruvian coast. From the mortuary caverns and graves numbers of objects have been obtained—maize, stone mortars, and spades, which help to prove that they were agriculturists—arrows and bows, and remains of dogs. Woven clothing of llama-wool was worn as a tunic, and over this a poncho, and a high cuirass has been discovered on the coast. There were also spindles and wooden needles furnished with eyes. In the neighbourhood of Salinas Grandes there have been finds of rudely chipped axes and polished celts with gems encircling the butt, star-shaped pendants and knives in metal. These are Stellar Cult remains.

Throughout the area of the so-called Diagnite the

Kakan language prevailed; the cult was Stellar, the same as that of the Andean people further north. The houses were built with walls of stone, and as traces of roof-beams have been discovered, the roofs were probably thatched; the houses were arranged in a rectangle or circle, about 1½ yards high, and formed in villages, nearly always on an eminence or hill; very often they contain one or two monolithic pillars, sometimes sculptured.

They made pottery, which had figures and pictographs on it, lizards and pigs, figures of birds and reptiles—Zootypes. They also buried their dead in the thrice-bent position. Frequently the head was removed. They used bows and arrows, and axes both of stone and copper.

They used a wrist-knife, like the Nilotic Negroes, and spear-throwers were widely distributed throughout America, precisely as we find at the present time in Australia and Inner Africa.

Their government was practically tribal or patriarchal. They were led by four independent chiefs; each of these had five ulmen or district chiefs under him. The chiefs were called Toki, who, as the word itself implies, carried a stone axe of a peculiar shape (see Fig. 31, on p. 247),[1] having the Horus Hawk Zootype Head with the All-seeing Eye.

The Arucanians and Aymaras are descendants of the old Stellar Mythos people here in South America, and it was the Arucanians who opposed the Incas in Chili, these fierce old warriors stopping further progress south on the part of the Inca's or Solar Mythos people.

The Arucanians are agriculturists. The Atacama are a pastoral people. They correspond to the two classes of the Masai, their customs, dress, and beliefs being the same originally. The two types still exist in Africa (see Masai, *supra*).

The true nomads clad themselves in skins.

The Tehuelche clad themselves in skins, used ostrich feathers for embellishment, sometimes nose-pins and lip ornaments; necklets of shells and beads were woven by the women, who were clad in aprons and cloaks, and sandals of sewn hide. A tall, round-headed race, making rude pottery.

[1] "The Secrets of the Pacific and Andes and the Amazon," by R. Enock.

Quetzaloatt was the presiding deity of the Great Pyramid of Cholula, the oldest and largest in Mexico. (This is the Egyptian Horus-Khuti.)

The Aztec Calendar stone is of Stellar Mythos origin and not Solar. The above meagre description has been taken from Mr. Enock's work, but many of the details of description are wanting in his book.

The Queche in Guatemala had the Popol Vich (Ritual of Egypt).

The Pilpils, separated from the Nahuatl, speak a very old dialect of the Mexican language of the highlands of Mexico, and have preserved forms still older than the original and classic Nahuatl.

When we turn to the Mayas in Yucatan we cannot doubt that we have here the story of a new civilization, or culture, founded upon or succeeding an older one, and this also applies to the Incas in Peru, who drove out and partly exterminated the Quechuas and Aymaras, who were Stellar Mythos people in South America.

The Quechuas and Aymaras were unacquainted with the principle of building arches, although the Chimus, a people of Peru of the region of Trujillo who spoke a different language, knew how to build arches. In the province of Sandia and Carabaya the Bolivian and Peruvian Indians speak nothing but the language of the Aymaras, and the men as well as the women wear their hair in long queues, like Chinamen and some Nilotic Negroes. The Cholo Indians of the Cordillera are the original Quechua Indians of the uplands—they are not true Indians.

The Temple Pyramids, which were orientated either South or North in the old Stellar Cult, gave place to the orientation to the East (Solar Cult).

Now let us turn to elucidate all the evidence we find in America, from those ancient monuments which are to be found for 4,000 miles or more from Arizona in North America to the south of Chili in South America, and in isolated patches across the Pacific in a north-western direction to Asia. These are the remains of these Stellar Mythos people from Old Egypt—Pyramids, "Mysterious

Courts," temples, sculptured façades, palaces, stelæ of great beauty, colossal images and fortresses, monoliths, pottery, hieroglyphic writings, etc., ruined and abandoned, or surrounded here and there by the wattle huts of half-savage Indians. Of the people of this old civilization how many remain?

It would be impossible in this work to notice each one, and decipher each and all details of them separately—because it would in the first place take many volumes and more time than I have at my disposal, and secondly, I have not obtained all the photographs or sufficient descriptions from those who have been fortunate enough to visit these places, but I will bring forward sufficient evidence, by decipherment, etc., of some of the principal objects, to prove critically that my contention is correct.

We find in Inca records stories of the discovery of mighty ruins, and the evolution of the several varieties of the llama species, two of which have never been found in a wild state. This requires a very long period of time, and can only be explained by centuries of settled life.

At Manta, in Manabi, was a celebrated temple to the God of healing, named Umiña, whose image was cut from an enormous emerald; on the island of Puna was another, equally famous, to Tumbal, the God of war.

On an island in Lake Titikaka (Titicaca) there are the remains of a sacred temple built by the Aymaras in honour of Huirakocka, in remembrance of the creation of the world by him; this was the supreme God of the Andeans, being "the Abyss of the Water," as Ea was amongst the Chaldeans.

On the island of Tumpinambaranas there are great ruins of these old Stellar Cult people.

This island is formed by an arm of the Madera River with the Amazon; it is 210 miles long, with an area of 955 square miles. Here we find these ruins in the very heart of South America. The above are not Inca remains, as is proved by iconographic figures depicted in stone here found, proving that this was once a great civilized island.

In the mythological history of the Chibcha, which is

common to all the cultured people of Central and South America (before the time of the Inca invasion), is the arrival of a white cultured hero, who gave the people laws and instructed them in arts and industries—

Quetzalcoatl of the Nahua;

Uracocha of the Peruvians;

Dabeciba of the Tamchi of Antioquia;

Tsuma of Venezuela, and known variously as Bochica, Nemtercqueteba and Xue, said to have come from the East. After the disappearance of this hero a woman came, none knew from whence, named Huitaca or Chie, whose teachings bore a very different complexion.

Here we have the tradition of the Lunar Cult following the Stellar.

Lunar Cult was practised in Ecuador and also amongst the Cañari.

In Quito there was a Lunar Temple erected by the Seyri.

In the annual report of the Bureau of Ethnology, U.S., 1881–82, there are notes by Cyrus Thomas on p. 63. He states: "It must be admitted that the Mexican or Nahua manuscripts have little or nothing in them that could have been borrowed from the Maya manuscripts or inscriptions. Hence, if we find in the latter anything belonging to or found in the former, it will indicate that they are borrowed, and that the Mexicans are the older. What is perhaps still more significant is the fact that in the Fejervary Codex we see evidences of a transition from pictorial symbols to conventional characters. For example, on Plate XXV in the Troans manuscript, under the four symbols of the cardinal points, we see four figures—one a sitting figure, similar to the middle one, with a black head, on the left side of the Cortesian plate; one a spotted dog, sitting on what is apparently part of the carapace of a tortoise; one a monkey, and the other a bird with a hooked bill.

"In the Maya manuscripts we find the custom of using heads as symbols, almost, if not quite, as often as in the Mexican codices. Not only so, but in the former, even in the purely conventional characters, we see evidence of a desire to turn every one possible into the figure of a head, and find this still more apparent in the monumental inscriptions.

"Turning to the ruins of Copan, as represented by Stephens and others, we find on the altars and elsewhere the same death's head with huge incisors so common in Mexico, and on the statues the snake skin so often repeated in those of Mexico.

"The long, elephantine Tlaloc nose, so often repeated in the Mexican codices, is even more common and more elaborate in the Maya manuscripts, and sculptures are also found in Copan. Many more points of agreement might be pointed out, but these will suffice to show that one must have borrowed from the other, for it is impossible that isolated civilizations should have produced such identical results in details, even down to conventional figures.

"One thing is apparent, viz., that the Mexican symbols could never have grown out of the Maya hieroglyphics. That the latter might have grown out of the former, is not impossible."

I have quoted Mr. Cyrus Thomas's article because here is one who recognizes that the Mexicans are older than the Mayas, that the Mayas still have some signs and symbols of the Stellar Cult people, but show a beginning of a change from the Zootype representations to that of the human, as we found in Lunar and Solar. "The Long Nose" (see Plate LXIX), etc., represents Taht-Aan, and the "huge teeth with gaping mouth" represents the Egyptian Hept-ro, i.e. Shu (Plate LXVI). Others I have dealt with elsewhere in this work.

The Serpents here mentioned are types of Renewal, either of the years or seasons or eternal life—or the God Tem, or the Great Earth Goddess Rannut.

The Jackal = Seer in the dark, or guide in Death = Anubis.

Big Nose = Ibis = Messenger = word or logos = Taht-Aan.

The Toltec nation preceded the advent of the Aztec; the latter made war on the former and drove them out of Mexico, down to South America, where colonies of the

Mexican Clay Figure.
That-Aan.

Clay Vessel, Peru. Shu.

Depiction of Shu.

Burial Urn, Mexico, showing Horus as a child of twelve years as under figure, and man of thirty years as upper figure.

Plate LXVI.

main stock had already been planted. These were both Stellar Cult peoples, hence the reason why we find that all the Central Americans and Stellar Cult natives of South America had calendars substantially the same in principle as the Mexican. The only difference is that one was an earlier exodus than the other from Egypt.

The Maya manuscript known as the Codex Cortesianus furnishes a connecting link between the Maya and Mexican Symbols. The reason is that the original inhabitants were connected with the Nahuas as Stellar Cult people, but when the Egyptians landed in Yucatan they imposed the new Solar Cult on these people, but did not obliterate the old signs and symbols still surviving here. The Solar people could not, however, make any impression upon the fierce old Stellar Cult Nahuas, or Mexicans; hence they spread down towards South America, and travelled along the line of less resistance between the two great mountain ranges as far as Peru.

We do not find any of the Solar signs and symbols amongst the Mexicans or Nahuas, because these people would have nothing to do with the incoming Solar Cult people, but drove them back. Hence we find in the Maya manuscripts the blending of the Solar and Stellar Cult signs and symbols, but there is nothing of the Solar in the Mexicans or Nahuas (see last Chapter).

As regards the route the Stellar Mythos people took when they crossed to America, the evidence points the way via Asia, because we do not find any evidence of the Stellar Mythos people in extreme North America. Going from north to south, the first positive evidences we meet with are the remains of the old cliff-dwellers in Colorado—Utah and Arizona—the Sierra Nevadas of California and Oregon, all on the western side of North America. From here to the north of Patagonia in South America there are found numerous remains of these old people, proving that ages ago they must have formed a dense population. Of course, it is possible that some may have crossed from the north of Europe via Greenland, and been driven south by a Glacial Epoch, never returning to the north; but if so, this wave of immigration could not have been a large one. They did not exterminate the descendants of the Nilotic Negroes which they met on their way, and left no evidence of cities, fortresses, colossal images, etc., which we find further south on the western side of North America. If built, these have been destroyed by the glacial epochs since. Therefore, I think the evidence still extant points only to waves of immigration coming from Asia and entering America on the western side; unless we except the very earliest, of which I shall treat later on. I have found no evidence that they travelled further than the north of Patagonia. Fresh evidence of course might modify this opinion. "How many waves" of these Stellar Mythos crossed it is impossible to say, but certainly more than one, probably many, from the characteristics of remains found at various places. In studying all that is extant at the present day, we must not confuse the two periods of civilization found here, which are sharply defined when we know and understand the key and are able to "read the writings on the wall." We must not confuse the ruins of old temples and cities, built by *these Stellar Mythos people who came first and were primary*, *with the Solar*, *who came last*, and spoke a semi-inflected language. The first Stellar people were here at least 300,000 years before the Solar, and *these Solar people did not come via Asia or via Greenland; they entered America on the east side in Yucatan.* There is no evidence that they ever went into North America, and only as far as Peru and Bolivia in South America (see later).

The Mayas and Incas were the Solar Mythos people, and the evidence left is very clear and distinct as to how far these extended to other parts of America. For whilst we find numerous and immense relics of the old Stellar people in North America, we find no remains of the Solar people. Their line of travel was to the south and west, the easiest route, where they found points of less resistance; so we trace them south and west until they met the Arucanians in Chili, who stopped their further advance south. To us there is no evidence that the Solar people landed on the coast of Peru and came north. They came from Yucatan and went south. Their line of communication was thus cut by the fierce old Stellar people, who surrounded them on every side (see later).

There is a difference in the architecture of the Stellar Mythos people and the Solar Mythos people, who followed these Stellar people in some places and endeavoured to "blot out" the whole of the primary cult.

In the Stellar we find iconographic carvings; in the Solar we do not. The Solar style of building, in polygonal blocks, was brought on from the Stellar and used by the Solar people. Monoliths of large size were common to both. *The Stellar never built an arch, always lintels.* The Solar people at a later date built arches, not at first (see Arches, later).

The principle of bonding the buildings was known to the Stellar people but not to the Solar people. Modern builders take no account of earthquakes. These ancients were far wiser. Shocks are of constant occurrence in the regions where we find these huge old buildings, and no doubt can be entertained that this danger was taken into consideration when they were being erected, and although these old buildings still stand after all these ages of time, I think they will be in existence when the present sky-scrapers and buildings of this generation have returned to dust.

The mural remains—great stone monuments, temples, pottery in general, and fortresses scattered throughout the country, are of different peoples and epochs, all easily distinguishable by the Zootype Gods and Goddesses and iconographic remains depicted.

After the Nilotic Negroes, the Stellar Mythos people must have advanced in immense waves from Asia to the shores of Western North America, and travelled south as far as the south of Chili, occupying on their way Mexico and the Central States of America. We have certain evidence up to the present time from undoubted Stellar Cult pottery and other remains, and also from the large ruined city found by Dr. O'Sullivan Beare in the forests of Brazil, to prove that all these countries were once occupied by Stellar Cult people. Agriculture was a large factor in the lives of these people. There is evidence of at least two different exodes of these Stellar people here: the first, of a comparatively primitive type; and the second, the people who fashioned the huge structures and planted the civilization of early America.

Many able writers, not only Professor Frazer and Dr. Brunton, but several other American authors, have advanced the theory "of the Autochthonous," which implies that man has been generated at several points simultaneously, and they state that "the inhabitants were indigenous to America." But I have, I contend, brought forward positive proof that man originated in Africa. There were no anthropoid apes in America, none of the Ape family higher than the Cebidæ, from which it is impossible to trace man, and it is impossible to accept Sir Harry Johnston's theory (*supra*). The proofs I have already given in this work establish the fact that the Pigmy was the first man, then we have the Bushman and Masaba Negro (who never left Africa), afterwards the Nilotic Negro of several types from the highest to the lowest, and after these the Stellar Mythos people.

The Pigmy is still met with in inaccessible forests of South America, but his instruments have been found in many places. The Nilotic Negroes are still here, in their highest and lowest types, and these I have proved to be identical with their African brothers, anatomically, physiologically, and in their religious totemic ceremonies. There can be no argument, scientific or otherwise, to refute this. I have given the direct line of human ancestry, with absolute proofs. *The Missing Link*, i.e. *the Bushman and Masaba Negro, is not found here*, and I now propose to unfathom the mystery, which has hitherto not been accomplished except by myself in "Signs and Symbols of Primordial Man," and I do so with proofs irrefutable. That America has been the scene of cycles of changes of humanity throughout very long periods of time and has shown great activity during its past—as proved by the mouldering ruins still extant—only proves my contention of the great antiquity of man, and the long period that the Stellar Mythos existed. In these I will include those scattered islands of the Pacific—Easter Island and the

Carolines—which still retain some of the structures, extraordinary images and walls, characteristic of the Stellar Mythos people. The teachings of the present Anthropological schools ignore the religious ceremonies and various cults found amongst the different peoples of the human race, because they cannot read Sign Language; yet these proofs in particular cannot be refuted by the statement that "variations in environment tend to develop culture along similar lines." The *Encyclopædia Britannica* has made this declaration from the Autochthonous Theory! (who was the writer?) and gives as an example: "No ethical relationship can ever have existed between the Aztecs and the Egyptians, yet each race developed the idea of the Pyramid tomb, through that psychological similarity which is as much a characteristic of the species man as his physique"! I have already proved in "Signs and Symbols of Primordial Man" the fallacy of the above, and that these Pyramid buildings and Mastaba tombs were copies of the old Egyptians brought here by the Stellar and Solar Mythos people, and the proofs are critically correct.

Not only in these old remains do we find a similarity to the Egyptian, but in physical and linguistic affinities, in analogous signs and symbols, devices of various kinds, the similarity of handicrafts, myths, and cosmogonies are proofs that all these could not have been evolved separately from their own surroundings. Moreover, we have the statement in the Ritual that the Egyptians traded with these people. Ignorant beliefs and false faiths derived from misunderstood symbology will not affect one single fact.

Mr. C. Reginald Enock, in "The Secrets of the Pacific," states: "In my opinion the terraces and statues of Easter Island, the Peruvian buildings of Caxamalca and Titicaca, the ruins of Angkor-Thom in Cambodia, of Brambanam, Boro Bodo, the Modju-pahit on Java, the Passumali monoliths of Sumatra, the great island Venice of Metalanim or Ponape, the canals and cyclopean walls of Lile, and the Langi and Druidical Hamonga of Tongatabu may be all, to use a homely expression, 'pieces of the same puzzle.'"

But it was the Toltecs, or old Turanians, or Stellar

Mythos people who, arriving on Mexican soil, built these colossal structures, truncated pyramids, divided by layers, like the temples of Belus at Babylon (which were in fact copied from the old Egyptians).

All these came across from Africa via Asia, which contains all the physical and mental prototypes of the race. Language, Pottery, Mythology, Religion, Dogmas, calendar and style of architecture, were all Egyptian. That they resemble those of Asia must be so, because all those in Asia came from old Egypt—so it was Egypt via Asia to America.

The illustrations given in this book of the natives of the Quechua district of Peru should be compared with the Tibetan and Mongolian faces and the Masai groups. These wanderers from Tibet might, indeed, have felt at home in Peru, with its remote towns on lofty plateaux, in the heart of snowy mountains; a land similar to their own, if any of those who left Tibet lived long enough to arrive there. In any case their children's children would have the traditions of the places of their wanderings since leaving, as well as their customs and religious beliefs.

Some of the cut and carved monoliths of the ruins of Tiahuanako are unequalled in size in any part of the world except Egypt. These existed in Peru long before the Incas—in fact, the Incas knew nothing of their origin. They state this, and it is recorded.

All the tribes of the Mexicans before the Toltecs to the Aztecs came down from the north. These came upon the American coasts from Asia to the north and travelled down south. The old Pueblo language, the Moqui or Hopi, is the same practically as the Nahuatl of ancient Mexico. There is also a great affinity between the Aztec and the Shoshones and Uti, and this is found as far down as South America.

The Zuni or Pueblo Indians' story of creation is the same as we find amongst the Masai group of Nilotic Negroes in Africa. Some of the pottery of the cliff-dwellers is identical with the figures and markings that we find amongst the Nilotic Negroes, with Egyptian Hieroglyphics and the Swastikas depicted thereon. These signs and symbols are common to the Utis, Navajos, Pueblos, Pimas, Apaches, and many other tribes. On the hills near Tacna in Tarapacs, between Peru and Chili, are "the remains of Hieroglyphics of enormous dimensions perfectly visible at a considerable distance, written in vertical lines, like the Egyptian and Chinese; eight leagues to the north-west at Arequipa may be seen engraved upon granite, on the heights of La Caldera, figures of men and animals, straight and curved lines, parallelograms and certain kinds of crosses and letters; these are all "pre-Inca."

The similarity between the objects of art of the pre-Aztecs (Stellar) and the pre-Incas (Stellar) is stronger than between the Aztecs and Incas (Solar) themselves.

CHAPTER XIX

FURTHER PROOFS OF STELLAR CULT IN AMERICA

I MUST draw my reader's attention to that excellent work recently published, "South American Archæology," by T. A. Joyce, Esq., Assistant in the Department of Ethnology, British Museum, and Hon. Sec. to the Royal Anthropological Institute. It is one of the most important works published on South America, and the contents will greatly assist the student to unravel the past mysteries of this Continent, the full description and fine photographs enabling us to decipher their meaning. I have quoted freely from this book.

Mr. Joyce states: "Along the coasts of Peru, Chili, and Brazil are found the remains of a very early population. These tribes were like the tribes now found in South Patagonia and Tierra del Fuego, a long-headed people burying their dead in the horizontal position on the side, facing south or north, using rough implements of bone, shell, and stone. These were driven south by the round-headed people who followed them from the north. The tradition of these roundheads is that they landed on the Venezuelan coast and spread south and south-westerly." The tribes in South Patagonia and Fuego are descendants of the Nilotic Negroes and use the Nilotic implements. There are no implements of Stellar Mythos people so far south, and up to the present no evidence that they even penetrated thus far. These tribes of "Nilotic Negroes" now living here were driven down and left as an isolated group. They are long-headed people. They bury their dead like the Nilotic Negroes in Africa. The round-headed people were Stellar Mythos people, who came from Africa via Asia: they bury their dead in the "thrice-bent position." The former probably came via Greenland and the latter via Asia.

Nilotic Negro implements, i.e., flaked implements (palæolithic), roughly chipped on both sides, have been found in Patagonia, Buenos Ayres, and many other parts of South America. Close to Corrientes, in Buenos Ayres, implements of a very primitive type have been found. These are plain oval pebbles, from one end only of which a few flakes have been chipped so as to give them a rough edge. They are Pigmy implements.

Mr. Joyce states: "From their appearance they might date from very early times, but they have been found in the most recent stratum on the surface, and cannot therefore be of any great antiquity."

As stated (*supra*) these are Pigmy instruments and prove that the Pigmies lived here before being driven away into the forests of Bolivia, where we now find them; how long ago the last one died here it is not possible to say.

Implements of polished stone, i.e. Neolithic or Stellar Mythos people's implements, are found almost all over the Continent, *but not in South Patagonia or Fuego*, although in Patagonia they have "axes"; these are of Palæolithic or Nilotic Negro types, and not Stellar or Neolithic. Those found in Chili are axes belonging to Neolithic or Stellar Mythos people, and some are beautifully polished.

The type of Nilotic Negro in Fuego is lower or more primitive than that found in Patagonia. They do not make pottery, whereas the Patagonians made a rude pottery, the same as is found in Chili. The Fuegians have no Hero-Cult; the Patagonians have.

The pottery found in South America shows that three distinct peoples inhabited this Continent. Mr. Joyce, who divides it into the following divisions, states there are—

1. Red-white-black type.
2. The black type.
3. The Inca pottery.

Amongst the black type of pottery found in Peru and

some other parts of South America is a vase in this form or shape,  called *Tekh*. This is precisely an Ancient Egyptian type, and also called in Egypt *Tekh*, which means "supplier of liquid" (Bun., p. 530).

The reason for the two vases combined in one is that it represented the two fountains of water supply, i.e. water of the Earth and water of the Heavens; it is of purely Egyptian origin.

A friend of mine has very good specimens of these.

In other words, we have the pottery of the Nilotic Negro, the pottery of the Stellar Mythos people, and the pottery of the Solar Mythos people all here.

The pottery found amongst the South Patagonians may be seen and characterized as that of the Nilotic Negro people.

The Inca Pottery, also shown in several plates of this book, was the Pottery of the Solar Mythos people.

The Bola Stones (p. 247), characteristic of the Puelche region, are similar to those found amongst the old Egyptians.

The Spear-throwers are the same as those found amongst the Nilotic Negroes in Africa and the Australians.

Mummification to preserve the corpus was practised amongst these people as amongst the Egyptians.

The bodies were buried in the thrice-bent position, sitting, or on the side, often wrapped in skins and other coverings; frequently numerous bandages, made of cotton, were wound many times round and round the body; false heads with eyes of shell, or silver, and sometimes wooden noses and painted faces, were added, just as was done by the old Egyptians. Dr. Smith is wrong in his statement that the Egyptians did not practise this rite until the Second or Third Dynasty, which was Solar. They did so more than 200,000 years before this, during the Stellar Cult period, and I say that Nilotic Negroes were the first to commence in a primitive way this process of mummification.

Throughout the country the various modes of burial found can always be traced to the Stellar or Solar Cult—or to the Nilotic Negro.

Ancestor-worship was, and is, practised; offerings of food and drink were made and left at the tombs.

Their belief that the dead were escorted to the other world by Dogs (p. 144), which here represent that Zootype for Anubis, is the same as the Egyptian. He was "the smeller of the way," and always accompanied the deceased through the underworld. (See Ritual and *supra* under Set.)

The two Uræi of the Egyptians were also represented and used amongst the Incas as a double-headed snake, and worn on the "belt" or head of the Kings of the Incas (Fig. 10), depicting Royal descent.

Arriaga states that the inhabitants of a certain village (in Peru) worshipped an ancestral "Huaca" in the form of a Stone Eagle, which was found, together with four mummies, said to be of its human sons, the parent of the Tribe. This Stone Eagle is the Zootype of Horus and the four mummies = his four Children or Brothers of the old Stellar Mythos (the Bacabs of the Mexicans, see *supra*). Mr. Joyce states that: "From the history we gather that certain of the Incas attempted to put down the Huaca worship. On the whole, especially in later times, they were broad-minded and were satisfied that the sun should be accorded the premier position." In other words, we have here the primary Stellar Cult well established and in existence before the Incas arrived, and these latter, not being able to eradicate it and impose the Solar Cult at once, were content to merge it into the Solar, as they had done in Egypt—by time.

Peru, like Egypt (whence the inhabitants originally came), was divided by the Incas into two divisions, an Upper and Lower, Hanan-Cuzco and Hurin-Cuzco, but the same division obtained under the Stellar Mythos people, under the names of Hanan-Chinca and Hurin-Chinca, equivalent to the Upper and Lower Egypt of the Egyptians.

The title for their King was the same as that of the Egyptians for their King, namely "Son of the Sun," during the Solar Cult.

The Stellar God in Peru which was pre-Inca, was

called *Con-Ticsi Uracocha*. He was the deity worshipped by the rulers of the pre-Inca empire. His name—

Amongst the Collas = *Tonapa* and *Tarapaca*.

Amongst the Quechua = *Pachacamac* (whose primitive name was *Irma*, another name identified with him on the coast). On the Uplands it was *Iraya*. Amongst the Arucanians his name was *Pillin*.

Amongst the Cara he was called *Lambayeque*.

The people of the Serra called him *Inti*, all different names for Horus I. Uhle's myth (S.A.A.) is the same as many others found in Central and South America, and must be referred to the precession of the Pole Stars, i.e. Horus and the Seven Glorious Ones. Wichama represents the Egyptian Set.

The legend of the Solar Mythos Incas can only be explained and understood through the Egyptian wisdom—Mama-Huaco = Hathor or Isis, who conceives by meeting the Sun in the underworld, or Amenta, and then gives birth to Horus, the Child (see Ritual).

Another name was Huaca, the clan-god of the ruling family of Cuzco, which Mr. Joyce calls a puzzling term, but it is not puzzling when you know the key that unlocks these writings.

Huaca is derived from the Egyptian Huh— Hu = Iu signifies, in Egyptian, the Lord of the Heavens and the Earth—ruler of the destiny of the world. The original was written Hu then Iu Iau, Iahe as the Son of Hu, which two were one. In later times the I was changed into Y. Another form of his name was Heru-Khuti = Light of the World, and we have this depicted here. The Jews used the word Iah = Jehovah, Phœnicians Iao, Hebrew Iah, Assyrian Iau, Egyptian gnostic Iaou, Polynesian Iho-Iho, Dyak Yavuah or Iaouh, Nicobar Islanders Eewu, Mexican Ao, Toda Au, Hungarian Iao, Manx Iee, Cornish Iau, Welsh Iau, Chaldean Iao-Heptaktes, Greek Ia and Ie, Iau in North America, Inti in South America. These were all the followers of the Cult of Aiu, or Iu, or Iau of the Egyptians—Earliest Solar = Atum-Iu.

I have explained this in other works of mine, which have evidently not come under the notice of Mr. Joyce. I trust this will explain the meaning clearly to Mr. Joyce, so that it may be no longer a puzzling term to him.

It is Stellar Mythos, here clearly depicted, which was the Cult of the people before the arrival of the Incas, who were Solar Mythos people, and then came the change from the one to the other—the names were changed, and one was brought on into the other, as in the Egyptian. (For further proofs see "Arcana of Freemasonry.")

Speaking of the earlier Peruvians, he states: "Another divinity which received worship was the Earth, which, under the name of Pachamama—'All Mother'—was regarded as the personification of fertility." He also says that each Ayllu claimed descent from a common ancestor, and this ancestor might be a rock, lake, river, tree, animal, or some *supernatural* personage, later transformed into a stone, beast, or bird. In Sign Language could anything be clearer or more definite to prove the Egyptian origin? The Ancestors here were the original Mother Totems.

Pachamama here is the old Mother Ta Urt of the Egyptians, and the "transformed" or the transformation I have already explained under totemic Sociology and Mythology. But what does Mr. Joyce mean by the *supernatural*, which term he uses frequently? May I suggest to him that there is no such thing as *supernatural*, but possibly he means *superhuman*, and supernatural is a printer's error; we will leave it at that, as so many fall into the above error, even my friend Dr. C. Haddon in his "Cambridge Anthropological Expedition," where he acknowledges it "as a printer's error."

The Chibcha, whose northern capital was Tunja, and who occupied the tableland of Bogota and the districts stretching to south and west, were old Stellar Cult people, and possessed a civilization which surpassed that of the Spaniards. Agriculture and trade flourished, closely built towns and villages with very numerous inhabitants and fine old Stellar Temples existed—ruins of these are still found in Columbia. They had a well-organized government, and were workers in gold and silver jewels and ornaments, mostly skilfully executed.

At Curicancha the golden glories of this Temple have often been described, the walls being built of accurately fitting rectangular blocks of stone, covered with precious metal and studded with jewels. The four shrines near the main building, one ornamented with silver (and dedicated to the Moon), another similarly decorated (dedicated to the Planets), a third (dedicated to the Thunder and Lightning), and the fourth, which was lined with gold (dedicated to the Rainbow), were built as shrines to the four Children of Horus. The Chibcha have—

Chiminigagua	= creative deity, the great Mother Apt.
Bochica	= cultured Hero and God of the Chiefs = Horus.
The Sun	= Sua = Shu.
The Moon	= Chia = Tefnut, or later, Hathor.
Bachue	= Patron of Agriculture, Horus as bringer of food.

These are the earliest Stellar Mythos "Great Mother and Gods," but much information is wanting, the above being only some part of their gods and legends. These Chibcha, Quesada and his followers put to the sword and destroyed. But there is sufficient proof that the original Totems of the Great Mother were brought on as in Old Egypt.

The people of Ecuador and Peru were analogous to the Chibcha, until a higher form of Stellar Mythos appeared, like that found at Cuzco (the Cara people and Canara to the south). The difference may be gauged by comparing that found on the coast with that of the highlands.

There is a tradition that a descent was made on the coast north of Manta by a people who came from the sea in "balsas," and that from the coast they travelled inland. They were called the Cara; and the "idol" of these immigrants was a green stone called "Lambayeque" = Horus of the Emerald Stone.

The Chibcha legend of themselves is: In the beginning all was darkness, until a Being, named Chiminigagua, created light and a number of great birds; these birds, acting under his instructions, seized the light in their beaks and distributed it over the earth. Subsequently Chiminigagua created the sun and moon. Shortly after this a woman emerged from a lake, called Iguaque, north-east of Tunja, bearing in her arms an infant boy: this woman, called variously Bachue and Furachague, came down to the plains, where she lived until the boy grew up. She then married him and bore innumerable children, changing her abode from time to time until the land was peopled. Finally she returned to the lake with her husband, and the pair disappeared beneath the waters in the form of snakes; all Egyptian mythology.

The Puruha plaited their hair into numerous little tresses (p. 61), kept in place by close-fitting caps, as the Kahama in Africa do at the present day. Their bodies are found buried in the thrice-bent position, and the various kinds of pottery, implements, and stone statues found with them are in every way analogous to those of the old Egyptians.

Mr. Joyce states: "With regard to religion, the beliefs and practices throughout Ecuador seem to have been much the same, to speak generally, as in Columbia. In the highlands the 'official' cult was that of the Sun and the Moon, and Moon-worship was found also among the Canari. In Quito were two chief temples on opposite hills, one of these, a square building of stone with a pyramidal roof and *a door facing the East*, being dedicated to the Sun. On either side of the door was a monolithic Pillar, said to have been used as a gnomon for calculating the calendar, and around the building were twelve short pillars, representing the months." This was a Solar Mythos temple and the two pillars represented Set and Horus, just as we find in all the Solar temples in Egypt. The twelve pillars would represent the twelve divisions of heaven, the twelve Zoo-type Banners, Camps, etc. (see *supra*). "The other temple was that of Set, a circular construction with windows of similar shape. These temples were erected by the early Scyri, but the cult seemed to have existed in the country before they arrived, since Sun-worship was probably not practised on the coast before the Inca conquest." Here we have a Temple with evidence of early Stellar. This was a temple of the God Set, such as we find in China and Mexico, in the chapter on the Chinese, and as these existed before the return of the Incas from the south again, it proves that the route of the Solar Cult, as is my contention, came from the north, i.e. Yucatan. These people travelled down from Yucatan as far as Peru, and probably their line of communication, which was comparatively narrow, was cut by the old Stellar Mythos people who surrounded them on every side, and that is why we find the Incas spreading north again after a period of time had elapsed, in order to open up communications.

The map given on p. 82 of Mr. Joyce's book will help to prove my contention, and the internal wars carried on between the old Stellar people and the Solar people further prove this, showing how tenacious the old Stellar people were, that the advantage of the wars for a time rested with them, but that eventually they were overthrown. The Solar people were surrounded by fierce fighting men with their old religious beliefs (Stellar Cult), and it was only by building strong fortresses that the Incas escaped annihilation; ultimately they succeeded because they were more advanced in civilization and the arts of war and were better armed; but they could not overcome the fierce old Arucanians in the south. Mr. Joyce states that—

"At Liribamba was a temple of the God of War, whose image was a pottery vase in the shape of a human head, into which the blood of prisoners was poured, before the Scyri abolished the practice. But the popular religion consisted in the worship of certain animals: for instance, the Canari believed themselves to be descended from a huge snake, which had its home on a lake above Sigsig and to which offerings of gold in the form of figures were thrown, just as in the Chibcha country. A similar snake-cult was found in the northern provinces, and the local worship of pumas, trees, and stones was common throughout the country. At Manta, in Manabi, was a celebrated temple to the God of healing, named Umiña, whose image was cut from an enormous emerald."

So much for the observation and explanation of their

Signs and Symbols by men who do not understand Sign Language.

The observations are good and of inestimable value; their explanations valueless where given. These were old Stellar Mythos people, and the figures, etc., were the Zootypes by which their deity and his attributes were represented. The Great Snake was a Zootype of the God Tem, such as we find amongst the Yezidis at the present day, and in the temples of Pa-Qerhet or Ast-Qerhet, that is, the house of the Snake-God Qerhet, in Egypt at Pithom, and the Great God of Pithom was Tem, and one of the forms or types of Tem was a huge serpent. The serpent was not worshipped, but was a representative Zootype of Tem. If Tem was the god and the serpent the symbol, obviously it was the divinity that was the object of the worship and not the symbol. Tem was a type of Horus. We find remains similar to these in North and Central America.

When Moses lifted up the Serpent in the Wilderness, his followers, who were mostly old Stellar Mythos people, understood. Moses was learned in both Cults, and tried to blot out the old Stellar and introduce the Solar, but he had to return to the Stellar Symbol to appease his followers at first.

The God of healing named Umiña was another form of Horus. In the Mexican his name was Ixtililton.[1] This is the Stellar Cult representation, and in the Solar he is portrayed as I-em-Hetep.

As Prince of the Emerald Stone his name in Egyptian was Her-uatch-f = The light of the world. Heru-Khuti was another name. He was known by association with the Great Pyramids, both in Egypt and Mexico (see "Signs and Symbols of Primordial Man").

In Mr. Joyce's book, "South American Archæology," on Plate VII, Fig. 3, is depicted Hept-ro, God of the gaping mouth, i.e. Shu, and Fig. 1 represents the Egyptian I-em-Hetep.

Unfortunately, Mr. Joyce has in some cases "confused" the "myths," i.e. he has confused Solar and Stellar Mythos

[1] See "Signs and Symbols of Primordial Man."

with the Nilotic Hero-Cult for the reason that he did not understand them and so makes no differentiation.

Mr. Joyce, speaking of the implements found in Ecuador, states: "It is a little surprising, therefore, to note that, in the area where such sculptures are relatively common, implements of stone are extremely rare, and that in the highlands they have been found in considerable abundance. These consist in the main of club-heads and axes, all of *polished stone*, of various patterns, none of which are peculiar to a given district." This only proves that the early Stellar Mythos people returned to the highlands as another wave of immigrants of higher culture arrived. These latter would have implements of iron (Ba metal), which by this time would have all oxidized and so we should not find them. That we find copper and "hardened copper" implements, the latter with which they could fashion stones, etc., is also accounted for, because these would only oxidize to a small extent, not enough to cause their entire destruction. They must have had iron and steel instruments, because the art of smelting and making these was known in Egypt from the time of the later Nilotic Negroes. The enormous specimens of axe-heads found are a puzzle to Mr. Joyce; he thinks they had "a ceremonial significance because they were too large to use." He is quite right; these axes were associated with "The God of the Axe." But many up to a weight of 50 and 60 lb. would be used for various purposes —these are found in Rhodesia, in Africa, and are associated with the Stellar Cult people (see my notes on Rhodesia).

They had the Spear-thrower, and slings for stones, as the Masai have, and their pottery was varied, much of a high class, with the same variations as we find in old Egypt. The pottery of the La Plata Island resembles that of the highlands of Ecuador, and we find the same resemblance to that of the early Stellar Mythos period in Egypt. We also see from Mr. G. Hey's expedition and description of Manabi and Esmeraldas that although there are certain similarities, great difference exists between their respective cultures, and the most important of these are as follows. Stone implements are rare in Manabi and relatively common in Esmeraldas; but on the other hand stone sculpture, which is highly developed in the former, is hardly found in the latter province, and the remains of stone buildings are confined to Manabi.

Thus we can trace how the wave of the second exodus of the Stellar people drove farther out and away from the most fertile parts those of the earlier exodus of these Stellar people, and that this second exodus was partially absorbed in some places by the Solar Mythos people.

"The myth of the history of the starting of the Peruvian Empire" related on pp. 78 and 79 in Mr. Joyce's book, is read through the Egyptian. The four brothers were the brothers afterwards denominated the four Children of Horus, and the four sisters were the four consorts assigned to them. (All issued from a Cave—a type of the mother, p. 85.) The Stellar chief deity in Peru, called Uracocha, an elementary nature-god and a culture hero = Horus (see names of the four Brothers or Children of Horus, and their names in different countries and various terminology, *supra*).

The fable of being turned into stone is equivalent to the precession of the Pole Stars as one takes the place of another, so the "types" represented are turned into stones; we find the same myth amongst the Mexicans and in other countries.

There is a tradition that many tribes came by sea and landed on the coast; some of these people were so-called giants, probably an exodus of Suk or Turkana or Jieng (men over 7 feet, see Nilotic Negroes). In the name of the people Tehuelches we have a word similar to the Egyptian Turkana.

The transference of the civil power of Tampu-Tocco shows a revival of Uracocha worship, i.e. Stellar, and a temporary eclipse of the Sun or Solar Cult. The religious persecution of the Stellar people by the Solar is well set forth in all histories of Peru, but probably the earliest arrival of these Solar people had been forgotten by the earliest chroniclers; it was so long ago that all trace would have been lost. The wars would have continued from the first.

Perhaps nothing demonstrates this better than the map Mr. Joyce gives on p. 83, which will speak for itself after reading the above, without further dilating on this matter. The line of march of the Incas to the south, along the Andes only, might probably have been caused by the geography of the South American Continent at the time of this early period; marching from the north down to the south between two ranges of mountains, they would to a certain extent be protected on each flank.

The Inca Empire ceased at the river Maule. The Arucanians were old Stellar Cult people whose individual freedom was a Stellar creed, and rose superior to all superhuman terrors; it therefore follows that when a people are prepared to perish rather than submit, they are unconquerable. They would not submit to the huge bureaucracy of the Incas, which this people had established in other parts of the country, and which in the end was the cause of their downfall when the Spanish arrived, and gained an easy victory over them. They were a lesson to all nations, especially to the present "Huns."

Near the southern end of Lake Titicaca, on the borders of Bolivia, are situated the ruins of Tiahuanaco, remains of the old Stellar Cult people connected with the Aymaras or the predecessors of these people (probably). I have taken particulars and descriptions from the works of Mr. Joyce and Mr. Enock. The latter states: "The ruins consist mainly of the outline of a great temple, shown by rows of upright monoliths, foundations, parts of stairways, a monolithic stone doorway, some colossal stone figures, and great stone platforms." A huge mound remains of what was formerly a truncated pyramid about 600 feet long, 400 feet wide, and 50 feet high. The stones in some cases are carved with hieroglyphics or low reliefs—this proves their pre-Solar origin.

These great stones of Tiahuanaco, unlike the Inca walls, are in some cases richly carved. The most remarkable of them is the monolithic doorway of Akapana, carved with a kind of frieze in bas-relief of figures, the central one of which has been taken by Peruvian archæologists to represent the mystic deity Huirakocha (Horus). Some of the lesser figures have human bodies, hands, and feet; some have human heads, others heads of condors (Zootypes); some wear crowns and carry sceptres, and they appear to be in an attitude of adoration of the central figure, which itself holds sceptres, carved with the heads of tigers, condors, etc. (Zootypes).

This central figure of Huirakocha (or Viracocha) was the supreme God of the Andean people, typical of the "Abyss of the Waters" as was "Ea" among the Chaldeans. Amongst other notable monoliths of this group is a stone image, about twice the height of a man, which stands upon the plain. One hand of the sculptured figure holds a fish against his breast—like those found in Easter Island.

(This is Horus as the fish-man. See Mexico and Assyria, later.)

At the Castle of Chavin there are large ruins with singular underground chambers and passages. The walls are of blocks of hewn stone, a number of singular carved monoliths—one of these, a beautifully carved stone, about twice the height of a man, has been transported to Lima. Another, in the form of a column with carved snake-heads upon it, remains in its place in a subterranean chamber. Chavin is of the pre-Inca period. (This figure represents one of the earliest forms of the God Tem, represented by the Serpent as Zootype. The so-called Great Serpent Cult (see God Tem) was practised amongst the Tusayan and amongst the "cliff-dwellers" of Arizona.)

Mr. Joyce's description of the ruins shows that there are two monoliths, one on each side at the entrance, about 16 feet apart, over 12 feet high (above ground), 6 feet wide, and 4 feet thick, weight estimated at 26 tons.

The central figure as shown (in Fig. 17) represents a being in human form with abbreviated legs, the head surrounded with rays terminating in puma-heads and circles. Similar puma-heads decorate the fringes of the tunic-skirt and sleeves, his belt, and the end of the engraved band which runs from each eye down his cheeks; on his chest he bears an ornament representing an animal with a fish-like body, curved into a semi-lunar shape and with a puma-head, supported on an intermediate object ♀ (the Egyptian Ank), flanked by bands terminating in Condor-heads. The object is repeated on the ornamental bands which run from his shoulder to his belt. In each hand he holds a kind of staff, the butt-end of which is carved to represent the head of a condor, but here the resemblance between the two ceases. The upper portion of that in the right hand is single, and a small condor rests on the inner surface of the extremity; that in the left is double, each hand terminating in a condor-head.

The figure stands on a kind of throne, also ornamented. The lesser figures (Fig. 18) are arranged in three rows on either side of the principal figure, and the figures composing each row resemble one another and those of the row opposite.

This figure is Horus I, with various attributes associated or attached to his name; the name is in the centre at the top, surrounded by two Uræi or two serpents, which represent the two feathers of Iu. There are four attributes (brothers or children of Horus, each one guarded by Uræi and with feathers on each side. Horus holds in his hands the symbol of sovereignty. It is part of the form of the supposed 20-day Sign of Dr. Seler in Mexico. The Hieroglyphic, Apt, is "the brow" of the god Iu.

I reproduce this figure from Tinogasta, Argentine.

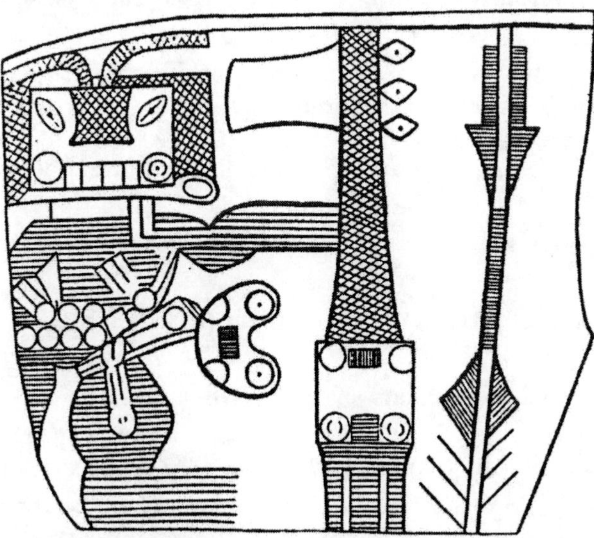

Con-Ticsi Uracocha, God of the Axe, from Tinogasta.

It is the God of the Axe in the Stellar Cult form. He is here called Con-Ticsi Uracocha, the same as the God of the Axe of Tepoxtecatl and the God of the Axe from Tepozteco (see later). It is the same figure portrayed on the Monolithic Gate at Tiahuanaco at the south end of the Lake Titicaca, Bolivia.

On his head are portrayed the Two Feathers . This gives us his name of Iu.

The Itheophallic depiction with the Fan  portrayed denotes a New Life—Spiritual, and the Ideograph for the House of Eternity is depicted on the Fan. The 6 with the 1 = 7 proves him to be the head of the seven Glorious Ones; we also have the eighth added, = Iu, as the Supreme One. Here, therefore, is portrayed the 6 + 1 = 7 and the 7 + 1 = 8.

The Axe is being held in his hand with the cutting blade towards him, with these  attached to the axe, denoting that he is the Primary of the Trinity; these three denote the Primary Great ones, i.e. Horus, Set, and Shu, the Primary Trinity.

The Axe is supported by the four children of Horus represented by  with the House of Earth and the House of Eternity between them (a double square); and the two poles or pillars of the north and south  —the two supports of heaven in one plane—are underneath.

On the right is the emblem or symbol of Sovereignty, Power, Majesty, and Might. This, therefore, is the God Horus as Iu—God of the North and South in Stellar form. There are no horizons here; these had not yet come into being, not until the Incas brought the Solar Cult. Then his name was changed to Pachacamac, and afterwards it became Horus of the Solar Cult as Atum-Iu.

The Quechua name Pachacamac was, therefore, the same God in Solar form as the Stellar Con-Ticsi Uracocha. Other names for the same in different parts of America I have given.

The fortress of Sachaihuaman of Cuzco, in South America, is also one of the great ruins left of the Stellar Mythos people. It consists of a series of four or more great walls from 12 feet to 25 feet high, forming terraces up the hill-side 1,800 feet long. The walls are built as great revetments with twenty salients at regular intervals, the masonry being formed of cyclopean worked stones, which in some cases are nearly 20 feet high, weighing many tons. But these temples and buildings of the Old Stellar people can always be distinguished and identified separately from those of the Solar people who followed after them, as I have shown.

Head-hunting was practised by many tribes in South America, namely the Muzo, the Colima, and the Panche. This, as can be seen (*supra*), was the custom of many tribes found in various parts of the world, originating amongst the Nilotic Negroes.

Amongst the Ahruaco Indians we find the same practice as amongst the Nilotic Negroes. The huts are round, with small entrance door, have not much in the way of furniture, but they use *wooden stools* like the Suk and Masai; also the man and wife live apart, that is, the huts are in pairs, one close to the other. They are agricultural, whereas the Goajiro are fierce fighting men and go quite naked.

The oldest known peoples in Peru, not including Nilotic Negroes, are to-day represented by the Aymara race of Indians.

Their ancestors built the beautiful temple of Pachacamac, and also the temple of Viracocha, whose massive ruins still testify to the great knowledge and skill of these old Stellar Cult people long before the Inca. There was another, that of Tiahuanaco, and all the ruins dotted about Lake Titicaca were raised by the same people.

The Guarani and Aymara Indians in some parts still wear the pigtail, which originated with the Nilotic Negro (see

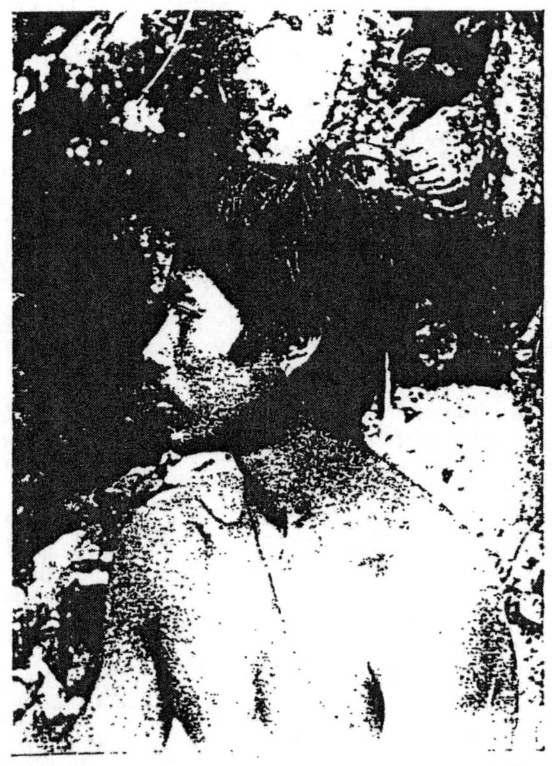

PLATE LXVII.

From Brazil, showing Japanese type, for which I am much indebted to the Minister of Agriculture of Brazil and Major Albert Levy.

PLATE LXVIII.

Kindly sent to me by the Minister of Agriculture of Brazil, to whom and to Major Albert Levy I am indebted for the photographs of the Indians of Brazil. I understand these have never been published before. The group represents the Indians occupying the country of São Paulo; they are being kindly treated by the Brazilian Government. They are probably degenerated Early Stellar Cult people.

photo of Guarani Nilotic Negroes and compare with the Japanese).

The Great Chincha Confederacy, ruled over by the Grand Chimu, was perhaps the greatest empire of the Stellar Cult people in South America.

We must now turn from the western part of South America to the eastern, and to what many have believed in ancient times to be a large island, Brazil and adjacent countries, which Mr. S. Landor has proved to be not so, but a huge, seething volcanic land, where we do not find any remains of the Solar Cult people—but non-Hero Cult and Hero Cult natives, who were followed here by the Stellar Cult people. The remains that have already been found are quite sufficient to prove that these people spread nearly all over South America, north, south, east, and west, as far down as Patagonia.

The evidence that remains proves that the Caral and the Arawak, who came from the north, gradually spread south, and occupied the Orinoco and Amazon basins; the Tupi-Guarani and Ges occupied the rest of the regions. The Tupi-Guarani, of whom I give a photograph (Plate LXVII) (for which, with others, I am much indebted to the Minister of Agriculture, Brazil, and also to Major Albert Levy of São Paulo, for all the Brazilian natives shown in this book), gradually worked down to the estuary of La Plata and then up the coast to the Amazon, and even beyond (or vice versa), coming in contact with many tribes still living, of greater antiquity than these, which have not up to the present time been described or visited by white men of the present generation.

At Lagoa Santa, in the province of Minas Geraes, a number of human remains were discovered, which were associated with the bones of extinct mammalia usually regarded as typical of the Pleistocene period. The human skulls which were there found have been shown to stand in relation both to those of the primitive tribes of the Ges family and to those found buried in the shell-mounds of the coast.

These shell-mounds are found scattered along the coast from the mouth of the Amazon to the most southern province of Brazil, as well as on the banks of the lower Amazon and Tocantins. They consist of shells mixed with earth, bones of fish and mammals, charcoal, and stone implements, as well as remains of human skeletons.

In the mounds of Santa Catharine have been found numbers of small diorite mortars, probably for grinding pigment, in the shape of fish or birds (Fig. 33 on p. 259, "South American Archæology"), extremely well finished and unlike any other objects discovered in the area under discussion. (These were made by Stellar Cult people.) In São Paulo finds have been made of peculiar bi-conical objects, Fig. 33 (*c*), probably arrow-heads or stone daggers, made from syenite and serpentine, which show equal skill in manufacture. Stone axes are common and no definite pottery has been found; it is a question, therefore, whether these belonged to the Ges family, whose best-known representatives of the present day—the Botocudo—are good workers in stone but are ignorant of pottery and do not practise urn burial, or whether they belonged to a High Stellar Cult people who have now disappeared. I produce here a photograph of the present natives around São Paulo (Plate LXVIII).

My own opinion is that the Stellar Cult people for some unknown reason have disappeared from these regions, whilst the Ges family—Nilotic Negroes—have retired to the forests and still remain.

For this reason—because throughout a large district we find a rude kind of pottery with simple decoration, in the south, but in the north; at the mouth of the Amazon, the art is on a considerably higher plane.

In Entre Rios and Corrientes, especially along the banks of the Parana, in the country between Buenos Ayres and Rosario, many traces of aboriginal settlements have been discovered in the shape of pottery, stone arrow-heads, mortars, implements of bone, antlers, and human remains. The pottery shows considerable homogeneity, and from two localities as far as Campana, a little to the south of Zarale and Goya, fragments which may be taken as typical of the district have been found ("South American Archæology"). These are portions of small vases, of rather thick ware, mixed with sand and well baked, with heads, chiefly of birds, in bold relief (the Horus Hawk of the Egyptians), string pattern, key pattern, zigzag, etc.

Further north, on both banks of the Parana, in Missiones and Paraguay, pots with similar impressed ornaments have been found.

Similar painted and engraved pottery has been dis-

The Egyptian representation of the Mesken, or birthplace. Ta-Urt giving birth to Horus—her young Crocodile, which is seen standing on its tail—or the birth of a new year.

Ta-Urt giving birth to her son, or the birth of a new year. This is the Mexican representation of the Old First Mother giving birth to her young Crocodile.

Steatite carving from the Lower Amazon. This portrays the Egyptian Great Mother, Ta-Urt, giving birth to her son, or the birth of a new year. (Drawn by K. Watkins.)

covered in the islands of the Parana delta. The ware found in South and Eastern Brazil also resembles that of Missiones, and it should also be mentioned that the vases from the cemeteries on the right bank of the Paraguay, right up to the Brazilian province of Matto Grosso, and also in the two provinces of Jujuy and Salta in the Argentine, are similar to the pottery of the upper Parana. Burial urns are found in most of the above districts — some with rudimentary human features in relief.

Finely carved "amulets" have been found in the Amazon valley, especially on the lower Trombelas. I produce here (p. 323) a figure of the great Mother Apt giving birth to her first-born, found in Brazil. Also a figure from a Theban tomb (drawn by Mrs. K. Watkins). The Old Great Mother Apt of the Egyptians, who in one form is a Crocodile, as portrayed here, has just given birth to her child Horus, Har-Ur, as shown by the young crocodile poised on end in front of her. It is the picture of the young child that was brought forth annually from the water by the mother, who was constellated as the Crocodile or Hippopotamus at the northern centre of the planisphere, as we see from the Hieroglyphics, [hieroglyphs], Mesken, the place of birth. I also reproduce here the Mexican representation from Mexican antiquities (p. 159). The comparison of these is critical evidence of their identity with the Egyptian.

These are all the works of the Old Stellar Cult people, and it is evident from these finds that in ages past the whole of this district must have been inhabited by the Stellar Cult people. It is a pity that the great ruins discovered by Dr. O'Sullivan Beare have not yet been revisited, photographed, and explained, as there can be no doubt, from his description, that this was one of the large cities with temples, etc., in South America.

It is a question if the Lunar Cult people did not at one time inhabit the North of Brazil, where in the area are found vases characterized in *human* form, with flat bases, a white lip and engraved, moulded, and applied and painted ornaments, of which the finest examples are found at Marajo, I am of the opinion that they were the works of Lunar Cult people, because of the human, or partly human, figures depicted.

The Stellar Cult people depicted by Zootype, prehuman.

The Solar in *human form fully developed.*

The Lunar people were those of the transition stage, i.e. partly human figures—which we find here, as we do in the Mediterranean basin, Europe, and the Nile Valley.

Petroglyphs and pictographs, engraved or painted on rocks, have been discovered in nearly every state in Brazil, as well as in the Guianas, and many of them bear a resemblance to those of Venezuela. It is impossible without seeing, or having good photographs of them to make a definite statement as to who made the above. Certainly either the descendants of the Nilotic Negroes or Stellar or Lunar Cult people. It has been stated that these are similar to those found in Africa, in the Mediterranean basin, and some parts of Europe.

Mr. Joyce claims that American civilization is indigenous, and the *Times* states that he is probably right, as he has such men as Dr. Brinton, Mr. Squire, Mr. Maudsley, Sir Edward im Thurm, and above all Sir Clements Markham, to back him—and states that "the latter points to the agricultural achievements of Andean man as evidence of his vast antiquity in those regions—the domestication of the llama and alpaca testify to this; the cultivation of maize and of several kinds of potato must have been the result of the careful and systematic labour of many centuries." *But these old Stellar Cult people had learnt all this before they left Egypt; their forefathers, the Madi and Masai, were the first to build, tame wild animals, and cultivate the land, bringing agricultural knowledge to a high standing. Even the old Hieroglyphic formed by the Masai for their house,* ⊔ *Per, I have found depicted on stones here. Agriculture and the taming of wild animals was African first, and their representatives or descendants built their houses as the Masai* (see *supra*). The knowledge of this was obtained from the Egyptians, passed on to Asia, thence to America.

The Lasso and Bolas, so frequently used in South America, were first used by the old Egyptians, as can be seen by the paintings still extant (see Maspero, "Dawn of Civilization").

The native boats found here in Peru and Chili are similar to those used in Egypt; these "balsas" are still used by the natives on Lake Titicaca, and the Quechua word for boat is derived from the Egyptian, also the word Huaca, or "holy place" of the Incas.

Delineated on these old ruins in South America I have found several pure Egyptian Hieroglyphics; the most important I consider is

Mra or Mr-tar; it is the name, or one of the names, for Egypt.

I have found this hieroglyphic not only in South America, but several times in Central America; also the following[1]:—

St—a sunbeam, sun's rays, daylight, splendour, to illuminate, to gleam.

Anx—life, living.

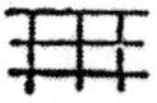

Hespu—name of countries, district, nome.

Ra—solar time, the sun, an hour, a day, yesterday.

Mu—to place.

Sxt—take, to net.

Sn—to open, pass.

Ha, ah, oh, time, day.

Hr—to open, to fight, to bind.

An—abode, name of Amen. (Horus was the eldest son of Amen, depicted as here, Bun., Dict., p. 102.)

Utbu (pl.)—wells, an ingot weight.

N—pool or tank.

Ta—the earth, the world.

Ht—a mace (taken from the Masai), order.

Am—Amsu, or the risen Horus.

which is a later form of  Amsu, or Horus in Spirit.

circumference of Paradise or Heaven (Pierret).

things—cake.

[1] See also "Signs and Symbols of Primordial Man," 2nd Edition, p. 264.

In the traditions of the Mexican and Central American races there is mention of a civilized nation said to have arrived in the country a very long time ago, namely the Toltecs, and that "they carried their books with them on their migration and were led by their wise men—the Amoxhuaque—who understood the books," i.e. their picture writings, which were the hieroglyphic writings—language, system of divination, their sacerdotal wisdom learnt by these Amoxhuaque, who were the High Priests from Egypt or their descendants.

These Toltecs are the people to whom the greatest culture in prehistoric Mexico is attributed.

They came from Egypt via Asia, and their empire or domain finally extended from the Pacific to the Atlantic. They probably crossed from the northern part of Asia and landed on the west coast of North America and travelled south.

The Great Pyramid of Cholula is the oldest and largest in Mexico, built by these Toltecs.

"Quetzalcoatl was the presiding deity," i.e. Horus I. Their God of War—Huitzilopochtli—was another form of Horus. Omechuatl, their creator goddess and mother of Huitzilopochtli, was Sebek-Nit, an early form in the Stellar, of Isis in the Solar; as Goddess of the Pole Star North she was Sesheta.

Their picture writings, metallurgical and textile arts, their carved serpents on walls which surrounded their temples, or teocatli, are all proofs, when deciphered, of the Stellar Egyptian Cult.

From Mr. R. Enock's book, "The Secrets of the Pacific and Andes and the Amazon," photo frontispiece and p. 105, I take the following descriptions, but it is to be regretted that he has not given fuller illustrations and details that one could decipher. "The famous sacrificial stone, made of trachyte" (on p. 105) "8 feet 9 inches in diameter and 2 feet 10 inches high, in the centre of which is a small cavity with a groove running therefrom, in which ran the blood of the victims, etc.," corresponds in all particulars to the one found amongst the ruins in Tunesia. In the Ritual it is called "The slaughtering-block of Set." It is a question as to whether these were human victims that were sacrificed. The Masai and Egyptians made sacrifices, but not human, except of prisoners; no doubt, however, exists as to offering prisoners of war as sacrifices; this is found depicted in mural paintings in Egypt, and it is possible that during earlier Stellar Mythos human sacrifices were made, but that they were subsequently discontinued, the Zootype of the "Great Mother" taking their place in the sacrificial offerings.

Mr. R. Enock in this book states: "About the fields surrounding Teotihuacan, hundreds of small terra-cotta masks and idols are constantly ploughed up; many seem to be moulded in likenesses. Some of them resemble the carvings and castings in stone and copper of the objects encountered in the tombs of the pre-Incas of Peru and also those found in Egyptian tombs." These are all remains of the old Stellar Mythos people, as their decipherment proves.

The Calendar Stone of the Aztecs is similar to those used by the early Egyptian Stellar Mythos people and by the Chaldeans, the latter people obtaining it from Egypt. It is of Stellar Mythos origin and not Solar. The central figure here depicted represents Tonatiuh, the Thoth of the Egyptians. Their Calendar reckoning was Stellar and not Solar, whereas the Calendar System of the Mayas was Solar, and their ornamental designs were similar to the Egyptian Solar.

The Aztecs were a wave of immigrants who followed after these old Toltecs, probably by the same route. But the Zuni or Pueblos Indians, who are the descendants of the "cliff-dwellers," are of an earlier exodus, and it is quite possible, taking the geographical features into consideration, that they crossed from Egypt via Europe, then travelled south, there being evidence which indicates this; they subsequently reached to the south of the Amazon in Brazil. Their stone axes and pottery are of a much higher class and better finish than those of the Hero-Cult Nilotic Negro which have been found here in South America.

Probably the Glacial Epochs which took place every

25,827 years drove them south, and they did not come north again, and that is the reason we find them south.

	CULTURE.
In America, we find in Alaska, British Columbia, Oregon, and Washington	No stone-shaping arts or remains of ruins. Descendants of Nilotic Negroes with and without Hero-Cult, but mostly Hero-Cult now.
In California, Colorado, Utah, New Mexico, Arizona	Hero-Cult Nilotic Negroes. The cliff-dwellers Stellar Mythos people only. Irrigationists.
In Mexico	Toltecs, Aztecs, etc. Stellar Mythos; some parts of Mexico Solar Mythos.
In Yucatan and Central America	Mayas, etc. Solar Mythos with evidence of older Cult, namely Stellar.
In Columbia, Ecuador, Peru, Bolivia, Chili	Incas (Solar Cult) and pre-Incas (Stellar Cult), and evidence of Lunar Cult.
In Brazil	Stellar Cult and descendants of Nilotic Negroes with and without Hero-Cult. Probably Lunar Cult.
In Patagonia	Palæolithic and Neolithic (Hero-Cult).
In Fuego	Palæolithic only (no Hero-Cult).

Is it reasonable to suppose that the huge continent of North and South America has lain incognito by the great communities of Europe, Asia, and Africa until the yesterday of Columbus—unknown throughout the vast ages that man has existed? Columbus reached America less than five centuries ago, and Eric the Red and his early Norsemen, according to Nansen, A.D. 983. Can we believe that the Egyptians, Chinese, and other people, so far advanced in knowledge and science thousands of years ago, had no knowledge of this land before that time? The evidence proves such an idea to be erroneous. America has been the scene of cycles and changes of humanity throughout ages of time; eras of activity (of which the remains can now be seen throughout the vast continent) have taken place; we are reminded of bygone people at every turn by ancient ruins now mouldering in deserts and forests, silent tombs of the dead past, still speaking eloquently of a high civilization. Whence came these people and what proofs have we? I cannot accept Dr. Brinton's opinion, which is analogous to Professor Frazer's and Sir Harry Johnston's and the *Times*. These opinions are theories without any foundation of facts, and I contend that the proofs I advance are unanswerable.

These people originally came from Egypt; the old Stellar Cult people travelling via Asia, not in one exodus only, but certainly there were two, and probably many.

The Solar people came direct to Yucatan and travelled west and south, but were comparatively limited, as stated before.

Arch.

The so-called Maya arch, which has puzzled many, is similar to those found in any Buddhist structure; many examples occur in Peru, and these were copied from the Egyptians.

Mr. Reginald Enock states ("The Secrets of the Pacific," p. 40): "The circular arch, vault, or dome is not found among the early American structures, nor any suspicion of it in prehistoric times, and if any relation existed between the ancient Mexican and the Egyptian this is strange, as the arch exists from earliest times as an Egyptian structure." Mr. Enock is not correct in this statement. *The Stellar Mythos people and early Solar never built arches—they had the lintel only.* The first true arch is found in a Fourth Dynasty Mastaba at Medum, *which was constructed during the Solar Cult.* Amongst the Druidical temples (which were early Solar) one always finds lintels; but at Uxmal we find that the arch has taken the place of the lintel. The arches we find in Central and South America are built in the same way and form as the Egyptian. The early Greeks also copied them from the Egyptians. All these arches are similar, one stone overlapping the other. This can be proved by comparing those of Las Monjas, Palanque, and those of Peru; and the "Treasure House" at Athens with the Egyptian Fourth and Fifth Dynasty arches. The "Mastaba" of Egypt was also copied by these people, and the date could not have been earlier than the Fourth Dynasty. The custom lasted only about one hundred years, from the reign of User-Ka-f to that of Meu-Kau-Heru.

So this gives approximately the date that the Mayas took the Solar Cult to America (Yucatan). They traded with the Egyptians direct and were called by the Egyptians Haûi-Nîbû, i.e. people beyond the seas. The translation is "People beyond the seas," or the people from behind, i.e. behind the setting sun. As the sun, setting in the west, went down into the seas, the Mayas would naturally be spoken of as people beyond the seas or behind the setting sun.

Along the sea-coast of Peru, before the Incas' invasion, these old Stellar people had perhaps attained the zenith of their power, extending over the domains of the Chincha Confederacy, and all the territory ruled by the Grand Chimu, as witnessed by their pottery, their manufactures of various articles, and their religious cult.

The agricultural achievements of Andean man were, we know, not higher than those of many of the Nilotic Negroes, but analogous to them, and this is an evidence of his vast antiquity in these regions. Their domestication of the llama and alpaca and the tradition of being a tall race of men indicate that they belonged to one of the Madi or Masai group, as we have seen (*supra*). The Nilotic Negroes tamed and domesticated animals, and cultivated all kinds of maize, potatoes, etc. All their boats and rafts are constructed after the fashion of the Nilotic Negroes. As regards language, *the earliest dialect preserved by the Aymara Indians,* which was the earliest Peruvian tongue, and extended largely over South America, was closely allied to the early Sumerian, which was old Egyptian. The late Sir Daniel Wilson and Hyde Clarke, two philological authorities, dealt fully with this undoubted relationship. In fact, the similarity between the Egyptian, Sumerian, Peruvian, Mexican, Chinese, and Tibetan has been proved.

CHAPTER XX

CENTRAL AMERICA AND MEXICO

We must now unlock the door leading to the hitherto hidden mysteries of the Mexican, Zapotec, and Maya remains of Ancient Cities, and of a great past civilization which, though dead and gone, leaves still sufficient evidence to enable those who can understand Sign and Symbol Language, and who know the Ritual and Cults of the Ancient Egyptians, to trace their history.

The Mexican Chronology was the same as the Egyptian record of time = twelve months.

The designation of the particular months differed in the sacred and civil Calendars. In the former, the different months are expressed merely as the first, second, third or fourth month of the particular season, the days being numbered like our own But for popular use the names were as follows:

The Wet Season: Season of Inundation. *Se.*	The Green Season: Season of Winter. *Pir.*	The Dry Season: Season of Heat. *Semon.*
Thoth	Tybi	Pachous
Paophi	Mechi	Payni
Hathor	Pharmuti	Epiphi
Choiak	Phamenoth	Mesori

Each cycle is of 1,460 years, the heliacal risings make the circle of the civil year, and as there is a corresponding series of settings performing a similar round, the two series in each cycle would make a double interchange (2 × 2,920).

Stellar time was 30 days for each month, making 360 days, but finding this wrong, in time they added 5 additional days, so that in Lunar and Solar time they reckoned 365.

(They also reckoned the great year as 25,827 days.)

The five days lasted as a Jubilee in the ordinary year and in the grand, or fourth year, there was a special festival of six days.

Hence every half orbit or passage from summer to winter or from winter to summer contained eighteen such decades of days, and these decades are represented, or headed, by a Solar Snake in the walls of the Temple of Denderah, which corresponds to the numeral *l* of Seler.

The Uræus, or symbol of the snake expressed in the "Polar Inch," was not only a symbol expressing the various serpentine curves traced by the motion of the earth and moon but also a symbol of time. Thus in Chapter 130 we read of a snake "70 cubits in his coil." Taking the well-known cubit of 20·6 inches and repeating it 70 times we obtain 1,442, which is proportional (within $\frac{1}{700}$th part) to the number of minutes of time ($24 \times 60 = 1{,}440$) in the daily rotation of the equator, or coil of the snake, so that it expresses our own division of the heavens into twenty-four hour circles, each divided again into 60 equal parts, or minutes of time.

Also "The Great Uræus" was a symbol used by the Egyptians for countless ages to represent time measured and the motion of the earth in reference to the celestial sphere.

The Polar inch was $\frac{1}{25}$th part of the length of the casing-stone of the Pyramid. It differs from ours by its thousandth part only; this Polar inch measures not only the axis of the earth but of the Earth's Polar diameter, a measurable space being contained in the former two hundred and fifty million times, and in the latter two hundred and fifty thousand billion times, as is shown in the Great Pyramid. Also in Ritual, Chapter 64: "I who know the Depths is my name. I make the shining cycles of the years and billions are my measurements."

The unit of the Polar inch was the secret to the architectural standards of Ancient Egypt and known to the high priests only; it is only by knowing this secret that the various units of measurement employed by the architects can be understood. The above was the cubit of 25 inches. The cubit of 20·6 inches was taken from the cycle of the equinox as a circle of 25,827 years, the radius equalling about 4,122 years, and, taking a century to an inch, the half radius gives the cubit of 20·6 inches. This cubit was the standard employed for the sacred "Tat," which measured the Waters of Life. Again, since the orbit of the earth is not a true circle, but an ellipse with the sun in one focus, there will always be one point in the orbit which will be in "perihelion"—nearer the sun than any other. This point is not stationary, but makes a circuit of the earth's orbit in about 114,000 years—whereof the half circuit gives us (at an inch to 10,000 years) the 57-inch cubit.

At the Temple of Hathor at Annu, or Denderah, there are fourteen ascents of the moon leading up on the fifteenth to the Throne of Thoth—The Lord of Measurements, and corresponding to the number of days between new moon and full moon. On the other side are portrayed eighteen boats, each led by a Solar serpent or Spiral, representing the eighteen decades which made up the half orbit. In the entrance hall rise eighteen enormous columns divided into three rows, each containing six columns, corresponding to the number of decades of days.

In two plates given in Mr. John L. Steven's work, "Incidents of Travel in Central America, Chepas, and Yucatan," published in 1848, there are two figures which represent the Egyptian Taht-Aan, the bearer of the symbolic Uat. He is portrayed carrying Horus in his hand and holding him aloft as the True Light of the World, and a symbolic likeness of a soul in human nature, that was begotten of Ra, the Holy Spirit, as the Father in heaven; precisely as we find shown in the Egyptian monuments.

We have found the same type in India, where we have the Great Mother Apt or Mut holding aloft her child Horus, here depicted by a drawing of the same, kindly given to me by Mr. A. H. Sutton.

Annu is the Eastern Solar Mountain where the sun rises or emerges from Amenta.

"O Divine Babe, who makest thy appearance in Annu." The picture of the Divine Babe lifted up into the upper world by two divinities is also shown and portrayed on the statue at Palenque, and was originally Stellar, as is shown here by a picture from India, where the Old Mother Apt is lifting up the Divine Babe.

Taht-Aan, as the Sacred Scribe, wrote the *Ritual*, the book which contains the Divine Word, and brings about the resurrection to the glory of Eternal Life. It is a book of the Mysteries, in which this picture shows part of the revelation, that is here seen dramatically enacted in Mexico (Ritual, Chapter 125), and it was not only here in the Central States of America that this was done, but throughout the whole world where the Stellar Mythos was established, all being taken from the Old Egyptian Stellar Mythos Cult, and nowhere else can the origin be found. Being so widespread, it must have had one origin, and that origin I have here given, with the reason for the same.

As further proof of its widespread adoption I give here two photographs of Taht-Aan, one from the Japanese Shinto Ritualism, which was the Stellar Mythos of the ancient Egyptians, the other from the Philippines, where Stellar Mythos Cult once existed (Plate LXIX).

These proofs are irrefutable and sufficient for any one who might doubt the statements I have made.

In the Papyri of Hunefer and Ani the deceased is judged not by the forty-two assessors of Osiris, but by twelve known divinities sitting on thrones, attributes of Horus I.

In this negative confession, however, amongst the forty-two assessors of the Osiris Cult we find twelve old Stellar divinities brought on—"Thou of the Nose," an allusion to one of the chief characteristic features of the Ibis, refers to Thoth, the god of Chemunnu; his name appears on the statue of Horus I, in the Museum of Turin, and on a very much more ancient altar in the same Museum. Aati, before it was appropriated to Ra (if it ever was), was one of the names of Horus as "the Cutter" or "Cleaver of the Way." The places spoken of in this chapter do not refer to the land of Egypt, or nomes of Egypt, but to the other world, and here again the Egyptian priests of the second and third part, in introducing "Chemunnu," the Paradise of the eight Stellar gods situated at the Pole, with Amenta of the Solar Mythos, which is situated "through the earth," have tried to bring in the Stellar and mix it with Lunar and Solar. The seventeenth chapter proves this: "The God of Serpent Face" is a Zootype of the god Tem. "O thou Horned one, who makest thine appearance at Sais," shows us the first of the Solar descent. It has been given as the attribute of Osiris, but as Osiris was not the first Solar god, but Atum-Iu, it must first have been one of Atum's attributes, and a prior name for him was Tem, son of Tmu. Therefore Horus = Tem, called the Serpent God.

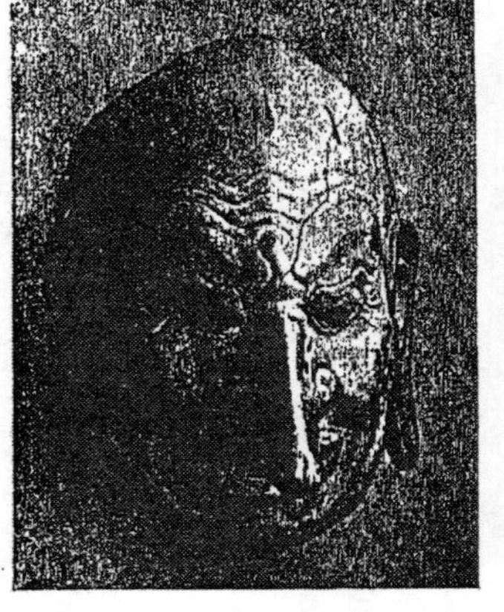

The long nose as a prominent feature in Shinto Ritualism: an ancient Japanese mask.

From "Epochs of Chinese and Japanese Art."

Ritualistic use of the nasal organ: a long-nosed wooden mask from the Philippines.

From "Epochs of Chinese and Japanese Art."

PLATE LXIX.

One of his names also, according to Plutarch, was KAIMIS, signifying "the seen" (Plutarch, C., 56).

The name of Horus as the grain god was in Mexican Cinteol; other names for Horus found here were Telpochtli—Tezcatlipoca. The red god of the Yopi was Tlatlanhqui, Tezcatl, or Tlatlanhqui Tezcatlipoca (the god who was dead in flayed human skin). In Mexican Zipe = to Horus-Amsu, and Ptah-Seker-Ausar of the Egyptians, amongst the Queches Tohil-C'abanil.

Professor Frazer's ideas of the "Corn Spirit Deity" in part v, vol. ii, "The Golden Bough," and the representation of the same by a Zootype, are principally taken from the Greek and Chinese, and he here enumerates how various peoples obtained their customs. He reads through the Greek, and the result is a vast labour of written nonsense; the ignorant Greeks never understood the Mythology or Ritual of Ancient Egypt, and what the priests told them they perverted. They could not speak the Egyptian tongue or read the hieroglyphics, but had to depend upon an interpreter. Diodorus of Sicily states that the Egyptians looked upon the Greeks as impostors who reissued their ancient mythology as their own history. Herodotus states (Eut. l): "Egypt has certainly communicated to Greece the names of almost all the gods"; (li) "With the above, the Greeks have derived many other circumstances of religious worship from Egypt." To understand these customs and folk-lore tales one must go back to where they sprang from, Old Egypt.

The corn god was Horus, and at Philæ the god Naprit—Corn Spirit—is represented with stalks and ears of corn springing from its mummy, near running water. This is Horus represented as a bringer of food in the shape of corn—a type of the eternal, manifested by renewal of food, produced from the element of water in inundation, i.e. an ear of corn near a fall of water. The ear of corn, green wheat-ear of the mysteries, which was held in the hand of Neith, or Isis in Virgo, and still survives in the Star Spica of this constellation, represents Horus, the child, as bringer or giver of food by the water of inundation or rising of the Nile, the food of Egypt being dependent on the periodical overflow of the Nile.

Professor Frazer falls into the same error in trying to interpret the reason why the pig, in one sense, was taboo or sacred, and in another was considered unclean and all that was evil. The reason is found only in the wisdom of the old Egyptians. The pig, as a Boar, was a Zootype in one form for Set, or the Great Evil One. In another form the Pig, as a Sow, was a Zootype for the Great Mother, Shaat, the many-teated sow. I have drawn attention to the meaning of this in my remarks on the New Guinea natives' customs.

Professor Frazer must study the Sign Language, Mythology and Ritual of Ancient Egypt, and the wisdom of their old priests, if he wishes to write anything that is the truth. His present labour will do much more harm than good, because he, as a Professor of one of our great Universities, is diffusing false and erroneous ideas (so-called knowledge). Osiris was not the original "Corn God," as the Professor states, nor was he the "Red-man," or even a type of him. The "Red-man" was a type of Set, and always of evil report, and the folk-lore tales that have been handed down always show him in this phase.

I will only refer to one more point in Professor Frazer's book "of errors" and that is the Aino, chapter xvii.

Let me put him right without troubling my readers with all that he has written, and let me assure them that the Aino were Nilotic Negroes with Hero-Cult, having the Great She Bear as their Mother Totem, tribally and astronomically (see S. & S.). Animals were never worshipped amongst the old people. These are only Zootypes, representing the Great Mother or Great Spirit, or attributes of the same; "Signs and Symbols" to portray these, when they had no language, as we have now, to express their ideas and beliefs.

Taking the "shadow for the substance" and mixing them up will not "help us to understand the wisdom of Ancient Egypt," nor give any true knowledge of the meaning of all that is to be found amongst the present remnants of people of a bygone age.

I think Professor Frazer should be classed as a simple Anthroposophist.

In the folk-lore of various races the human soul takes the form of one of the Egyptian Zootypes—the snake, mouse, hawk, etc., each of which was an Egyptian Zootype of some power or soul of nature, before there was any representation of the human soul, or ancestral spirit in the human form. But these tales are by no means a faithful reflection of the world as it appeared to the primitive African mind.

The picture seen on the pillars walling the entrance to the cellar at the Pyramid of Tepoxtlan, part of a huge glyph of the chalchinith, or "green precious stone," represents Horus as "Prince of the Emerald Stone" or "The Light of the World"; and, as we have seen, the emerald stone was an emblem used by the Zapotecs and Mexican and Peruvian people to represent "Rays of Light" or Diffuser of Light, this would correspond in every particular with Egyptian Heru-Khuti, spoken of in the Ritual in connection with the Great Pyramid of Ghizeh.

On p. 305, "Mexican Antiquities," the central figure represents Horus with the god Bes over his head, the same as we see on many Egyptian plates; for instance, Maspero gives a good representation in the "Dawn of Civilization," of "Horus strangling Serpents," where we see Horus standing on two crocodiles strangling two serpents, and over the head of Horus is the head of the god Bes. The anatomical features of these pottery figures are similar to those of the Masai Group. The lobes of the ears are pierced and hang down as those of the Masai do. The two Crocodiles are sometimes represented as Zootypes for Set and Horus. The goddess Nit, only another name for Apt—"the heifer born from the midst of the primordial waters"—is sometimes represented with two Crocodiles as her first two children, and on the monuments these are depicted as hanging from her bosom. They are the two brothers Set and Horus, born of the Great Mother.

That the primary Stellar Cult—with Set as primary God, and afterwards with Horus as primary God—was brought to Central America is proved by the picture reproduced on p. 340 from the "Mexican Antiquities," p. 292, Fig. 66.

We have here a temple of Set; his name is depicted in hieroglyphic form on the spire, and he himself is shown in front of the temple.

Dr. Edward Seler, on p. 292, "Mexican Antiquities," after giving the description by the interpreter of the Codes Telleriano-Remensis, has mixed Horus with Set and

Stellar Temple of Set.

placed him here, but afterwards he states: "The above description makes it plain that this figure must be considered a deity of the earth of the *hollow interior of the earth and the mountain wilderness, who has nothing to do with the light, pure upper regions.*" He further states: "We seek in vain for mention of this deity and for statements concerning his worship in the works of the historians who lived in the capital of Mexico in the midst of the Mexican-speaking people, and who therefore drew their information chiefly or exclusively from Mexican traditions. Neither Sahagun, Duran, Motilinia, nor Mendieta mentions this god. On the other hand, we have reliable information that in the territory with which we are here concerned, Zapotecs, and, indeed, among both the kindred nations of the Mextics and Zapotecs, he was known, and received special veneration." This was the primary God Set, the twin of our two brothers, who are sometimes spoken of as twins. He reigned or was the primary God for 52,000 years in Egypt, and then when the South Pole Star sank, he was deposed as primary God. In one form he became the God of the Underworld, or God of Darkness. The "middle earth" or Amenta had not yet been founded, and so Dr. Seler is quite correct in describing him as "a god of a mountain wilderness" (the great void or underworld), *but not of the hollow interior of the earth as yet;* this did not come into existence until the time of Ptah of the Solar Cult, and Set was primary God of the Stellar. Horus had taken his place as far as Mexico and the Mexicans were concerned, and he had been forgotten.

Dr. Seler's wind god (p. 133) is the god Set, but not the wind god, which was Shu. Speaking of this god, Dr. Seler states that "his *temples were built in a circular form.* The cap which he wears is cone-shaped. His head ornament a snail-shell." This cap is well shown in the figure — the two brothers born of the same Mother Apt contending for the mount (full description of this is given in S. & S.).

Amongst the Zapotecs the principal pair of gods were the twin brothers, Xolotl and Quetzalcoatl = Horus and Set, and the traditions amongst these corresponded in every particular with the Egyptian; the deposition of Set, as primary God of the South, was followed by the raising of Horus as God of the North.

The legends contained in chapter iv, book 5, of "Origin de los Indios," by Fray Gregario Garcia, are precisely similar to the Egyptian, i.e. two brothers born of the old Mother Apt: two of the first three transformers, and the first two elemental powers divinized into gods—light and darkness. These two brothers Dr. Seler calls "9 snake" and "9 cave."

These are Xoloth, i.e. Set, and Quetzalcoath, i.e. Horus, brothers born of the old mother Apt—having no father, "bulls of their mother" in mythology. I have given the decipherment of Fig. 64 already, and Fig. 65 represents Horus "the Nile God, bringing food and water of life" (see plate, Maspero, "Dawn of Civilization," p. 39), and not the goddess Talaclquani, expressed by a man eating his own excrement. The proof of the remains of Stellar Mythos, with Set as primary God amongst the Mextics and Zapotecs, is still found here. His principal temples were at Zoopaa or Mictlan, the holy city of the Zapotecs, and Nuundecu or Achiotlan, the holy city of the Mextics. Later, the change took place in Egypt, and Horus became primary God and was carried, as we have shown, over the greater part of America.

According to Father Burgoa's statement, in the above codex, "There was among those altars one of an idol, which they called Corazon del Pueblo, which received great honour. The material was of marvellous value, for it was an emerald of the size of a thick pepper-pod (capsicum), upon which a small bird was engraved with the greatest skill, and with the same skill a serpent coiled ready to strike. The stone was so transparent that it shone from its interior with the brightness of a candle flame; it was a very old jewel, and there is no tradition extant concerning the origin of its veneration and worship." This, as I have pointed out elsewhere, was one of the names of Horus, Her-uatch-f, i.e. Prince of the Emerald Stone.

His "Corazon del Regno," a deity of the earth, to whom earthquakes were ascribed, was Shu—another name was Pitao-Xoo, who was known also to the Tzentals as Votan. Dr. Seler is quite correct in saying that he is the third god in the calendar, because the primary was Set, then Horus, and third Shu, the three Kamite originals, and the original Trinity. Shu had two forms: in one he lifted up the heavens in the day, in the other he lifted up the sky at night or in the underworld. He was also the god of breathing power or force, or the god of the winds, which will account for the Indian tradition here found "of his building a house in the underworld, or underneath, with the breath of his nostrils."

In Fig. 67, p. 205, of "Mexican Antiquities," we have Horus, D, and Set, C, depicted from the Mendoza Codex and the Sahagun manuscript.

A *does not represent the sun here setting, as Dr. Seler states; it is the Pole Star South going down, and disappearing below the horizon, with the House and Symbol of Horus in the ascendant North;* B is a symbol of Horus, God of the North and South, with his five attributes attached below "the five cardinal points" of Dr. Seler! It is the old tale of the Egyptians plainly portrayed in symbolic language. There *was no Sun worship here, it was all Stellar.* Sun worship did indeed obtain among the Mayas, but not among the Zapotecs. The following is Father Burgoa's description of a very ancient sanctuary at Teotitlan, in the Valle de Oaxaca-Xa-Quie, "fingiendo haver venido del ciclo, en figura de ave, en una lumiosa constelacion." It cannot be admitted that this luminous bird is to be regarded merely as a particular conception of the sun, as Dr. Seler would have it. The true interpretation is that it was the "Hawk of Horus." A proof of this is seen on p. 299, Fig. 70 (*a*), where he—Dr. Seler—has drawn a relief exhibit. Here we see besides the jaguar (which took the place of the lion of Egypt), the special local deity, a man whose face is held by the jaws of a bird; that is, the god who came down from heaven in the form of a bird. In this exhibit we have the three feathers on his head, which denotes his name Iu, and the fragments here found of an iconographic nature also prove that it was Stellar Cult. His Xipi found here, which he calls "the flayed one," was Horus of the resurrection—Amsu of the Stellar Cult corresponding to Ptah—Seker-Ausar in the Solar—of the Egyptians.

Dr. Seler's "figure of a blazing star" (which he wishes the hieroglyphic "ce acatl, 'reed,'" to represent, *but which is absent*) is the representation of the Egyptian "Orion," and is shown by his "God of the Night," who guided the soul through the underworld before Amenta was formed (Anubis).

Coqui-Xec, Coqui-Cilla = Zapotec;
Tlanizcalpan Tecutli = Mexican

= "Lord of the Morning."

The Lord of the beginning—the Lord of the Morning Star—is the Egyptian Horus, or in this case, Heru.

He was the annual bringer of food and drink before there was a Sun-god (i.e. he was pre-Solar), when the stars were the annunciators of the coming times and seasons to the waiting, watching world. Both Orion and Lupus (which was represented by the Hare) are two southern constellations.

Pije-Tao and Pije-Xoo = Zapotec;
Quetzalcoatl = Mexican

—the Lord of the Wind—is the Egyptian Shu.

Pelle-Nij = Zapotec;
Citlalpol = Mexican

—the Bright Morning Star; Sothis = Heru or Horus (see *supra*).

Ometecutli;
Omecuiatl

= Horus and Set as the Lords of Duality.

These occupy the first place and are represented as the deities dominating the beginning, in the first divisions, as Thoth and Set, of their calendars. Their initial sign, as I have proved, is a "Crocodile." As already pointed out, the commencement of the Great Year, or the termination of the Great Period, was always marked and symbolized by the two Zootypes, those for Set and Thoth.

July 16th was accepted as the beginning of the Maya and Mexican year, the same as in Egypt.

The ancient temple of Tepoxtlan contains the representation of Horus I and Set, as well as many other Egyptian deities in the Stellar Cult.

Dr. Seler wishes to trace the Mexican word "Amoxtli"

back to the Maya root!—because he does not understand how the Sign of the Crocodile = the Great Lizard, was the sign for the first day of the year, and how it fits in with the rainy season of Mexico—and why this word *Amoxtli*, which he translates as "lot beans," or "book," and finally, after a long argument, suggests it means "beans and corn," should be associated with this first day of the year. The answer is:—

Anoxtli, or Amoxtli, is equivalent to the Egyptian Taht-Aan, the Recorder of Truth, who was always associated with the Great Water Lizard, as the two signs and symbols indicating the commencement of the year (the Great Year). It was so in Old Egypt from the first, and that is the reason why we find the portrayal here, as well as in all other parts of the world where time has been recorded. The old Zapotec name for the Crocodile was Tlaloc, or Ce Cipactli.

In studying these American ancient remains there are several points one must emphasize to form a correct conclusion.

There is no single, uniform type among what is known as the *Maya Antiquities*.

The manuscripts form an independent group.

The relief representations from the ruined cities of Yucatan form a second group.

The clay images a third.

Remains of the different groups are alike in many particulars.

The architectural remains in Yucatan originated first with the ancient inhabitants, who were Stellar Cult people and were then partly obliterated by the Solar.

These architectural remains bear a striking resemblance, especially in the bas-reliefs, to Mexican antiquities such as we find amongst the Stellar Cult people, which we do not find to the same extent in the Maya manuscripts and in the clay figures.

The type of the representations in the codices and the clay figures agrees with that found in the antiquities of Palanque and Copan, but even here there is a difference which shows that "later knowledge" has been effective enough to alter the original.

The Mayas did not deform their skulls (heads) artificially, as did the inhabitants of Copan and Palanque; these latter were not Mayas. Their mode of writing was also different. The peoples of Copan and Palanque were old Stellar Cult peoples, and therefore wrote in Ideographs and Hieroglyphics, whereas the Mayas were Solar Cult people, and their writing was the Hieratic of the Egyptians. That some of the characters used by the Mayas are found to be similar to those discovered in Copan and Palanque only proves that some of the old Stellar Ideographs and Hieroglyphics were brought on into the Solar by the Mayas, when they were in the transition stage, as were the Egyptians (see later).

For the understanding of the differences that we find, and the solution of the whole problem, there is only one key, which I have given to all who wish to unlock and read the past history of these countries.

Stellar Mythos in its lowest and highest forms was brought to America, North and South, by people who originally came from Egypt.

What does this show and prove?

Stellar Mythos being much older than Solar Mythos, it follows that all the civilization of North, Central, and South America was originated by people who believed in and practised Stellar Mythos—a cult identical with the Stellar Mythos of the Egyptians; identical also with that of which traces are found in the ancient ruins in the Caroline and other islands, and which we find practised to the present day among certain peoples of Asia.

Signs and symbols discovered on the remains of many an ancient city undoubtedly substantiate the contention that, prior to the advent of the Lunar and Solar cults, the practice of Stellar Mythos was universal—and to a great extent homogeneous—throughout Asia, as well as North, Central, and South America and Africa. Moreover, in civilized Europe remains are still left to us which prove that this cult was also in vogue here before the Lunar and Solar cults, and that these people of the old Stellar Mythos attained to a comparatively high state of civilization.

CHAPTER XXI

STELLAR MYTHOS PEOPLE IN ASIA

THE earliest exodes of the Stellar Mythos people were probably all along the Mediterranean Basin. Then to the East and North-East, the earliest Botiya or earliest Turanians. These spread all over Asia and thence to America. Some of whom were the Kashshi, Elamites, Guti, Shuti, etc., who were wandering Nomad tribes, and all the Mongolian races and the Scythians. Those of the Mediterranean Basin spread up through Europe, the Southern countries first, then gradually going North.

Another early exodus was that of the Sumerians, who left Egypt carrying all their codes, learning, cult, and laws with them, as witnessed by the Stelæ of Hammurabi and other evidences of proof brought forward here. They were before the Chinese, or the Chinese were some of those who travelled on as far as China and then settled down, probably the same exodus, as there are so many similarities between them.

All spoke an agglutinated language, and their priests wrote the hieroglyphics, which in time have been altered (see Chapter on China).

Traces of the early Turanians are still found in many parts of the world.

The old Turanians probably developed from the original ancestors of the Madi and Masai groups in Egypt, and may all be classed as members of the Turanian or Mongolian family. Their anatomical features are identical, and all spoke dialects of one family of languages, viz. : Old Egyptian.

There were early and late exodes of the old Turanians,

the first having Set as primary God, and the later Horus I as primary God.

The present Mongols, who still inhabit a large part of Asia, can tell you nothing now of their ancient greatness.

Before the first century they were a united body of the old brotherhood of the Stellar Cult; their "Tengri" is the Horus God of the North and South, and at that time they were well organized, fierce old warriors like the Arucanians of South America; the ancestors of each were of the same stock, one body remaining in Asia, the other crossing over to America via Asia and then going South, but, as in South America, the Solar Cult crept in, and was one of the causes which assisted to destroy the nation. Now we find these Mongols to-day, much as we find the Arucanians of South America, downtrodden, over-taxed, poor pastoral people.

Climatic changes may have been another cause for their degeneration and downfall, by robbing them of their pastoral wealth; and the various glacial epochs which have occurred no doubt had some great influence upon them from the time they first came into this country until the present. But, like the Stellar inhabitants of Peru, the introduction of the Solar people and the degeneration of their Cult here in Asia, has been the principal cause of their downfall and destruction as a nation.

The Old Chinese, who were of the same Stellar Cult brotherhood, and at a very early time were no doubt part of an exodus of these Mongolians, all having certain lands allotted to them under the old laws, could not interfere with the Mongolians as long as they belonged to the Stellar Cult brotherhood. The Law of Tabu prevented this. As soon, however, as the Mongols allowed the Solar Cult (in a debased form) to come amongst them, the Chinese, who had resisted the invasion of the Solar Cult people and their doctrines, were astute enough to see that the Lamaserais, under this degraded form, absorbed all their best young men, would gradually destroy and weaken the Mongolian power to protect the land which had been formally allotted to them, and who now professing a different cult, there was no Law of Tabu. So they encouraged them in their downward course until the Mongolians became practically serfs to them.

At the present day, however, the Chinese appear to have fallen into the same error, and dissensions have arisen, internal wars have taken place, with the result that Russia is now pushing her arms forward to embrace the whole of Mongolia.

The *God of Heaven* of the Uriankhai, who were the Ancient Tuba, was called *Tengri.* Uriankhai is an Egyptian word.

Uri-
ank-
hai } signifies "The eldest or Great Priest, living assistant of Horus."

Mr. Douglas Carruthers, in his most interesting book "Unknown Mongolia," has given some certain and most interesting proof that all these old Mongolians were Stellar Cult people, divided up into tribes with "heads" over them, just as we find in South America, and amongst the Nomes in Egypt.

He calls the ancient inhabitants "Tubas" or Yeneseinans, who worked in metals and sculptured Iconographic figures. The remains of huge monoliths are still found here, and their rock drawings are characteristic of these people. On p. 54, vol. i, he gives a photograph of two of these monoliths, as well as a carved stone image with folded arms on p. 66.

In the Northern Minnusuish region, tumuli and monuments are found distributed over a wide area, from the South of Siberia along the Urals as far as the Volga. He states that their number in the Russian Altai, in Mongolia, and in the Uriankhai country is astonishing.

The shape of these giant mounds, according to Adrianoff, máy easily be divided and distinguished into two groups.

Those of the Kemchik valley, the Upper Yenesei and the Russian Altai he groups together, and entirely separates those of the Abakan and Minnusuish Steppes. The first type have a cobbled surface and are always *surrounded by a circle of stones.* The second type are earthen mounds surrounded by large stone slabs in the *form of a square.*

Mr. Carruthers states that "all slabs" *without exception faced north and south*, with the narrow edges east and west. Their position was so exact in this respect that he could always observe the points of the compass so long as he was in sight of a "grave mound."

"Some of these stones measured 10 feet in height, 3 feet in breadth, and 8 inches in thickness; others reached 13 feet in height.

"Although the total number of stones surrounding a particular 'grave' was generally even, the stones were placed without any apparent method; I counted six on the east side and six on the west side, three on the north and three on the south—a total of eighteen. Other examples give twelve and twenty, made up in like manner.

"In many tumuli were found skulls, death-masks of beaten gold, stone, bronze, gold, silver, and iron ornaments, trinkets, mining tools, and implements. Many of the tumuli were situated on the hills and surrounded by circles of stones singly. They contained no graves. The largest were generally placed singly, *or in couples, on the top of a hill*, or high mound, and contained nothing, i.e. no remains of anything were found on exploration."

These latter are the remains of their temples. I do not wish to pass any criticism on Mr. Carruthers' remarks and theories; the contents of this work will show how erroneous theories are as regards their religious ideas. What I do wish to emphasize is that here in Mongolia we have objective records still left of the past great race of the Stellar Cult peoples, both at the time of the reign of Set and later at that of Horus. *The circular temples*, or remains of the same, surrounded by these monoliths (twenty-four in number), represent heaven in two divisions of twelve each, the same as we find throughout the world where those Stellar Cult people went. Later, followed the square or double square temple of Horus—some of these monoliths have disappeared, and that is why the correct numbers are not always found. The skulls found here also prove my contention, the oldest being of the *Dolichocephalic* type, *followed later by the Brachycephalic type*, which is the type found predominantly amongst the present

Mongols. These Yeneseians were mentioned by Herodotus, fifth century B.C., and their customs of burial were like those of the Scythians.

Mr. D. Carruthers has written a most interesting and important work, "Unknown Mongolia," but there is a field of study here for the future student to follow, which can be explained by the evolution of the human race as I have set forth in this work. I have merely touched upon this field as a guide to point the way for others to follow, since more details are required and more time than I have at my disposal. But the evidence here is the same as we find in Egypt, South and Central America, and should be a warning to our present Government and people generally, who are rapidly following in the footsteps of these countries.

The Mongolians still lead a nomadic life, and cattle-raising is their principal industry. At present their religion is Lamaism—a debased form of Buddhism which arose out of the Solar Cult of the Egyptians, but, as stated above, they were originally Stellar Cult people. Their prayer was Om-ma-ni pad-mé Hŭm, i.e. Glory of Padma-Pani (the Lotus-bearer, i.e. Horus). The Korean people also belonged to the Stellar Cult, but to-day they are a mixture of this and Buddhism.

The Japanese at the present day are composed of several types, mixed or merged into one another. There is the Mongoloid with his aquiline nose, oblique eyes, high arched eyebrows and bird-like mouth, cream-coloured skin and slender frame, and the Malayan cast of countenance, high cheekbones, large prognathic mouth, full straight eyes, with a skin almost as dark as bronze, and robust, heavily boned physique.

In the north are the Hero Cult Ainu, with luxuriant long hair and long hands, and all the anatomical features of the Hero Cult peoples: Nilotic Negroes.

The remains found, of old Stellar Cult temples—skeletons buried in the thrice-bent position, facing north or south, with characteristic amulets in these tombs—the old signs, symbols, and glyphs, universally similar, found in many parts of the world, and the ivory tablets and Stellar Cult Implements discovered at Nagada, all point to the long-past ages when the Stellar Cult must have existed, and the many exodes that must have left the old Mother Country in different states of progressive culture, from the time of the end of their totemic Sociology to the time of the builders of the Great Pyramid and other huge temples and structures (all built by these people), found in many parts of the world. All their Osteo-anatomy corresponds, and may be classed, as far as this goes, as the "present type of man."

Professor Max Müller said, "There is a continuity in language which nothing equals," and language in this case bears out the other evidences. Gallatin was the first to draw attention to certain analogies in the structure of Polynesian and American languages, which pointed to a past relationship; but we can go further than that. The ancient inhabitants of Asia Minor were the Sumarians, and so remote was their arrival in Mesopotamia that no approximate date can be given, except that they were Stellar Cult people. The earlier their civilization is traced, the more certain is it that they came from Egypt after totemic Sociology, 600,000 years ago at least. The finest remains in the ruined cities of Babylonia are attributed to them. They claimed to have arrived on the shores of the Euphrates from another land, somewhere in the direction of the south-west. The next certain thing respecting them is that they were members of the Turanian or Mongolian family, thus absolutely contrasting with the later invaders. The third and extremely interesting point about the Sumarians is that their language was closely allied to the Peruvian tongue, by which is meant, not the Quechua, the language of the Inca (Solar) conquerors, but the earlier dialect preserved by the Aymara (Stellar Cult) Indians. The late Sir Daniel Wilson, a philological authority, dealt fully with this undoubted relationship. Hyde Clarke, another very discriminating authority, has classified languages into two main classes—the prehistoric and protohistoric; in the latter he tabulated the following tongues: Sumarian, Peruvian, Mexican, Egyptian, Chinese, Tibetan, and Dravidian.

The birthplace of the Mongolian and Turanian races was Egypt, and nowhere else can it be found. The earliest of these people left Egypt before the Chinese (see Chinese). The Botiya language being of an older form of hieroglyphic than the Chinese—who probably were the third to follow after—these former would drive out and exterminate all the Nilotic Negro people they found when they conquered countries, taking their women, with the exception of a few who would remain in isolated places, as we find them to-day.

No weight can be attached to the statement of Sir im Thurn's "Indians of Guiana," pp. 156–7, which I here quote :—

"It has been estimated that within the 15,000,000 square miles of the whole continent there are nearly 500 distinct vocabularies and 2,000 dialects. Yet there is one great and important feature common to all these diverse languages . . . *and absent from the language of the rest of the world*; and that is that, though the vocabularies of the languages differ, their structure is the same and is peculiar. *The structure of all, and only of these languages, is poly-synthetic.* This community of speech is a strong, though not absolutely certain, indication of community of race. When, however, the bodily structure, and to some extent the customs, of these groups of Americans are examined, it appears that in these points also, with considerable differences, there are yet features which are on the one hand common to all these groups *and are on the other hand unrepresented elsewhere in the world.*"

One part of this is correct, but that which I have italicized is not correct. Until he takes into account *the Inner African roots and understands* the various cults and formation of language from the Old Egyptians, as set forth in this work, Sir im Thurn is not in a position to form any rational opinion. The correct term for the above is *Megasynthetic Language.*

You cannot judge from language alone the people whose origin and evolution you wish to find out. Anatomical features, cults, and modes of life and industries must all be taken into account, as well as the old African ideographs, *which have never been studied by the philologists.*

All the above have representatives left in Old Egypt,

whence all these Stellar Mythos people came, after the Totemic people, who had left, some of them, hundreds of thousands of years before.

The Sumarians and Chaldeans were an exodus from Egypt during the first 52,000 years of Stellar Mythos, because Bel, El Shaddai, i.e. Set, was their primary deity, as witnessed by a passage which reads: "for Nannar, *the powerful bull of Anu*, the son of Bel—his king Urbau," etc. (Maspero's "Dawn of Civilization").

Nannar, the mighty bull of Anu, the son of Inlin Bel, is Set. He was the bull of his mother, and "father of all the Gods" (Egyptian Ritual). He was the first-born of the old Mother Apt of the Egyptians (see *supra*).

It is of no use, therefore, to try to compare the buildings of the Sumarians (early Stellar people) with the buildings of Egypt, built by the later Stellar Cult people, or those built still later by "the Solar"; you must compare these with the oldest Circular Temples.

We have no doubt that, *at the time of the exodus of these people, the Egyptians built in a somewhat similar way*; this fact is borne out by the style of the buildings and walls erected by the Madi and Masai (see Nilotic Negroes).

Old Egypt was progressive, but those who went out into other lands, after they had left, kept to the same style and cult as they practised at the time of their leaving, whilst the old Egyptians would pull down and destroy all these old buildings and replace them with those of the solidity and form we now find. A further proof is that although we know that Set was primary God for 52,000 years, yet with the exception of his statues in hard stone, which were built at that time, few remains now exist; on the contrary, nearly all the present-day remains were built at the time of Horus I, or during Lunar or Solar Mythos, which was much later.

Therefore it is useless to compare the temples of the Chaldeans and the Sumarians with the Theban, as Maspero wishes to do.

The temples of the Chaldeans were built on the same plan as we find that the early Stellar Mythos people built in Central and South America, during the time that Set was primary God. And the temples dedicated to Horus, which we find in Central and South America, were the same as we find in Egypt at the time that Horus was primary God there.

All the gods of the Euphrates correspond with those of the Nile at that time, and were, like the Egyptian, depicted in Zootype form—the Elemental Spirits divinized. The Chaldeans, who were a College of Priests afterwards connected with the Babylonian Empire, were celebrated for their extensive learning, and it is said that they took their origin from Zerutusth, an Egyptian priest; but in all these exodes there can be no doubt that a full body of priests accompanied each great exodus. The priests only were the learned men, and commanded the people generally.

Belus came originally from Egypt. He went, accompanied by other Egyptians, to Babylon; there he established priests—these are the personages called by the Babylonians, Chaldeans. These Chaldeans carried to Babylon the science of astrology, which they had learnt from the Egyptian priests (Larcher, "Herodotus").

Pliny calls Belus the "inventor of sidereal science" (N.H. 6, 26), and Belus as the elder Bel was a form of the Egyptian Bar, a name of Set. As Diodorus also relates (1, 28, 29), the Egyptians claim to have taught the science of astronomy to the Babylonians, and declared that Belus and his subjects were a colony from Egypt. Belus (the first Bel) being identified with Bar = Set, this means that the colonizing of Babylonia from Egypt was during the reign of Set, or at least in the time of the primordial pole-star of the South. In Astronomy the status of an-arch-first depended on being formed in time, and Set was first as Bull of the Mother, or the male hippopotamus with the female. The founder of astronomy therefore was the establisher of the pole, as Set or Set-Anup, in the South. The Book of Enoch says that, previously to the Noachian deluge, Noah saw that the earth became inclined, and that destruction approached. Then he lifted up his feet and went to the ends of the earth, to the dwelling of his great-grandfather Enoch (chap. 64). The "Ends of the Earth" was an expression for the two poles—the Dwelling of Enoch being equivalent to that of Set at the southern pole.

The Assyrians in their funeral rites in all respects imitated the Egyptians.

In the Stellar Mythos we find the "Khuti" or "Glorious Ones" succeeding the elemental powers, with Horus as the eighth and supreme, and these the Babylonians converted into evil spirits, not understanding, or having forgotten, the original, the old Egyptian wisdom.

In comparing Egyptian Mythology with the Babylonian, the types that represent evil in the latter had represented good in the former.

The old Great Mother of Evil, called the Dragon-horse in the Assyrian version, was neither the source nor the product of evil in the original. The serpent goddess, Rannut, as renewer of the fruits of the earth in the soil, or on the tree, is not a representation of evil. The types of good and ill were indiscriminately mixed, pre-eminently so in the reproduction of the old Great Mother as Tiamat. Originally she was a form of the Mother Earth—Ta-Urt of the Egyptians; the Great Mother, variously named Tiamat, Zikum, Nin-Ki-Gal or Nana, was not originally evil. She represented source in perfect correspondence to Ta-Urt, Apt; and Rannut of Egypt, howsoever hideous, was not bad or inimical to man, the mother and nurse of all, the mother of gods and men, who was the renewer and bringer forth of life in earth and water. Likewise in Egypt, the seven Anunaki were spirits of earth, born of the Earth Mother in the earth, but they were not wicked spirits.

The elements are not immoral.

In the Cuthean legend of creation we are told that the greatgods created "warriors with the body of a bird," and "men with faces of ravens." "Tiamat gave them suck." "Their progeny the mistress of the gods created." "In the midst of the (celestial) mountains they grew up and became heroes," and increased in number. "Seven kings, brethren, appeared as begetters," who are given names as signs of personality ("Babylonian Story of Creation": "Records of the Past," N.S., vol. i, p. 149). The seven children of the Great Mother Apt, as Egyptian, were produced as two plus five.

Set and Horus, twins, were the first born warriors or fighters, portrayed as two birds, the Black Vulture or Raven of Set, and the Golden Hawk of Horus.

The Set and Horus twins were succeeded by five other powers, so that there were seven altogether, all brothers, and all males or begetters. These seven constituted a primary order of gods as *fellow males*, who were the "Nunu" of Egypt, and became the Anunas or primordial male deities of ancient Babylonia; but these seven nature powers evolved in the Egyptian Mythos were the offspring of the Great Earth Mother first, Ta-Urt, and then divinized as children of the Great Mother Apt; not the progeny of the evil reptile, Apap.

These the Babylonians changed into the seven evil spirits or devils.

They were native to the nether earth, but were not wicked spirits. The Ritual (Chapter 83) speaks of them as "those seven Uræus deities who are born in Amenta"; elsewhere they are called the seven divine Uræi, or serpents of life and renewal. The serpent type is employed to denote the power—the good serpent, not the evil reptile, Apap. The spawn of Apap in Egypt are the Sebau, numberless in physical phenomena and never portrayed as seven in number. There are no seven serpents of death, no seven evil serpents.

The Evil Spirits are mixed up with the good, and the Hebrews follow the Babylonians in confusing the Uræus-serpent of life with Apap, the serpent of death, but Apap never was a spiritual type and was never divinized, not even as a devil. The Babylonians and Akkadians turned the old Great Mother into Tiamat as the evil serpent Apap.

Conclusive evidence of the way that changes were made in the appropriation of the prototypes and their re-adaptation to the change of fauna, and likewise the evidence of later theology, can be shown in relation to the primordial Great Mother who is Tiamat in Babylonia. One of her typical titles is the "dragon-horse," and as the Egyptians had no horse it might be fancied at first sight that such a compound type as the dragon-horse, which also figures in Chinese Mythology, was not Egyptian. The Ancient Egyptians had no horse, and their dragon was a crocodile. The hippopotamus was their first water-horse—as male, i.e. the water-bull—as female it was the water-cow.

Now the old first genitrix, Ta-Urt, when represented as a compound figure is a hippopotamus—that is, a water-horse—in front, and a crocodile—that is, the dragon—behind. The dual type of Tiamat, the dragon-horse, is based on the Crocodile and Hippopotamus, which are to be seen combined in the twofold character of the Great Mother Apt, *and these two animals were unknown to the fauna of Akkad and Babylonia.*

Thus as Babylonian they are not derived directly from nature, but from the Mythology and the Zootypes that were already extant in Egypt.

The Kamite elemental powers were the powers of the elements represented by Zootypes. They might sometimes be fearsome, but were not baneful. The Babylonians have mixed these seven elemental powers with the "Sebu" of the Egyptians, and in the darkness of their ignorance confounded the Great Mother as the Crocodile (zootype) of water with the Apap reptile of evil, assigning the seven elemental powers (which were afterwards divinized) as the progeny of the Great Serpent, Apap.

The Hebrews followed and did the same; they confused the Uræus—Serpent of Life—representing Rannut (zootype serpent), the Mother of life and giver of food in fruits of the tree and earth, with the Serpent of death, darkness, disease, etc.

As in Egypt, the heavens were first divided into north and south.

Bel was assigned to the Tropic of Cancer.

Ea was assigned to the Tropic of Capricorn.

The seven supreme gods were the seven Glorious Ones of the Egyptians.

Like the Egyptians, the high priests were seers, and they had prophetesses "who told the message," in other words, these prophetesses were trained spiritual clairvoyants whom the priests, or seers, used in order to communicate with the spiritual world. Maspero and other Assyriologists have made the same mistake here as they have done in Egypt, and called all *the attributes of the one God different gods and goddesses,* whom they have mixed up indifferently.

Ea or Ia.—Anu, Bel, and Ea were all three one, as a trinity, as were the Egyptian Horus, Shu, and Set.

The pictures set forth in Maspero's "Dawn of Civilization" represent Shamash as Horus of the double horizon in one, and as Atum in the other. (See Chapter, Solar Cult.)

Their primitive forms can only be read through the Stellar or early Solar Cult of the Egyptians.

The story of Ishtar and Dumuzi is the same story of Isis and Osiris in Solar Cult in its primitive form. Stellar Sebek-Nit and Horus.

At the time of the Babylonian Empire we see the gradual change from Bel or Set, to Merodach or Horus.

Merodach was regarded as the son of Ea, as the star which had risen from the abyss to illuminate the world—Horus, Orion, Kehui, and to confer upon mankind the decrees of eternal wisdom. He was proclaimed as Lord—bîlu—*par excellence,* in comparison with whom all other lords sank into insignificance, and to this title was added a second, which was no less widely recognized than the first: he was spoken of everywhere as the Bel of Babylon—Bel-Merodach = Horus—before whom Bel of Nipur, = Set, was gradually thrown into the shade.

Thus we see the same assimilation here as we did in Egypt and the changing from the deity of the South Pole Star to the North, proving also that there must have been a fresh exodus from Egypt, bringing this on to the Babylonians, and that the first 52,000 years of the Stellar Mythos had now been completed.

The interpretation of "Shamashnapishtim being saved from the Deluge by Ia or Ea" must be read through the Egyptian wisdom; it was at the end of one great year of precession of the Pole Stars, before Horus came into being as the primary God (see *supra*).

Ia or Ea as "a smith and fashioner of metals" is the Egyptian Horus-Behutat (see Nilotic Negroes, *supra*).

As a scribe and physician he represents Iu-em-Hetep.

Sin was the Egyptian Thoth.

Shamash is Horus of the double horizon, in one form as Atum-Iu in the east, and Atum-Ra in the west, a Solar God of the earliest part of the Solar Cult.

Zu = Shu.

Shutu, the South Wind or God of the South.

Ramman was the God of the Axe (see Solar Chapter).

Their goddesses were the same as the Egyptian under other names.

Damkina, the lady of the soil, was Egyptian under the name of Kep, or Ta-Urt, and the Inner African Nzambi, and NKissi-Usi, Peruvian Mamapocha, Finn Ukko, Esquimaux Gigone, Khonds of Orissa Tari-Pennu.

It was Bel-Merodach, = Horus, the divine Bull (Bul of his mother and the second born of Apt), who fought Tiamat—the great evil one. The twelve constellations were combined during this time of the Stellar Mythos, representing the twelve divisions of heaven north and twelve divisions for the south. It was not Solar Mythos here at this time, as Maspero would have us believe; he has, like a good many others, "made a very bad mixture," which I would recommend no one to take. This was pre-Zodiacal.

The Babylonian Tiamat, who waged war against Marduk, the champion chosen of the gods, was held to be the incarnation of all evil. This monster was 300 miles long and had a mouth 10 feet wide, moved in undulations 6 miles high, and was 100 feet round his body; this corresponds to the Great Serpent of the Nile (see *supra*) and to the Warraminga of the Australian Aborigines (see *supra*)—a copy of the Egyptian Apap. The tales are the same in each instance, and he is the same as "the Leviathan" of the Hebrews—the serpent of many twistings and folds (see Job xli., Jeremiah li. 34, Isaiah xiv. 29). He was hunted for slaughter by Gabriel, and with the assistance of Yahweh was slain by him. Gabriel here is the counterpart of Marduk, Yahweh taking the place of Aushar or Anu as the head of the gods. Although the Hebrews obtained their hell direct from the Egyptians, they took their "Leviathan" probably from the Babylonians, but these latter probably obtained it direct from the

Egyptians at a much earlier date; although the evidence is not quite conclusive, I believe this to be the case, from records found. If it was not so, then the Hebrews brought it direct from Egypt, but the tale goes back to the time of totemic Sociology, and the Sumarians left Egypt thousands of years before the Hebrews did. These have mixed the good and bad together.

In "The Dawn of Civilization," on p. 657, we see Atum seated on his throne with the Ank of Life in his right hand. The three hieroglyphics above and in front of his head are two stars of eight rays, as also the emblem of the moon, and there are four of these stars below (Stellar Mythos Cult symbols brought on).

The two stars with the Lunar symbol above indicate two divisions of heaven, symbolizing that he is Lord of the North and South.

The disc has sometimes four, sometimes eight rays, indicating wheels with four or eight spokes respectively, and Rawlinson supposed that these two figures indicate a distinction between the male and female power of the deity, the disc with four rays symbolizing Shames, the orb with eight rays being the emblem of Ai, Gula, or Anuit (on the Religion of the Babylonians and Assyrians in G. Rawlinson, "Herodotus" 2nd edition, vol. i, p. 504). But it was not so. The four rays depicted heaven in four divisions, and the four intermediate wavy lines indicate the subdivision into eight, heaven in eight divisions, and Horus was the God of this Stellar Pole, Paradise, and Atum is here represented as the Father in the earlier Solar Cult. They have mixed up the later Stellar, or rather carried on some of the signs and symbols of the Stellar into the earlier Solar.

The wing disc, here depicted by the two cords and being held up by Hu and Sau, was attached and given to Horus.

These people had the earliest Solar with signs and symbols of the Stellar Cult mixed in the same way as we find amongst the Mayas.

Maspero's two "separators," Nergal and Ninib, were the two forms of Shu, i.e. Shu and Shu-Ahner. "The line of union between heaven and earth" was uplifted or raised by Shu. He is seen seated on a throne or square with two figures kneeling, the early forms of Isis and Nephthys; and this is supported by the waters of space, in which the four stars represent the four children as the four supports of Horus.

We see also the Solar representative of the Stellar in the papyrus of Ani. Horus, the son, is presenting Ani to Atum, with the wife of Ani following. She is arrayed in celestial garments as Atum is, she having preceded Ani and then gone to meet him.

Atum is presenting the Ank of Life to Ani.

The Egyptians would blot out all they could of the Stellar and convert what they could into Solar, and that is why we find so much mixture of this in the Ritual of Ancient Egypt, but all the facts I have brought forward in this work prove that the whole of their doctrine had been originally worked out in the Stellar.

According to the legend related by Berosos a divine fish-man, Oan or Oannes, who dwelt in the Persian Gulf, came from thence to teach the Chaldeans all that they ever knew, when, as it is said in the native tradition, the people wisely "repeated his wisdom" (W.A.I. 16, 37–71). In Babylonia the instructing fish-man was represented by Ea, whose consort was Davki or Davakina, the Earth-mother, corresponding to the Egyptian Great Mother, one of whose names was Tef. Among the chief deities reverenced by the rulers of Telloh was one whose name is expressed by the ideogram of a "fish" and an "enclosure," which served in later days to denote the name of Nina or Nineveh (Sayce, Hib. Lectures, p. 281). This same ideogram, i.e. a fish and enclosure, in the Egyptian hieroglyphics signifies An

to teach, to appeal, to show, as did the fish-man.

An, in Egyptian, is a name of the teacher, the scribe, the priest. An was the fish in Egypt; An, with the fish for ideograph, is an ancient throne name that was found by Lepsius among the monumental titles on a tomb near the Pyramid of Ghizeh (Bunsen, "Egypt's Place," vol. ii, p. 77). Horus-Sebek was the earliest fish-man known to mythology; he was the Great Fisherman of the inundation, typical of food and water and constellated during the Stellar Cult, and symbolized as the Great Fish by a Crocodile, which was portrayed as a figure of force in his capacity of Solar God. He was thus brought on from the Stellar to the Solar Cult and reapplied to the power that crossed the waters as the Solar Horus of the Double Horizon—Horus the fulfiller of time and law, the saviour who came by water, by blood, and in the spirit. Horus the fish and the bread of life was due, according to precession, in the sign of the fishes. A new point in the evolution for the religion of Ichthus in Rome is indicated astronomically when Jesus or Horus was portrayed with the sign of the fish upon his head and the Crocodile beneath his feet. This is illustrated in the Catacombs. Bryant copied from an ancient Maltese coin the figure of Horus who carries the crook and fan in his hands and wears a fish-mitre on his head. This was Horus of the Inundation, who was emaned from the water as a fish and by the fish, but who is here depicted in a human form with the fish's mouth for a mitre on his head (Bryant, vol. v, p. 384). Thus we see that the "Fish-man" is portrayed in various forms and found throughout the world, all emanating from the Egyptian Horus—in fact, is Horus of the Egyptians in the various types adopted and carried on from the early Stellar, through the Lunar and Solar Cults, and into the Christian doctrines.

As regards the cuneiform character of their writings, if you turn to p. 727, Maspero's "Dawn of Civilization," there are fragments of a tablet on which some of the primitive hieroglyphics are explained in cuneiform characters. These hieroglyphics are undoubtedly of Egyptian origin.

Oppert was the first man who pointed out the foreign origin of the cuneiform syllabary, and attributed the honour of its invention to the Scythians of the Ancients, but the Scythians were also a Stellar Mythos people of old Egypt. On p. 726, Maspero has shown how the old Egyptian hieroglyphics were converted into the cuneiform, although he does not acknowledge it.

These old Stellar Cult people, who went out via Asia, never attained an "alphabetical form," because they had left old Egypt before the Egyptians had evolved their alphabet, *and always wrote in glyphs.*

Their system remained a syllabary interspersed with

ideograms, but excluded an alphabet. It is the Polysynthetic of Sir im Thurn's "Indians of Guiana," and may be classed as the "Encapsulating" or "Megasynthetic language." Therefore we know approximately the time that these people left, viz. at the period that this stage of writing had been reached, but before the alphabet had been formed. We are in the stage when the first prefixes had been introduced to determine definitely the meaning of the hieroglyphic. That stage was the early Stellar Mythos period when Set was primary God.

The *old Sumarians* wrote the hieroglyphic of the Egyptian. The Semitic is an advance on the pure ideographic. Maspero states: "It may be almost affirmed that most of the grammatical processes used in Semitic languages are to be found in a rudimentary condition in Egypt. One would say that the language of the people of Egypt and the language of the Semitic races, having once belonged to the same group, had separated very early, at a time when the vocabulary and grammatical system of the group had not as yet taken definite shape. Subject to different influences the two families would treat in diverse fashion the elements common to both. The Semitic dialects continued to develop for centuries, *while the Egyptians' language, although earlier cultivated, stopped short in its growth.*" But the last sentence is wrong; *the Egyptian was the only progressive language.* From ideographs we see the growth first of the affixes, then of the suffixes, after these, of linear alphabetical forms, etc. The old Sumarians left Egypt when only the ideograph was used, in the early part of Stellar Mythos, when Set was primary God. The Chaldean and Semitic were people who came after, when the Egyptians had formed their affixes, and in the Babylonian time we see the Cult of Set being replaced by that of Horus. The old Egyptians still progressed to the higher evolution by forming and adding prefixes, followed by alphabetical form, and finally from a monosyllabic language developed the agglutinated, and ultimately the reflected, to which latter the Semitics never attained. But these did not go out to Asia; from Egypt they went to the Mediterranean basin and then to Europe. Hence we find two different "treatments" of the original. That is the reason why we find all the old Stellar Mythos people writing hieroglyphics and speaking an agglutinated language, both in Asia and Central and South America. They had all left Egypt at this stage, and being cut off from their old Mother and subjected to different influences, "would treat in a diverse fashion the elements they had when they left Egypt." All these Stellar Mythos people throughout the world, wherever they went, can thus be identified by their anatomical features, by their cults, by their language, their buildings, and their arts.

The Solar people who left Egypt long after were of a higher type in every way, i.e. anatomical features, cults, and language, like the Druids, Mayas, and Incas (see Solar Cult), thus proving the progressive development that had taken place in old Egypt. Nowhere else on the face of the earth could this have been.

Maspero states that: "The Children of Defeat, in Egyptian Mosû batashû or Mosû batashet, are often confounded with the followers of Set, the enemies of Osiris. From the first they were distinct," but this is not so; Maspero, like all Egyptologists, failing to trace the Egyptians from their totemic Sociology to Stellar, Lunar, or Solar Mythos, has mixed these up together.

Apap was another form of Set, used as a term under a different cult. He does not recognize totemic Sociology as the first form, any more than the present Anthropologists and Ethnologists recognize any difference between totemic and non-totemic people, or the first people who commenced to form Egypt into "Nomes." He knows nothing about the change from this to Stellar, then Lunar, and afterwards Solar. He puts "Osiris" as the Chief in the latter Cult, anywhere he is able to fit him in the former cults—when he was not in existence, and in all, he has made such a mixture that no one could possibly understand the real meaning of anything. He is just as bad as the Greeks. His conception of Ra is on a par with Osiris. He thinks that Ra *was first*, even before Nu and Nit or Shu and Seb, but Ra was a Solar type, the others Stellar, a difference of at least 350,000 years in time; and if one wishes to understand the Egyptian wisdom one must study how they brought on all these old gods, the attributes of "the One God," converting them into other names and adding to them other attributes, plus all the old ones possessed. This Ra was Solar, but he was the Solar representative of Atum, Stellar; likewise he was the Solar of the Stellar Amsu in Spirit form. At first Set was a brother-warrior with Horus and a God of the South Pole (primary God); it was not until Horus became God of the North Pole and assimilated all the attributes of Set that Set was deposed and associated, as in one type, with the great Apap, evil serpent. In another type, after dis-association with Horus, he was "Apt-Uat," the opener of the way in the underworld, "the King of Darkness."

In the Babylonian legend concerning the generation of mankind attributed to Oannes by Berosos, the beginning is with hideous beings in the Abyss, which are described as human figures mixed with the shapes of beasts. "The person who was supposed to have presided over them was a woman named Omoroca," who indeed is The Great Mother, who at first was Mother Earth. "Belus came and cut the woman asunder," which in totemism is the dividing of the one woman, or the type, in two. At the same time he destroyed the animals in the Abyss. Thus the pre-human period was succeeded by the Matriarchate and the two female Ungambikula, who in the Arunta tradition cut and carved the rudimentary creatures into totemic men and women. Then Belus, the deity, "cut off his own head," upon which the other gods mixed the blood with the earth and from thence men were formed. Thus the source of life, or a soul of blood, was changed from the female to the male deity, who in the Egyptian is Atum-Ra or Tum, the image of created man, or man who was created from the soul of blood—that is at first female, and was afterwards fathered to the male. This is represented in the Ritual (Chapter 17). The God, as Father, takes the Mother's place, and the Matriarchate terminates in the mythology of Egypt. Tum is described as giving birth to Hu and Sa as the children of him who now unites the Father with the Mother as one divinity, or one divine person.

CHAPTER XXII

THE CHINESE PEOPLE

ARYANISTS assert that Chinese is the parent of all Asiatic languages. To refute this assertion let us determine the origin of the Chinese people.

It is stated in a very ancient Chinese manuscript called Pih-Kea-Sing, the date of which is said to be 3000 B.C., that the Chinese originally came from the north-west as colonists, and that the whole number of them did not amount to more than a hundred families. The names of these families are still preserved in this manuscript, and are to be found in China at the present time. They came over the heights of Kwan-lun towards the borders of Hwang-ho, and subdued and exterminated in succession most of the barbarous clans of Nilotic Negroes whom they found in the country. They had considerable knowledge of the arts necessary to social life. They could write, but their writing was hieroglyphic (Egyptian). They knew the course of the stars, and had the same astronomical knowledge as the ancient Egyptians, the same ideas of symbolism, a similar Stellar Cult, and the same cycles in time. Their Y always had the same meaning as the Egyptian ☥, the Hebrew T, and the Christian †. On the monuments of Yu and Mount Thrae-shan, in the province of Shantung, we can still see the original hieroglyphic writings, which are Egyptian characters. Egyptian hieroglyphics have also been found on some very ancient Chinese coins. The deity of the Pole Star was known to the earlier Chinese and Japanese as the supreme god in Nature, who had his abode on the Great Peak or Mountain; this god the Chinese called Tien-hwang Ta-Tici, god of the Pole Star ("Religion in China," p. 109).

Shang-Ti, the supreme ruler, was the highest object of worship. His heavenly abode, Toze-wei, was "a celestial space round the North Pole," and his throne was indicated by the Pole Star (Legge, "Chinese Classics," vol. iii, pp. 6, 34, and "Chinese Repository," vol. iv, pp. 9, 194). This is the most sacred and ancient form of Chinese worship; a round hillock represented the altar on which sacrifice was offered to him. In the Archaic Chow Ritual (Li) it is said that when the Sovereign worshipped Shang-Ti he offered upon a round hillock a first-born male as a whole burnt-sacrifice.

Both the mount and the first-born male are typical. Set-Anup was the first male Ancestor. The hillock is an image of the mount. This deity was also known to the Chinese as the "Divine Prince of the Great Northern Equilibrium," who promulgated the laws of the silent wheels of the heaven's palace, or the cycle of time determinable by the revolution of the stars (De Groot). This was the Stellar Mythos of the Egyptians. The silent wheels of the heaven's palace were Ursa Minor, $6+1=7$. The Temple of Prayer in Peking (on Plate LXX) is a good example of the oldest form of temple that was ever built. As elsewhere stated in this work, the primary temple was in the form of a circle, and afterwards in the form of the double square. All these circular Temples were built during the earlier part of the Stellar Cult, and dedicated to Set, the primary God.

The altar here is entirely composed of white marble in the form of a circle, built in three tiers, rising one above the other, and quite open to the sky. There are steps at each of the cardinal points leading to the summit. This is one of the oldest Stellar Cult temples, and is dedicated "To the Great Nameless One," "Shang-Ti," the one God or supreme ruler.

The three tiers here represent the primary Trinity. The carved Dragon and Phœnix are representations of the Egyptians, as explained.

In countries where the later Stellar or Solar Cult predominated, or in the northern countries which were affected by the Glacial Epochs, these temples would be destroyed, and only the remains of two circles of stones would remain, which we find both in the British Isles and elsewhere. It was such a one as this that St. Paul discovered amongst the so-called "heathen" Athenians (Acts xvii. 23).

Sir C. Alabaster, who had much insight into China, says: "Going back to the earliest historic times in China I find a clear evidence of the existence of a mystic faith, expressed in allegorical form and illustrated by symbols. The secrets of this faith were orally transmitted, the chiefs alone pretending to have full knowledge of them."

The exodus of the Chinese from Egypt was during their Stellar Mythos, and at a time when their hieroglyphic system was in vogue.

How, then, can we determine the age of the Chinese, and at what period they left Egypt? By three certain or definite facts. We know that it was during the time of the Egyptian Stellar Mythos; further, that from the formation of this temple (round) the first exodus must have been during the formation of the Stellar Cult, when Set-Anup was primary God of the South, i.e. the first of the Chinese came out of Egypt about the end of the first 50,000 years of the Stellar Cult, and a later exodus went out of Egypt soon after Horus was primary God of the North, because although they here possess the temple of the first formation, which was built during Set's reign, or during the time that he was primary God, they have changed their Pole from South to North. This can be demonstrated by their doctrines still existing, and by their traditional history of the past and their traditional myths.

The old Chinese had no Lunar Mythos and no Solar Mythos. Their extant traditions and present beliefs are of the Egyptian Stellar, connected with Set, and then of the early part of Horus I, and, as we know, China is a country which has cut itself off for thousands of years, and allowed no intercommunication; it has always carried on its original beliefs and mythology, altering comparatively little from its beginning to the present day. Some later incursions, however, probably at the time of the Hindus, introduced some Solar.

Again, these Chinese must have left Egypt at the time the Egyptians had their hieroglyphic system of writing and before they had evolved their hieratic; this is demonstrated conclusively and critically by the present writings of the Chinese language.

We know that the hieratic was perfect in the third dynasty in Egypt, and this must have taken a very long time to be evolved from the ideographs, because we find that the Egyptians first commenced to add prefixes and suffixes to these. Then we have the transition stage which Dr. Evans calls the "Proto-Egyptian," and after this the hieratic was formed, followed by the demotic.

The Chinese left Egypt at the time that the pure ideographs were in vogue, but, instead of continuing to progress in the same way as the Egyptians, they treated their characters with explanatory keys, a system most awkward in itself and tending to cramp the mind with mere conventional and fortuitous forms, and no advance could be made in evolution. The same applies to those old Stellar people who travelled across Asia to America. As Professor Maspero states, "the progressive Egyptians showed the same artistic mind in the representation of physical objects, and applied it to those of a metaphysical nature—that is, in actions and objects representing certain invisible phenomena, impressed upon the human mind by its contact with the external as well as the internal world. For example, the word 'night' the Egyptians represent under the idea of the starry heavens by the Symbol of Heaven united with that of the stars. The drawing of a road with trees on each side signifies movement and progression. If we take the eye and the two legs, we can discover in them the two principal ideas through which the Egyptian writing appears to have advanced, viz. the principles of homophony and determination. The principle of determination consists, first, in the distinction between the individual and the genus, and then in the explanation of the image of the first by means of the accompanying image of the second."

In the original Chinese, like the Egyptian, every syllable is a word and every word a complete root. They are identical with hieroglyphic writings. These facts, I believe, have never been brought forward before, but must be taken as another proof that those who believe in Asia "as our original home" are absolutely wrong. The suggestion that Turanian and Botiya were formed and originated from the Chinese, and that the Semitic and Iranian came from the Khametic, is devoid of critical argument.

Aryanists have to acknowledge that the stage anterior to Semitism is Khamatic (i.e. Egyptian). This was contemporary with Iranian and Turanian, previously to which was Botiya (long before the time of Menes); after all these came the so-called Aryans, a term which to me has no meaning. But all these languages and their forms were late in Egyptian life. Egypt had existed and passed through thousands of years to form their Mythology—Astro-Mythology, Stellar Mythos, Lunar Mythos, and the Solar doctrines, because at the time of Menes they were at the zenith of the Solar doctrines. Menes it was who founded Memphis and raised a great temple in honour of the god Ptah, who was a god of the Solar Mythos. How long before this Egypt was in a high state of civilization it is impossible to say, but certainly many thousands of years, because we know as an historical fact that at the time of Menes Egypt was united into one kingdom, for Menes was King of Upper and Lower Egypt. As to how many kings reigned before him as such, there is no record, except the six mentioned in the ivory tablets found in his tomb. But we do know, also as an historical fact, that before this, Egypt was divided into Nomes, over each of which a prince reigned, with one Principal over all, although we find no certain dates of reigns on the monuments earlier than Menes; names of kings, on the other hand, are found on contemporary monuments which must date at a very early stage. Abydos was built and founded before Memphis, and Thebes was built and founded before Abydos. At Abydos, Solar doctrines were in full force.

Again, those who regard Asia as the primary birthplace of man have taken no notice of the aboriginal inhabitants; or, if they have, have given no explanation of these aboriginals, whom we still find existing in various places, have given no explanation of the different and progressive forms anatomically and physiologically, have no explanation to offer with regard to their Mythology, their language, their Totems and Totemic Ceremonies, which are found amongst all. Only Professor Frazer, Sir H. Johnston, and Dr. Brunton say that they all sprang from their own surroundings and developed all these Totemic Ceremonies, Mythos, and Totems from the same Polygenism, and are Autochthonous!

How is it conceivable for any one with such knowledge as we possess in the present generation to arrive at such an absurd theory? It is only from want of knowledge that the beginnings of ancient peoples do not offer to the student of history any trace of foreign origin, or any direct affinity, with the corresponding institutions of other peoples or tribes, in these parts as well as in America. The fact of merely isolated, or general features in these beginnings having come down to us, which have been rescued from the forgetfulness of posterity and the destruction that has taken place in subsequent ages, is no argument for Professor Frazer and his school to base their theory on. The fact that the earliest religious or sacred institutions give a prominence to the special peculiarities of the peoples in question, which were formed when they lived in groups of families, or tribes, and that they are identical in various tribes in various parts of the world—anatomically, physiologically, and totemically—and that the explanation and decipherment of these various social rites can be found in the African Nile Valley and in Egyptian Ritual, is critically a positive proof against their arguments. The affinity between their Mythology, Totemic Ceremonies, and language, anatomy, and physiology, and those of the earliest Egyptians, is unanswerable.

Neither do we find the Bushman or Masaba Negro in Asia nor in any other part of the world except Africa. They form the connecting links between the Pigmy and the Nilotic Negro, and cannot be left out of consideration, because this is a positive proof that those who believe in the origin of man in Asia or the Lemurian Continent are absolutely wrong in their conjectures.

The Pigmy was the first exodus from old Egypt and the Nilotic Negro the next; the Bushmen were developed from the Pigmy, also the Masaba Negro. The Bushmen went South and were never allowed by the Nilotic Negro to come North; so no exodus of these ever took place, and no type of these can be found or traced outside Africa.

That different races and tribes speak different languages is unanswerable, but this is explained by all primarily speaking the original monosyllabic language, or root word, and then separating and developing independently; many new words being coined and added to the old Egyptian root or monosyllabic word, which, however, can still be traced in all lands. An evolutionary change, the consequence of the law of development, is prepared in one generation, it progresses in the second, and is adopted in the third. The father dies using words which he knew as a boy; the son is independent of the old words and goes with the stream of new formations which time has adopted; the grandchild receives these new words and formations as already established, and in time a new language springs up in consequence of the forcible doctrine of the nation, etc. It is, however, impossible to state the duration of a language when a nation or tribe remains in its own country, cut off from those around, or have little inter-communication—like the Chinese—except by studying the origin of that language in use. How little has the language of Brittany changed in the last forty or fifty generations, in spite of French words having been introduced!

Supposing we wished to find out the position of Anglo-Saxon in the general system of Germanic languages, and by that means, its place in time, and its importance in the internal development of that branch of language, we should have to discover when and where it branched off from the language of the home country, and next its first appearance in an independent shape, i.e. different on the one hand from the old mother tongue, on the other from the language of the country, which in its own home either grew out of the common mother tongue or the language of the previous home. It is certain that the Anglo-Saxon grew out of the emigration of the Angles and Jutes, Saxons and Frisians, who about the middle of the fifth century sent colonies to England. But as regards the second question, Anglo-Saxon was an independent tongue, in both these respects, about the middle of the eighth century. It was the same with the Icelandic, which was originally nothing but the language spoken in Norway when, about the end of the ninth century, a number of Norsemen went to the Northern Island. The Edda contains the record of that ancient Norman tongue, and the writings of the fifteenth century exhibit to us the Icelandic exactly as it is spoken and written at this day, different from the Edda language, as well as from the new Scandinavian idioms, which had grown up in the meantime, but still much nearer than they are to the common mother tongue. Also owing to the immigration and dominion of German races from the fifth century downwards, we find the Romance languages, Italian, Provençal, French, Spanish, Portuguese, in neighbouring countries, each grown into such an independent shape within six centuries that none of the above-mentioned people could understand each other, or the Latin, still less the Wallachian, which had been formed contemporaneously in Dacia by the military colonies of the Romans.

Therefore in tracing the origin of language we must go back to the origin of man—the Pigmy, Bushman, and Nilotic Negro of Inner Africa and the Nile Valley, where we find the Gesture Signs and Ideographic Symbols—the root and origin of all language.

Egypt alone furnishes authentic and critical proof of this monumentally and by the papyri still extant.

First, Signs, Symbols, and Ideographic characters represented a word, and every word was a sentence, and the meaning of every word purely objective, yet undistinguished. The particular meaning of a phrase depended on its relative position; the relation itself was not specifically expressed, for it is the non-objective that the mind supplies to things. But a time came when the mind became conscious of possessing this power. It then coined words as the special expression of that creative act by which it became Lord of the objective world and Speaking Man.

The first language therefore was monosyllabic.

Every syllable is a word, every word conveys an entire proposition, and therefore a sentence, and every word is purely objective.

The second language or next progressive stage of Language was Agglutination, or the Agglutinated Language, or the rendering of the objective and subordinate into the unity of a word, as a part of speech in which the noun and verb remain unaltered, the mark of vitality being the unity of the tone or accent of the word, i.e. the mere agglutination of principal and subordinate stems, and then the formation of affixes (prefixes and suffixes), the stems and their adjuncts being reciprocally affected.

Dr. Bunsen states that "the inflected, or the Flexional Language, commenced with alphabetism and progressive advance in phonetism and flexional syllables. The suffixes and prefixes and endings as pure formative syllables attached to the main root, the further extension in which the root is affected, especially in the inflexion of the personal relations of the verbs: there, and in the expression of the copula, the personal pronoun predominates, not the verb substantive, and after conjugation and copula are expressed by the verb substantive; unfettered by subordinate and therefore one-sided formation until it has risen to the most perfect syntactical arrangement."[1]

The first or Ideographic or Monosyllabic is expressed by a simple sign or symbol, a hieroglyph, of some object, *and every word must be looked upon as a complete root with*

[1] The Semitic and Indo-Germanic branch of language is a copy of the Egyptian. Their verbal root of nouns remains the same and unchanged in all cases—both singular, dual, and plural—the numbers being formed by affixes and prefixes. This was the original, and Egyptian. But the difference is this, that the Egyptian differs from both of these languages in declension and the full use of pronouns.

In the Egyptian *they are declined by their initial instead of their final*, the initial referring to the gender of the noun, or object possessed, with which they agree, the final being of the same gender as the possessor.

We see here in this respect that they are just the reverse of the Indo-Germanic pronouns, which change according to their final syllables.

a gesture sign and this ideographic symbol, which may be a noun, or verb, according to its position in the sentence. Consequently it is a not yet individualized stem. Therefore there can be no question about affinity of grammatical forms, for these are not purely formative words and not grammar beyond the syntax, i.e. beyond the law of *architectonic* arrangement of simple words. The first stage, then, must be that of uniting several ideographs by the unity of tone, or accent, into the unity of a word as a part of speech, and therefore into a noun or verb, but one particle does not affect the other, it is merely agglutination. In these agglutinative languages it is a fundamental law that the stem which is to be more closely defined by those agglutinations is not affected by them. (The prefixes and suffixes affect the stem, but they themselves have no independent signification as single words.) In other cases they retain their full meaning (flexional).

One language formed or developed itself out of another without any violent influences, the new one being as unintelligible to those who speak the old language as to those who speak the living offshoot of it.

If one studies the development of language and sounds in a child, it will give a good idea of how earlier man learnt his language and what the primary and later letters or sounds of letters were. Any average child a few months old can only utter certain sounds—Ma, Da, Ta, etc., but as the brain develops and he learns, other difficult sounds and words are uttered; from two and a half to three years of age he will be able to pronounce all the alphabetical letters except C, G, J, X, and W. These he will be able to pronounce last, when about three or three and a half years old: but will not be able to aspirate the "h" until a little later; so primitive man must be looked upon as a child able to pronounce certain letters or sounds only, but as his brain developed and evolution continued, he gradually became possessed of all the various sounds known up to the present day, though not in a few months or years, as with a child. Time with him must be reckoned by thousands of years; *it must be reckoned by evolution*, from the Pigmy to the white man, in contradistinction from baby to boy.

Reasoning from the above facts will help considerably to determine the time the various exodes left Egypt, as we can demonstrate.

1. The Pigmy has no written language—he speaks a limited monosyllabic one, illustrated by a Sign Language and signs, gestures, and symbols. The Masaba redoubled these, as Ma-Mama, Ba-Baba.

2. The next exodus—the Nilotic Negro—has no written language, but a further developed monosyllabic one, illustrated by gesture, signs, and symbols.

3. The Chinese, Japanese, old Botiya, and all this exodus, had a written language which was either pure hieroglyphic or hieroglyphs with prefixes or suffixes.

4. An exodus of a transition stage between agglutinated and reflected—Dr. Evans's Proto-Egyptian or alphabetical form, like the Moain and other script found in Egypt, Mediterranean basin, and other parts of Europe.

These people, or remnants of them, can be found in many parts of the world, and this is one point in helping to trace the time of their exodus from Egypt and the corresponding types in various countries at the present day, although taken alone this evidence is not so conclusive as the determination of their Totemic Ceremonies and religious cults, anatomical features and arts. (Further proof in "Signs and Symbols.")

What I have written here with regard to language and religious doctrines applies equally to that of the early Koreans, Japanese, and the Subbas or Mandogs, i.e. "The Ancients of Mesopotamia," who are still followers of the old Egyptian Stellar Mythos. Among all these old Asiatic nations, the Pole Star is in their Mythologies an emblem of stability, a seat or throne of the Power which is the highest god—Set-Anup, or Horus, in Egypt, Sydite in Phœnicia, Anu in Babylonia, Tai-Yi in China, Avather or Zivo in Mesopotamia, Ame-No-Foko-Tachi-Kami in Japan, and various other names for the same being in different parts of the world.

The Turanian languages, which are agglutinated languages, form an advance in evolution upon the older Chinese.

The most ancient Tibetan is still spoken by some tribes in Asia and by the Indians of North and South America.

Here we find the pure old Egyptian Ideographic Hieroglyph, with suffixes and affixes; and so we can trace the origin of the Chinese and old Turanian. The Phœnician script was not obtained from the East.

The cuneiform letters of Assyria were derived from treating the old Egyptian Hieroglyphics in a different manner from the script as found in Crete and the Mediterranean basin, although both originated from Egypt (see later).

The Egyptians called the Phœnicians Kephtiu, that is, Cretians (H. R. Hall, A.B.S.A., viii, p. 163), and the Greeks Phoinikes, signifying red man. The Cretians are painted red in their frescoes (A. Evans, "Scripta Minoa," p. 94).

The oldest Minoan script is similar to, and in fact taken from, the Egyptian Hieroglyphic signs when they were converting them into "linear script" (see later).

From these came the Turano-Finnish and Aryo-Germanic, also the connection of the Turanian and Iranian, and in no other way can it be traced or explained; the Egyptian alone furnishes authentic proof of the origin of the Botiya, the Chinese, the old Turanian, and the identity of the Semitic and Iranian, as well as the Ægean, Canaanitish and old Greek script.

In the Rig-Veda, the habitation of the one god is placed in the highest North, "beyond the seven Rishi"; these by some supposed to be represented by the stars of the Great Bear; but it is not so: these seven Rishi —Urshi, or Divine Watchers—were grouped in Ursa Minor, "*the stars of which constellation never set to the Egyptian.*"

These were the chief of the Akhemu under Horus I, the God of the Pole Star North. And although the Hindus have the Rig-Veda, which is a sacred book to them, this was carried on from the Stellar Mythos and merged into the Solar, because the Hindu people were an exodus from Egypt after the Chinese and departed from Egypt in two different exodes, the first exodus in Stellar Mythos, and the second exodus in Solar Mythos. Unlike Moses, who was versed in both doctrines, they have not tried to obliterate the old, but to merge it into the new—whereas Moses tried to obliterate all the old Stellar doctrines and only carry on and practise the Solar, which was new. In all past histories we find that new religions endeavour to obliterate every trace of their predecessor. The Chinese have no reminiscences, any more than the Egyptians, of the great catastrophe which we know by the name of the flood of Noah. Africanus, quoting Manetho and Eusebius in Syncellus, in the Armenian version of Eusebius, mentions 13,900 years of reign before Bytes, "and then other reigns." I do not give the full text of this because it is a stupid rendering of what these men did not understand, but the length of time is probably correct. This also proves the approximate time of the Chinese exodus. Sesostris is said to be the originator of the divisions into castes, which will give the time of the later exodus of the Hindus. This proves my contention that there were two exodes of the Hindu, one without caste, with Stellar Mythos, and the second when the division into castes was established and Solar Mythos was in vogue.

Prior to Menes the historic age cannot at present, at any rate, be computed in years; it must be in epochs—of one great year, i.e. 25,827 years, or the time of the precession of the Pole Stars; or a part of one great year—a seventh, i.e. the change or time of changing of one Pole Star to another, roughly about 3,000 years. This is not an arbitrary reckoning.

With regard to Hieroglyphic language and its progressive formation, the following points must be considered as to the ages of the different people who left in the many exodes.

1. The writing (Hieroglyphic) was purely a representation of an object, a word, a syllable, a sentence, and original root Ideograph.

2. The first progressive divergence from this was a system of affixes to this Hieroglyphic.

3. The next stage was one of suffixes.

Therefore the first stage may be stated to be purely

monosyllabic, and the next two stages by evolution were the agglutinated.

The next stage in the development in Egypt was linear alphabetical signs, and then the hieratic, *which is not found amongst the Chinese or Turanians.* It was found at a later stage than the exodes of these people, amongst those that went to the Mediterranean basin and travelled over Europe.

In their books the Egyptians stand forth pre-eminently as a people of reminiscences, of monuments. Their sacred writings evince considerably more historical cultivation than we can see among the ancient Persians, judging from the Zend books; and the Greeks had only traditions which they obtained from the Egyptians without understanding them. The echo was caught by the curious and inquisitive Greeks, but the real substance they failed to find. Herodotus perhaps made greater and more reliable research than any other man, but he lived 484 B.C., which was very late; moreover, he was unacquainted with the Egyptian language.

I have not entered into the late history of China, and the influence the Moslems had upon the Chinese, because Mr. Broomhill, in his well-written work "Islam in China," 1910, and Mr. E. R. Huch in "The Chinese Empire" have set this forth fully, although I cannot see eye to eye with these authors in some of their statements.

In Egypt, "The Book of the Dead," or Ritual, is the only one we have found of the Sacred Books—forty-two in number—and even in this we find old and new blended together in the new Empire. In the Turin Papyrus the characters are written in pure monumental hieroglyphics, which proves its antiquity, in contradistinction to those written in the hieratic character.

The tombs of the first Ming Emperors in China, and the ruins found at Mycenæ and Torques, are precisely similar to those of the Kings at Thebes.

Mr. I. Hanlay has made the following statement, drawn from Chinese historians and geographers: "Fourteen hundred years ago America had been discovered by the Chinese and described by them."

They stated that land to be about 20,000 Chinese miles distant from China.

About A.D. 500, Buddhist priests repaired there and brought back the news that they had met with Buddhist priests, idols, and religious writings already in the country (Solar Mythos).

The description in many respects resembles that of the Spaniards 1,000 years after.

They called the country "Fusarry," after a tree which grew there, the leaves of which resemble those of the bamboo, whose bark the natives made clothes and paper of, and whose fruit they ate.

The above almost proves that from the Stellar Mythos there must have occurred some great event to cut off all communication with America, and further that the Chinese must have landed on the west coast of Central or South America, where the Solar doctrines were in existence.

I give here a photograph of "A Royal Relic of Ancient China" (from *Man*, 1912, No. 27): the Chinese representation of the old Egyptian Hare-headed Sceptre. This Sceptre was the Hare-headed Symbol of the resurrection (Egyptian), and I here reproduce a Hare-headed Sceptre found in Egypt by Jean Capart, which will be sufficient to prove the identity of the Chinese. It is the "Tem" Sceptre of the Egyptians (Plate LXXI).

In the same issue of *Man*, p. 51, there is given a page of ancient Chinese writings; many of these are pure Egyptian hieroglyphics, whilst the others can be recognized as such with some variation in the characters drawn.

Professor Sayce's article in *Ancient Egypt*, Part iv, 1914, proves very clearly that the Chinese knowledge of making the biscuit or egg-shell ware was obtained from Egypt, and that the centre of manufacture of this was at Meroe. The documents brought back from Western China by the Pelliot expedition prove conclusively that trading voyages were in vogue at a very early period between China and Egypt.

All words of the Coptic which are not Greek are Egyptian. Nothing is Coptic but the Christian arrangement of the Demotic, alloyed with Greek words under the Ptolemies, which came from the old hieroglyphics.

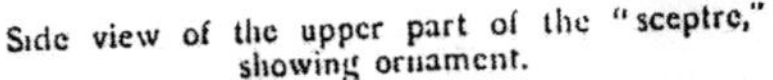

Side view of the upper part of the "sceptre," showing ornament.

The "sceptre" viewed from the other side, showing part of the genealogy incised on the shaft.

A Royal Relic of Ancient China.

Hare-headed sceptre, found in Egypt.

Plate LXXI.

CHAPTER XXIII

FURTHER EVIDENCE OF STELLAR CULT

I AM convinced that the so-called proto-Egyptian writings were simply the earliest attempts to form writings into "an alphabetical form" and an improvement upon the ideographs. They are a commencement of the old hieratic.

Eusebius, Plato, and Tacitus all state that the Phœnicians did not claim to be themselves the inventors of the art

Comparison of Alphabetiform Signs (Evans).

of writing, but admitted that it was obtained by them from Egypt. Therefore the Egyptians were the inventors of the alphabet. Moreover, the Phœnicians were traders, later than the Minoan. They possessed no art or literature of their own, and their history was written in later days by men of other nations.

The signs discovered by Dr. Evans in the Eastern Mediterranean are Cretan. I reproduce here Evans's comparison of alphabetiform signs.

Now one can tell approximately at what date these people obtained these from Egypt, because here is the hieroglyphic name of Ptah (No. 21 in the following list).

It must have been after the Stellar and Lunar Cult, because Ptah had no existence then, and before the Osirian Cult, over 30,000 years ago; it was, therefore, at the beginning or early part of the Solar Mythos, after the Lunar, perhaps 100,000 years ago, that Ptah came into being, and it shows how far the Egyptians had progressed in their writings up to this time. It was written by Solar Mythos people, being of a later formation than the Chinese or Tibetan, or of any of the people of the Stellar Mythos in America or elsewhere, and after Atum-Iu and Atum-Ra in the earliest Solar.

As an example I will give here some of the decipherments of the hieroglyphics found by Dr. Evans.

1.  = Pa, P, also  (Pierret).

2.  = indicates also the number 60 *dans les bas-temps* (Pierret).

3. = Pa, a house.

4.  = Dominion of the North (Pierret).

5.  = Dominion of the North and South, also (Temptu Pierret).

6. 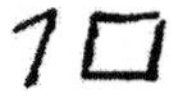 = Sep, time, turn.

7. 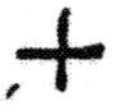 = Am, with, by, resident in, to eat; one name of Amsu.

8. 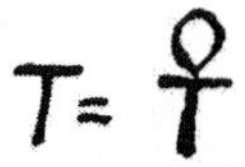 = Ank, living, life.

9. 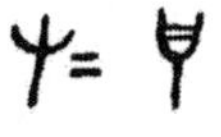 = Ha = moderate, to endure, the head, abode.

10.  [glyphs] = He descended, he ascended, living.

11. 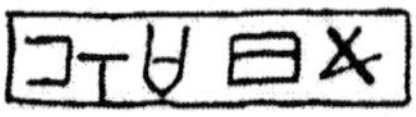 [glyphs] = Guide and also Judge, the living } the Living Guide and also Judge.

12. 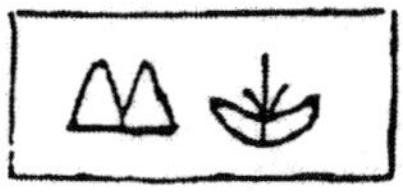 [glyphs] = The Egyptian double horizon. (Amsu-Horus, born or risen.) Lord of the Double Horizon. Therefore early Solar.

A fan denotes a new Spirit born.
Atum-Iu or Amsu-Horus.

13. 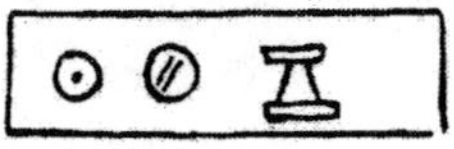 [glyphs] The Sun, light, division of time. Sxti, to take, to meet, Mastaba, i.e. burial-place.

14. 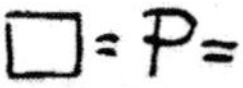 [glyph] = P = the.

15.  [glyph] = He descended, He ascended.

16.  [glyph] = Su, living; also S in Sebtu.

17.  [glyph] = Kes. To tie, to attach, to address, to spin.

18. 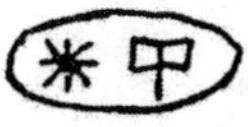 [glyphs] = The God of the Axe. The Eightfold One, i.e. Horus as the highest of the seven = Horus as God of the Axe in Lunar or Stellar form.

19.  [glyph] = Horus, God of the North and South.

20.  [glyph] = Horus and Set, Gods of the North and South, also the Khui Land.

21. [hieroglyph] = Ptah, the Great God of Earth and Heaven.

22. [hieroglyph] = A Nome, a district, a field, a vineyard.

23. [hieroglyph] = Meux, utensils, or Neh, to go abroad, a foreign people.

24. | = u, one, indefinite article, masculine.

25. [hieroglyph] = Mra or Mr-tar, Egypt, labyrinth, land.

26. [hieroglyph] = Neb, Lord.

27. [hieroglyph] = Lord of the Universe = Horus.

28. [hieroglyph] = Se, the backbone, the back, to cut in pieces.

29. [hieroglyph] = P = The. Bes = impurity, to prohibit (to prohibit the impurity).

30. O O O = Corn, wheat, crops, nourishment.

I can approximately give the date at which these people left Egypt by the decipherment of the hieroglyphic denoting the name of the god delineated on the stones and other objects.

In the Royal Tomb of Nagada not only knives, scrapers, and awls of stone were found, but a more diligent search later revealed in this tomb blades and chisels of copper; the same occurred at Abydos, where at first no objects of metal were found, but later excavations produced important metal objects, among which is a plate of bronze which M. Maspero gave to Angelo Mosso to analyse. The analysis shows copper 96·00, tin 3·75 per cent., therefore before the time of the First Dynasty bronze was known and worked into thin plates, as this piece proves (Fig. 26, "Dawn of Mediterranean Civilization," by Professor Angelo Mosso). Silver was also found here. Bronze was made with an alloy of 9 per cent. of tin, and great vases were cast from this (see Montelius, "Die Chronologie der altesten Bronzezeit in Nord-Deutschland," and Mosso's "Dawn of Mediterranean Civilization).

Four pieces of copper leaf are known from predynastic times, and a considerable number of small objects, such as needles, fish-hooks, bracelets, and rings, which have been catalogued with indications of the place of discovery by Dr. Meisner. Therefore no invading race brought into Egypt the knowledge of copper, nor did another people come in later, bringing with them the use of bronze, as many have stated. Both discoveries were made in old Egypt and carried out from there by the Stellar Mythos people.

Dr. Petrie published a diagram showing the copper objects which have come to light at various depths in the excavations made by him in Egypt.

First come the pins for fastening clothing, then fish-hooks, chisels for carpenters' work, flat axes, and daggers. The analysis of one of the objects of metal found gave copper 98·60, tin 0·38, and zinc 1·55 per cent., so that this metal resembles brass.

Flat axes of copper discovered in Sicily and other places in Europe are similar to those found in Egypt, and short daggers found in Crete and other places are identical with those found by Dr. Petrie at Abydos.

The priority of Egypt is absolute as regards copper and bronze, both as to the early date when these were worked and the perfection of craftsmanship.

Much has been written on the dawn of Mediterranean civilization. There are many remains of the first exodus of the old Stellar Cult people from Egypt, who went forth to colonize, found here. From the Mediterranean Basin these people, or another exodus from Egypt, gradually spread over Italy and other European countries. All the objects of art that have been discovered in these countries, pottery, implements, etc., are found in corresponding strata in Old Egypt.

The black, drab, or red clay vases, the rectangular incisions made with a stamp similar to those of the Neolithic period, excavated in Crete and other places, are found also at Dakkeh in Nubia.

The pottery found at Phæstos, where the vases are decorated, not only outside but also inside, and others having the whole surfaces covered with zigzag parallel lines found in Italy, black pottery with incised decoration filled with white, and the Cretan made of black clay well polished, with conically shaped bottom and lip, besides many other styles of pottery discovered on the continents of Europe and South America, all originated in Egypt, as may be seen by studying those found in Egypt by Drs. Flinders Petrie and Weighall.

The beautiful pottery found by Paolo Orsi at Stentinella and Matreusa, in Sicily, and that at Pulo, near Molfetta, excavated by Angelo Mosso, is identical with the pre-Dynastic pottery of Egypt, also the characteristic decorations of the vases of the first Sicilian period described by Professor Orsi, as discovered in Crete and other Grecian islands, are identical with the Egyptian.

Before Dynastic times the Egyptian potters placed their private marks on the pottery they made; this proves that the pots were not always made at home, but that there were special factories whose goods were distinguished by their marks.

At Tordos in Hungary, in Greece—at Knossos and Phæstos—in Spain and Caria, makers' marks, similar to those discovered in Egypt, were found on the pottery; also on pottery in the Museum at Taranto discovered by Professor Viola. As I stated before, the Nilotic Negroes were the first pottery makers, and it is only in the Valley of the Nile that we can find the origin of this art.

The recent finds prove that a uniform culture existed in the whole basin of the Mediterranean, which lasted for thousands of years, and that this culture had been brought there from the Nile Valley, or Old Egypt, during the Stellar, Lunar, and Solar Cult periods. Old Egypt was always progressive—not only the flint flakes of the Pigmies and the roughly chipped implements of the Nilotic Negroes, but recent excavations and researches have yielded the most perfect examples of the flint knapper's art known, and flint tools and weapons more beautiful than the finest that any other country can show have been found in Egypt.

It was during the Stellar Cult that the Egyptians appear to have made the greatest strides in civilization and culture, which cult was the longest in duration, lasting over 300,000 years. Pottery of this period is of remarkable perfection; they also carved and worked in all the metals. Their skill in manufacturing glass[1] indicates a knowledge of chemistry.

That the Mediterranean civilization was not obtained from the Libyans (except as carriers of the same to certain places), and that the Egyptians were primary, may be seen from Eusebius's extract from Manetho, which will prove that the Libyans were subject to the Egyptians during the First Dynasty at least, because he states that "*they revolted during the Second Dynasty.*"

As Egypt was then famous for the sciences and arts, the Greeks, who were beginning to emerge from barbarism, travelled thither to obtain knowledge, which they might afterwards communicate to their countrymen. Some of these I will mention—Orpheus, Homer, Musæus, Melampus, Dædalus, Lycurgus the Spartan, Solon of Athens, Plato, Pythagoras of Samos, Eudoxus, Democritus of Abdera, Œnopus of Chios, Herodotus of Caria (by birth a Dorian), etc. The latter states (Eut. lviii) that "these religious ceremonies are in Greece but of modern date, whereas in Egypt they have been in use from the remotest antiquity."

Angelo Mosso, in his "Dawn of Mediterranean Civilization," has exercised his mind as to the origin and meaning of the Mother Goddess being represented as "two different forms," "the fat woman or steatopygous" and "the normal woman," figures of which have been found in Egypt (Fig. here reproduced), also by Mosso at Phæstos and by Tsountas in Thessaly; Piette, in the South of France, found nine, one at Malta of the fat woman, reproduced, one at Vho, near Cremona.

He further states that the models of these women could

[1] Some authorities on Egypt dispute this.

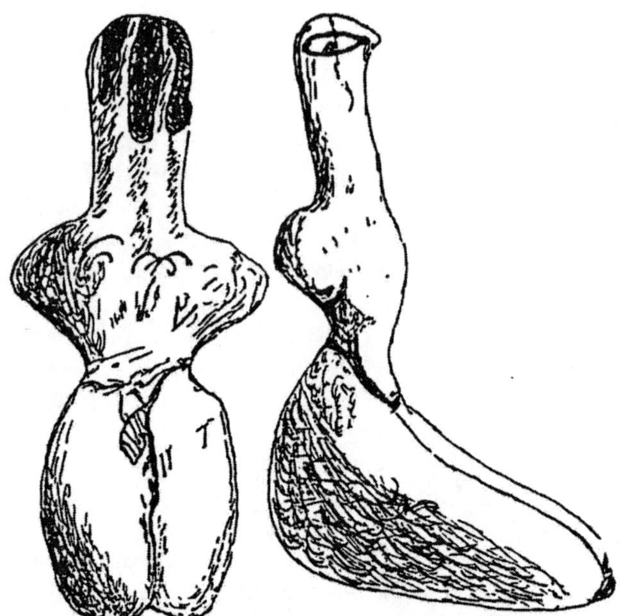

Egyptian figures formed during Lunar Cult, showing Steatopygous form, made of unbaked clay, with hieroglyphic characters on them. (Drawn by K. Watkins.)

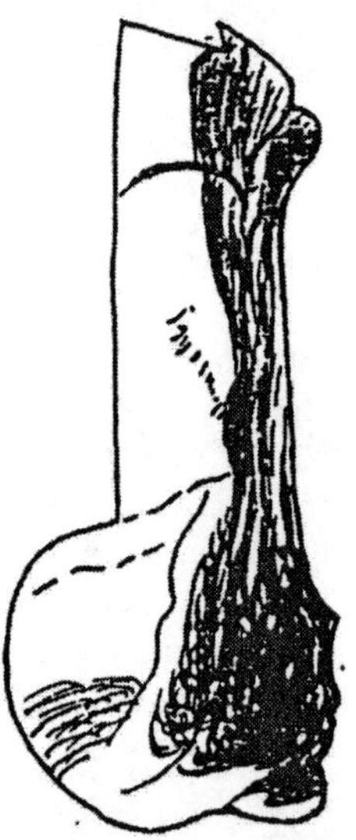

Early Lunar form of steatopygous female representing the Great Mother. Found at Vho, near Cremona. (Drawn by K. Watkins.)

Type of a fat woman found at Malta, and made of local calcareous stone, probably early Neolithic. (Drawn by K. Watkins.)

not be either the Hottentot or Bush women, on account of the labia minora being normal in these figures, whereas in the Hottentot and Bush women they have an extreme development, and hang down between the thighs for at least 15 cm. In this he is quite right.

The Bushmen were developed from the Pigmy and the Hottentot from the Bushmen. They went South in Africa and never came North, and therefore could not be models for the Nilotic Negroes or for the Stellar people who left Africa and came North. But the Pigmies did come North and went all over the world, and these were known both to the Nilotic Negro and Stellar and Lunar Cult people as "their original ancestors."

These sculptured or moulded female figures representing the "Mother Goddess" in the Lunar Cult are, therefore, the representations of Primary women, the Pigmy, just as the Solar people represented Horus, or Ptah, the first Solar God in human form as a Pigmy (see *supra*). So we see these Lunar Cult people began to represent their Goddess in human form, in place of the Zootypes of the Stellar, but inasmuch as these figures are by no means perfect, we can thus gauge the progress of their faculties. Possibly these figures were formed during the early Lunar Cult; personally, I believe they were. Amongst the Pigmy women there are two types—"steatopygous" and "non-steatopygous"; also, as we find represented on some of these figures, one type "hairy" and the other not. On Plate LXXII is a photograph of an extreme type of steatopygous woman, and, as I have stated (*supra*), there still exists a tribe of Nilotic Negroes with this hypertrophy of the gluteal region, the cause of which I have mentioned. (These types still exist in Central Africa.)[1]

These people knew that the Pigmy was the first or original Homo; hence the type taken for the Great Mother in human form, which was primary, the same as the Solar, who came after, portraying the Pigmy as this first God in human form. It was during the Lunar Cult that man began to model the human form, and replace the Zootype representations, the pre-human form, of the

[1] Some authorities on Egypt dispute this.

PLATE LXX

TEMPLE OF PRAYER, PEKING.

On the staircase is the Carved Dragon and the Phoenix.

Stellar Cult people for the Gods and Goddesses, but as we see in these cases, examples of which we find in many countries, these models were anything but perfectly accomplished.

The resemblance between the most ancient female figures in France and the so-called Neolithic figures of Crete and Egypt is very noticeable, and the fact that these are women without arms, flattened and steatopygous, cannot be accidental. The dress and girdle, the Egyptian mode of hairdressing, in some of the so-called Neolithic statues in France, the finding in the tombs the same red colouring of iron with pebbles and shells, which were employed in pulverizing it for use in colouring the skin, the same hieroglyphics or signs in the Mediterranean writings, and many other similar things, prove that there were the same general developments of religious customs, industries, and art throughout France, Spain, Portugal, Switzerland, Italy, and the Mediterranean Basin generally, as we find in Old Egypt. These developments spread North.

When excavating at Nagada and Gebelen, Dr. Petrie and Mr. Imbell found figures of birds exactly like those found in Liguria in the cave of Pollera and the Caverna delle Arene Candide, by Don Morelli.

The figures from Upper Egypt are of marble, quartz, or bone. Similar figures are found in South America and many other parts of the world, and Angelo Mosso is quite right in stating that "pre-history now furnishes indisputable proof that there was a community of religion, of customs, of arts and of industry, so that the people of Europe, of Asia, and of Africa" (and he might add Central and some parts of South America and Pacific Isles) "appear as brothers and members of one family." Angelo Mosso also states that "the independence of Minoan religion from the religion of Egypt is one of the glories of the Mediterranean civilization. The domestic cult of the penates and the absence of a dominant sacerdotal caste are two of the characteristics of Minoan civilization, *but the later forms of religion were less ideal and passed from the cult of nature and beauty to that of man-like divinities*." Here our friend is at fault, not understanding the progressive evolution that took place in Egypt from the totemic Sociology to the Stellar Cult—then Lunar (both forms of which have remains here), and then to the Solar, to which the latter part of his statement must refer. The so-called Horns of consecration found in Crete, Italy, and many other parts of the world, have been deciphered and dilated upon in another part of this work, as well as the Swastika—mother goddess of Crete, surrounded by lions, trees, axe, etc., and as here portrayed. The modification undergone by the religion of the Cretans is shown by the fact that the double axe rarely appears painted on the vases of the third period of the Middle Minoan epoch. It is, however, frequently found on the vases of the second period of the same epoch, and disappears in the last period of the Minoan epoch, the reason being that the oldest found was Stellar Cult, and Lunar, and finally Solar—different exodes of the Old Egyptians peopling these regions.

Again, if we take the Dolmens of Africa, which extend as far South as the Soudan and are common over North Africa, these have been found to extend as far North as Scandinavia in Europe. Dr. Zinck, of Copenhagen, attributes importance to the fact that objects of bronze and copper are found in the Dolmens of the South, whilst only stone implements and weapons are found in the Dolmens of the North. This was due to the progress of Egyptian civilization and the exodes from South to North and not from North to South.

The pottery found in these Dolmens and the Egyptian hieroglyphics found on some of the Monolithic Stones are sufficient proof of my contention, but as another proof I may mention the bell-shaped cup (Fig. 142, D.M.C.) discovered in Sicily, which is characteristic of the epoch of the Dolmens in North Europe. Similar cups have been found in the valley of the Po. Sophus Müller says: "Types of utensils, weapons, and ornaments may be preserved almost unaltered for a considerable time, especially when they are transmitted to fresh regions." Angelo Mosso states: "The Bell-shaped Cups must have belonged to a period of small sepulchral chambers of great antiquity;

PLATE LXXII.

Extreme type of steatopygous woman, from South Africa. Sent to me by Major William Jardine, South Africa, for which I return my thanks.

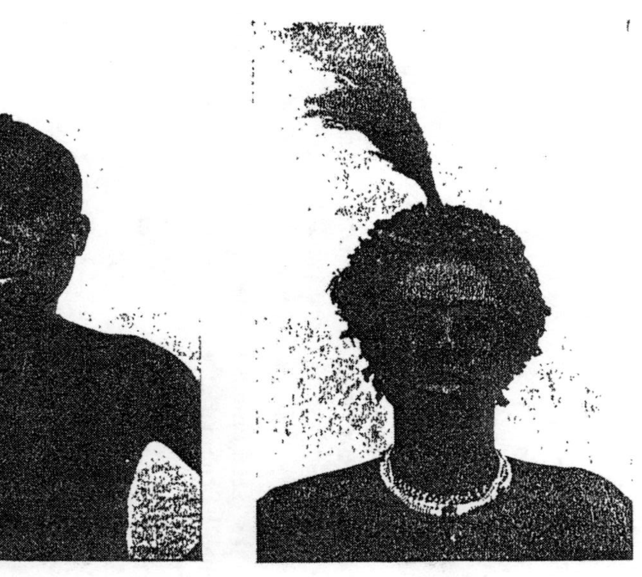

"Chripeta" native hunter. An "Angoni" maiden.

PLATE LXXIII.

I take this opportunity of thanking Major William Jardine, of South Africa, for these photographs of types of natives (Nilotic Negroes) living near the cave, and for his kindness and interest in my work.

in Denmark they are found in the chambers of the Giants, which must have been derived later and enlarged from these primitive chambers" (D.M.C.).

The numerous small bronze statues and terra-cotta figures, also some formed of clay or stone, representing animals—Sow, Bull, Dog, Bear and Birds—found in the Mediterranean Basin, throughout Europe, and in various countries, and used in religious customs, were made during the Stellar Cult period, and are the Totems of the Nilotic Negro, now divinized and used to represent symbolically the God or Goddess and various attributes, as, for instance :—

> The Sow was = to Rerit.
> The Bull was = to Osiris (also to Horus).
> The Dog was = to Anubis.
> The Bear was = to the Great Mother.
> Birds—Hawk = Horus and Vulture = Neith, and very early type of Horus.

Bull's Head amulets were found by Jean Capart in Egypt, and may be compared with those found in the Mediterranean

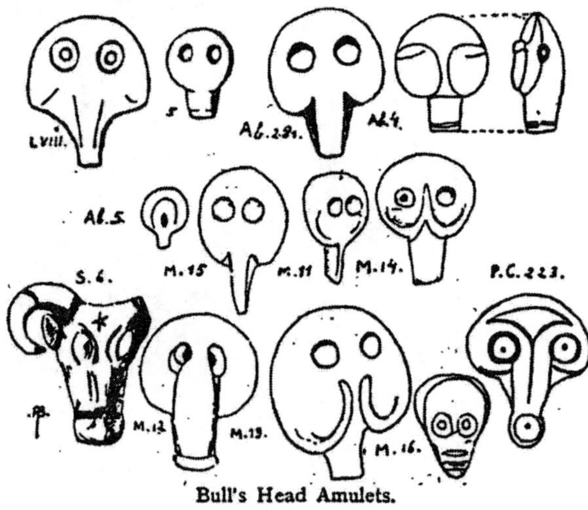

Bull's Head Amulets.

Basin (see "Origin and Evolution of Freemasonry connected with the Origin and Evolution of the Human Race").

We must always remember that new religions endeavour to obliterate every trace of their predecessors.

Professor Flinders Petrie, speaking of amulets, states :—

"The value of amulets found in Egypt lies in showing what such things were like in the Fifth Dynasty and *before.* Hitherto nearly all that were known were those of the Twenty-sixth and Twenty-eighth and a few of the Twelfth Dynasty.

"Several types used in the Fifth disappeared in later Dynasties and others superseded them.

"The 'Uza,' eye, differs in form, having two projections below, but not the rounded cheekpiece.

"The clenched hand and open hand are rare later, but common in the Fifth.

"The Hornet, Leopard's Head, and Jackal's Head are unknown later.

"But the commonest amulets of later times, the Heart, Headrest, Feathers, Crowns, Tat, etc., were apparently unknown in early times." I am of opinion that Feathers and a form of Crown were known from earliest times, but not the Tat or Heart.

It appears, however, that we can distinguish two different periods by studying some of these rock and cave paintings and carvings. I am of opinion that the earliest were formed or painted by the Nilotic Negro with Hero Cult, and others by the Stellar and Lunar Cult people. The Abbé Breuil and Don Juan d'Aguila made discoveries of rock paintings near Cogul, in the province of Lerida, representing a dance in which nine women are dancing around one nude man; which, from the description, one would say was a Lunar Cult dance of Old Egypt. On the rock carvings in the high maritime valleys of Liguria are found many figures and linear conventional signs made with tools of metal.

Rock carvings like those of Finland are formed of deep furrows made with a rather blunt tool.

Similar designs are found carved on rocks in Switzerland, and Señor José Fortes discovered many in Portugal similar to these.

Rock drawings are common in North as well as South Africa. That these carvings are made upon high rocks where it would be inconvenient to stand for the work proves that they were considered of great importance.

Dr. Evans admitted the resemblance of the signs carved on the rocks of the Alpes Maritimes, and especially those of Fontanalba and Lago delle Meraviglie, to the linear script of Crete, whence he concluded that "in the prehistoric age there already existed linear script common to a great part of Europe."

In the caves of the Pyrenees at Niaux, in France, paintings and carvings are found upon the walls in which pictographic inscriptions in red and black may be recognized among the figures of deer, horses, bison, and fish, drawn with great exactness. The "Bushmen's" pictographs are of a distinct class and only found in the South African district.

I here reproduce a photograph (frontispiece) of "Signs and Symbols and Hieroglyphics" cut on the rock walls of a cave (with decipherments of many). This cave was discovered by a shooting party in Portuguese territory, north of the Gambia, and I am much indebted to Major Wm. Jardine, of South Africa, for sending it to me. The immense number of glyphs here depicted is quite astonishing, but, owing to the action of water and atmosphere, much obliteration has taken place in many cases.

The remains, however, show that these were sculptured by different people. The most primitive were done by the Nilotic Totemic Negroes, and the others by Stellar Cult people. It is interesting to note that some of the characters are similar to those found in the Mediterranean Basin, and some are similar to those glyphs found in America, whilst others are pure Egyptian hieroglyphics; others again are similar to those found amongst the Australian natives, and from Mr. S. Landor's description the same as he found in Brazil.

The discovery is a very important one, and I regret that I am unable to visit this cave to investigate further and decipher those not clearly shown in the photograph. There is much "overlapping" in some parts, one glyph having been cut partially over an older one. Major Wm. Jardine states that the cave is situated near Missale, in Portuguese East Africa, where the territory joins Nyasaland and North-east Rhodesia. It is some sixty miles direct south from Fort Jameson and on the western bank of the Voobue River. He encloses the photos of two natives who live there to-day, here reproduced (Plate LXXIII).

Decipherment of some of the Signs, Symbols and Hieroglyphics.

1.  = Heaven raining.

2. = Ta—the world.

3. = P, the, or 60.

4.  = Water; ab, to thirst; sa, to drink.

5. = Drop of blood, parts of the body.

6.  = Sa, an egg (several of these).

7.  = Ten or Xta, a tomb; hri, to fear, hate, to be afraid; tna-t, half—Amsu. Also applied to the dams of the Nile (see Ancient Egypt).

8.  = Ak, bread; sus, kind of bread; xa, food.

9.  = Names of countries.

10.  = Mn (several of these).

11.  = Xi, sieve (many of these).

12.  = Sti, a sunbeam; ht, daylight, in brilliancy; hai, light; Zu am, a beam of light, or Zu, glorious.

13. ⊙ = Ntr-str, incense, a disk of the sun; atna, light; Ra, a day, etc.

14. = Px = P + m, to prevail (many of 11, 12, 13, 14) (to meet) (Kh, thing). S + t = take, to net.

15. = Pool; atr, river.

16. = Suh, to bind; nuh, a cord.

17. = S + l, to net.

18. = Marka, metal armlet?

19. = 10 (several of these).

20. = Serpent (Apap).

21. = A hand, a palm, a palm measure (many of these, ancient form).

22. = Kar-ntr, divine subterranean region, Hades or Amenta.

23. = T, a cake (a great many of these).

24. = Iu (God).

25. = Iu, Lord of the Universe (many of these)

26. = Ket?

27. = P (many of these).

28. = Offering, bread. ? Altar offerings.

29. = Offering, bread. ? Altar offerings.

30. = Offering, bread. ? Altar offerings.

31. = Heaven in Seven Divisions (? eight).

32. = Altar (many of these).

33. = Tank of flame (many).

34. = Altar offering?

35. ⊙ = Sep or Amsu (many of these).

36. = Sep or Amsu (many of these).

37. = Sep or Amsu (many of these).

38. = Ta (many of these).

39. = Heaven raining.

40. 8 = Division of heaven into North and South.

41. = Altar offering?

42. = He + *Hades*.

43. = Probably basket trap for fish.

44.  = Mn ?

45.  = Altar offering ?

46.

47. 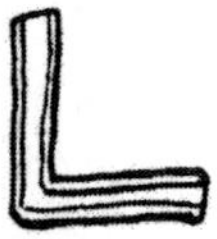 = Keb ? X, b = Keb.

48.  = P ?

49.  = Early form of Swastika.[1]

50.  = To tie in a knot, to bind up, to be ?

51.  = Concealer of the waters?

52.  = Ts, to carry, to go out, to transport.

53. 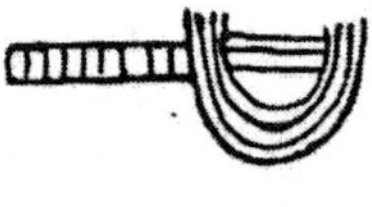 = ?

54.  = Crown of glory.

55.

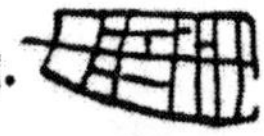

56. 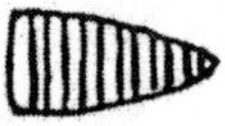 } = Probably represent basket traps for fish.

[1] Dr. Drummond found a figure depicted on a stone in Scotland precisely similar to this.

57. ☥ = Ank, living, life.

58. + = Am, with, by, resident in, to eat, name of Amsu.

59. = Keb.

60. = Sep.

61. = Sa.

62 = Sa = the backbone.

63. = The pelvis.

64. = Tank of flame—when the sun set in the west.

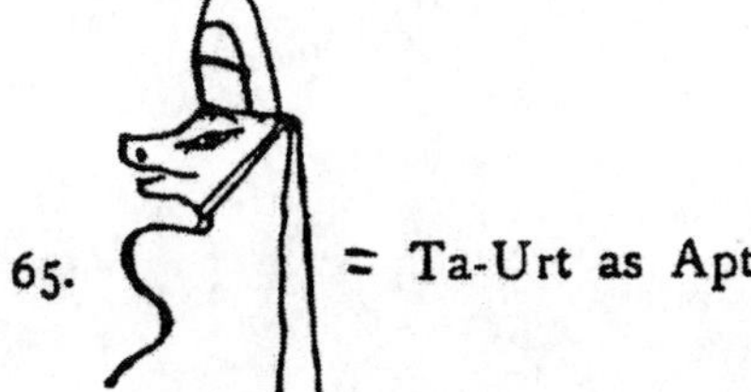

65. = Ta-Urt as Apt

66. = Head of Anubis.

67. = Aba - Xu. Greatest Light, or Most Glorious Light.

68. = Un? Heaven raining.

69. = Keb.

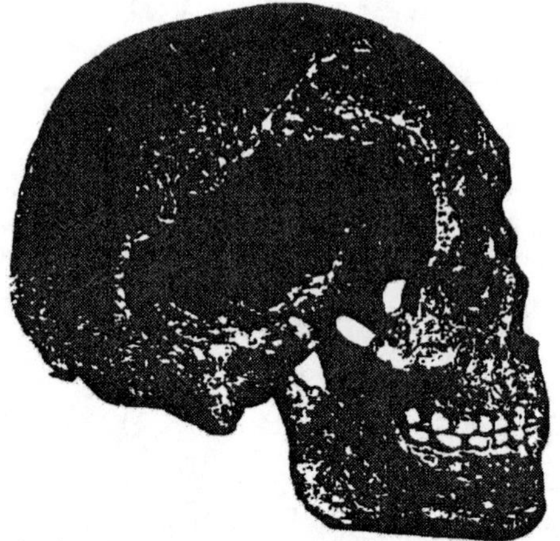

The skull of an early Stellar Cult man, which would resemble, when living, somewhat the type on the newly found piece of mammoth bone.

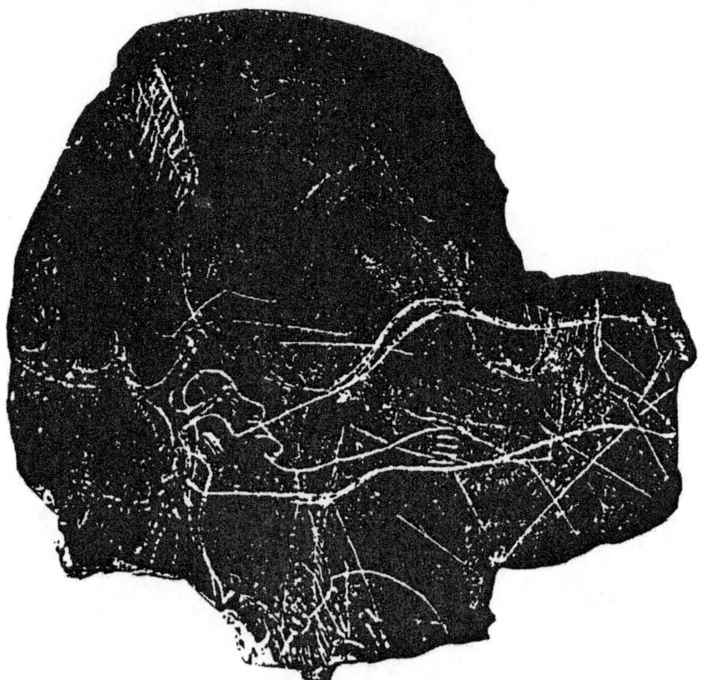

The human face engraved on part of the shoulder place or pelvis of a mammoth by a man of the Stellar Cult period; discovered in the Columbière Shelter. (Actual size.)

PLATE LXXIV.

70. [glyph] = P.

71. [glyph] = Ak, bread; Xa, food?

72. [glyph] = Metals.

73. [glyph] = A nome, a district, a field, a vineyard.

One can compare some of these with those "offerings of cakes and loaves" shown at bottom of Plate XXVIII of "Deshasheh," in "Memoirs of the Egyptian Exploration Fund," by W. M. Flinders Petrie, D.C.L., LL.D.

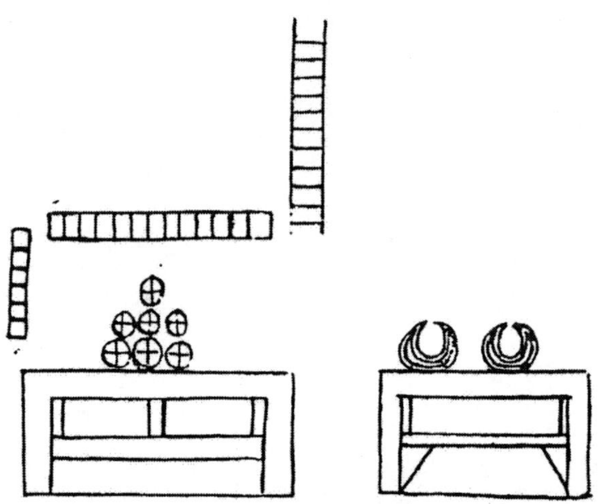

The rock carvings found by Mr. Savage Landor and described by him in his book "Across Unknown South America," vol. i, pp. 340–1, are evidently those of the Early Stellar Cult people: "concentric circles with central cross, ovals, triangles, squares, the Egyptian Cross, series of detached circles, generally enclosed within a triangle or quadrangle, were frequent. The circles were generally divided into four or eight sections, diameters intersecting pretty well in the centre of the circles. An oval with a number of dots inside, also hands and feet." The above description is good evidence that these were formed by Stellar Cult people. It is a pity that Mr. Landor could not have taken a photograph, as Major W. Jardine did, so that one could decipher these.

The drawings found in the *Grotte des Fées*, in the Gironde (A), in the Font-de-Gaume cave at Les Eyzies, Dordogne (B), and that in the Colombière shelter—so

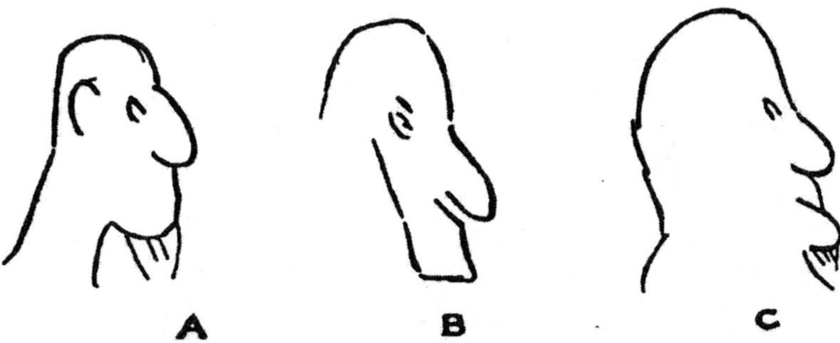

called drawings of the "Aurignacian Man"—are the works of "Stellar Mythos Man."

Sir Gaston Maspero has in his "Studies in Egyptian Art" attempted to define the chief schools of Egyptian sculpture, and particularly draws attention to the Theban school, in which the old Egyptians scrupulously depicted every detail in a true and lifelike form as then seen, whereas the Memphian "softened these details" when ugliness might have been portrayed, i.e. "he would not have failed to harmonize the lines and soften the colour."

In speaking of "the Cow of Deir-el-Bahari" he states: "Neither Greece nor Rome has left us anything that can be compared with it; we must go to the great sculptors of animals of our own day to find an equally realistic piece of work."

It was the Egyptian art market which supplied the Phœnicians with materials and works of art, religious and otherwise, to teach in foreign countries and the Mediterranean Basin.

He states that "the oldest monuments hitherto known in Egypt are of an art so fine, so well determined in the main outlines, and the country revealed so ingeniously combined a system of administration, government, and religion, which must have taken a long accumulation of centuries behind them when risen from barbarism into such a high degree of culture."

The term, or name, of Egypt, and of Egyptians, is not Egyptian. The Egyptians called themselves Romitu or Rotu, and their country Kimit, or Kam = the blackland.

In the "Dawn of Mediterranean Civilization" (Fig. 145) there are illustrations of three "pintaderas" from Mexico, now in the Archæological Museum of Turin.

They are stamps made of terra-cotta, and were used to impress patterns on the skin.

The handle is shown in profile in the central figure. When the Spaniards arrived in South America they saw that the indigenous population impressed designs like these on the skin, upon the face, in bright colours, and they called these stamps "pintaderas."

These stamps have been found in the Neolithic caves of Liguria and in hut foundations in Upper Italy. Don Morelli discovered some in the Caverna delle Arene Candide, and they have been found at the Pulo Molfetta. Dr. Schliemann found two at Troy, and they are found widely diffused throughout Europe.

Dr. Issel found them in Columbia and other parts of America. They are common in Lower Austria, Morive, Hungary, and throughout the Mediterranean Basin.

Dr. Evans found them in Egypt (Fig. 149 and Fig. 150, as may be seen in D.M.C.). These are from Tholos Haghia Triada, and the Egyptian are in the Turin Museum. They are all of either the Stellar, Lunar, or Solar periods, and of Egyptian origin.

The art of writing began in Egypt and was carried by the Stellar Cult people to all countries wherever they went, in the ideographic, hieroglyphic, lesser or mixed form, probably throughout the Mediterranean Basin first. On the monuments in Crete and other places in the Mediterranean, we find that these writings have been obtained from a more primitive or an earlier source than the two linguistic parent stocks (so called), the Aryan and the Semitic. The discoveries in Crete alone prove that the art of writing existed before the presumed penetration of the Aryan races, or of its influence in the island. The primitive script found here in the Mediterranean Basin, and other places to which these old Stellar people travelled, was altered after it had left Egypt. In the Egyptian, the writers drew the outline of the object of which they desired to convey the idea, in the artistic temperament of the race, with increasing skill of sculptors and writers. Outside of Egypt these hieroglyphics became degraded from the original forms on account of the difficulty experienced in copying them with the stylus on clay tablets; thus they lost the original portrait, finally retaining only the faintest resemblance to the original model. As in the primary, so it was in the transition script. The exigencies of haste and the unskilfulness of scribes soon changed both its appearances and its elements; the characters, when contracted, superimposed on, and united to one another with connecting strokes, preserved only the most distant resemblance to the persons, or things, which they had originally represented.

To obtain the key to read the script which is found in many parts of the continent of Europe, and throughout the Mediterranean Basin, formed at the time of the Stellar and Lunar Cults, one must seek for the original in Egypt. Much of it has already been found there, as Dr. Evans has proved. It was the intermediate step of the progressive evolution of the writings of the Egyptians, or the transition stage from the pure old Egyptian hieroglyphic to that of the hieratic, but much altered after having left Egypt, just as the pure hieroglyphic had been altered by those people who had travelled north-east, namely Assyrian, Babylonian, Chinese, and Japanese, and those old Stellar people who went to America, the latter's glyphs being altered from the original hieroglyphic of the Egyptians by themselves after they had left Egypt, which was before the time linear writing had been evolved.

Thus we do not find "alphabetical" writings amongst these early Stellar people, but only glyphs or hieroglyphics, which now differ so much from the original that they are not easily recognized.

The signs used by seamen to enable their vessels to be recognized from a distance, as Professor Angelo Mosso puts it, giving illustrations (D.M.C.), are the Zootype representations of Horus—as Horus the Fish Man

; the symbol on the fish proves this.

As I have shown (*supra*), this was one of the symbols used to identify him as depicted on the ivory tablets found at Nagada. These seamen used the symbol as a mascot—"Horus the Fisherman," or Horus who was the Saviour by water (see *supra*), the giver or bringer of life by food and water.

CHAPTER XXIV

EVIDENCE OF STELLAR CULT IN AFRICA

Mr. H. S. Cowper, in "The Hills of the Graces," a record of investigation among the Trilithous and Megalithic sites of Tripoli, has given some photographs and drawings of sculptured remains found ~~on~~ ancient ruins in Tripoli, which are certainly not all Roman. Some few of these I have deciphered, and here give their meaning and interpretation. They are unmistakably of Egyptian origin at the time of early Stellar Mythos. One or two, however, are not quite clear, part of the ideograph sculptured having been obliterated, evidently by time and the elements, but *those there portrayed are all Egyptian hieroglyphics,* and some of them are found in many parts of the world, connected with Stellar Mythos and otherwise.

1. Two circles, depicting heaven in two divisions, North and South.

2. A cross within a circle, the same as was found on the Stelæ at Palenque, on the Dolmens in Brittany and in Cornwall, as I have shown elsewhere in this book = to ☥.

3. 
The double triangle called *Sbaau.* The Abode of Stars or the subdivision of the Celestial World—*Stellar Cult.*

4. P, the, or 60.

5. T, me, thou, they, to embrace, to bring together, also negation.

6. Ht or Ut, a mace, order. The stone club of the Masai.

7. Ta, to seize, to tear, to bear away, to rob.

8. Rain, clouds, etc.

9. Ar, make, do.

10. Sem, total, combine, join.

11. Am, in, belonging to, Amsu.

12. Khen or Shen, brother, sister, two male and female spirits united for eternity.

13. Hr or U, great, to open, to fight, to bind.

14. Two eyes (Maa, to see), also a representative symbol of the two Pole Stars, Merti (Egypt).

15. The two earths, Ta.

16. Sut or Set.

17. Sen, open, pass.

18. = to ☥ the taw = Ank—life, living.

19. Circumference of Paradise, or heaven (Pierret).

20. Two-thirds (Pierret).

21. Am or M (Pierret).

22. Hespu, district, nome.

23. Per, a house (Masai house, see "Nilotic Negroes").

24. Tn, to revolt—conduct.

Of course one cannot read what the original text on the monument was, because so much is obliterated, but there is quite sufficient evidence to prove my contention (as *supra*). These ancient remains have the same characters as we find on those in many other parts of the world belonging to this remote date—huge blocks of stone set up with perfect accuracy. The forms of the temples here are precisely the same as we find in Ancient Egypt, Central America, at Kronos, in the temples of Golgos and Paphos, and apparently of Stonehenge, and in Kilat Siman in Northern Syria, with pedestal in the form of a double square, one above the other, in the centre of the building. (This will give Mr. Cowper the reason why these stones were grooved and made so as to form and dovetail into two separate blocks to form the double square.) Sometimes there were three, as Evans found at Knosos.

There are no arches, only lintels here.

Mr. Cowper draws particular attention to what he calls "dot" markings, making seven, which are found abundantly amongst the ruins, "making seven squares." That is symbolical of the seven Pole Stars (Ursa Minor) or the Seven Glorious Ones. It is also a tribal mark amongst many Nilotic Negroes and South American Indians. Mr. Cowper is under the impression that these temples were dedicated to Baal, as he states that "it was worthy of note, also, that the dual form seen in the two jambs of the Senams has a number of parallels in Ancient Rituals," and mentions Egyptian temples, Solomon's Temple, and Melkarte at Tyre. In this he is quite right, for Baal was Set, the Egyptian primary God of the South Pole Star, before Horus I of the North, and these two pillars represent Set and Horus. There is quite sufficient evidence in the numerous remains found here to prove that in long ages past a dense population inhabited this part of Africa, and from all the evidence found in this most interesting book one cannot but come to the conclusion that these were *early Stellar Mythos people*, who in later times perfected their wonderful buildings found throughout the world, as, for instance, the Great Pyramids, and those old cities in America and elsewhere, the evidence here pointing to the fact that these temples were either open or only "covered in" with wood, etc.

Mr. Cowper also states that they were not orientated; neither were the temples of Golgos and Paphos, nor those at Kalat Siman. Personally, I am unable to say if this was so, or if "some mode" of orientation had been observed. It is therefore probable that the temples were built at the earlier part of the Stellar Mythos, at the time that Set or Sut was primary God of the Pole Star South, when the orientation would be South, or if built at the time of Horus I it would be North.

The Arabic names here are also very significant, viz.:—

> Senam el Nejm, which means the idol of the Star = Horus.
> Senam el Thubah, which means the idol of Sacrifice.
> El-Aref, which means the place of divination.
> Ras el Id, which means the hill of the Feast.[1]
> Jabel Thubin,[1] which means the mount of Two Sacrifices.

The evidence found here points to the fact that this was a place of sacrificial offerings—whether to the first Great Mother Earth, or when she was divinized as Apt, or to one of her children, it is not quite possible to say, but probably the latter; and the sacrificial stone found here is described as similar in many ways to the one now in the Mexican Museum.

Personally, I feel convinced that we have here the two different remains of the primary Stellar Cult temples, when Set was primary God, and also a later form, when Horus was primary God. I form this opinion, which is quite conclusive to me, from the *forms of the altars* he has found and depicted on p. 149, Fig. 39, "Altars at Senam El-Ragud and Ferjana." *The first is circular;* these were built during the time that Set was primary God. I have shown in other parts of this work examples of two round temples (one Chinese and one Mexican). The other is square. These *square altars* were erected when and where *these temples*

[1] The Egyptian Hieroglyphics are 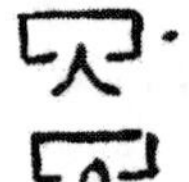

were built in the form of *a double square, first when Horus was primary God and never during the time of Set.* Therefore we can form some idea of the great ages of these remains, and we can always distinguish the temples of Set, which were round—at first, and orientated south—from the temples of Horus, which were in the form of a double square.

On p. 180 Mr. Cowper reproduces a design on a Babylonian seal which he states is similar to one found on the Senams here.

The interpretation of this figure is Horus, God of the North and South. The two poles or pillars are those of Horus and Set, the two Gods of the North and South. The cone with  (star with eight rays) is a symbol of Horus, God of the North.  above represents the circumpolar Paradise, and the  (two only shown, other two not here shown) = four houses of the Children of Horus. The four supports below, enclosing the two poles and the tree, or cone, represent the four supports of heaven, or four Children of Horus supporting heaven, which is symbolized by the tree and cone or mount. The priest here has the hair of the Suk, the same as Ramman of the Chaldeans (see *supra*).

(For much of the following information I am indebted to the Scientific Association of Rhodesia.)

Other Ruins still found in Africa.—Many yet, I believe, have not been discovered, but if we take the ruins found in South Africa, we observe two distinct classes of ancient architecture and construction apart from those of any of the decadent periods. On critical examination it must be admitted, after having examined some two hundred ruins, that—

1. There are two principal styles of architecture and construction, each possessing its own peculiar characteristics.

2. The original buildings of these two types of architecture and construction occupy distinct areas in the mediæval country of Monomatapa.

3. The first period buildings are found only on the east, south-east, and south of a line drawn from Zimbabwe to the Matopos, with the exception of a few ruins overlapping on the western side of this line, as, for instance, some ruins at Khami. This area extends to a point on the Sabi River within Portuguese territory.

4. The original buildings of the second or terraced type of architecture are to be found to the north, north-west, and west of this line, extending from and including Nyanza (if not as far north as to include the ruins of Mount Fura district), down in a south-westerly direction to the western border of Bulalema district, south-west of Bulawayo, including on its way the large number of important terraced ruins of Upper Insiza, Khami, and Bulalema.

5. The second period architecture is also to be found in the shape of extensions of, and additions to, the first period buildings in the area in which the original buildings are only those of the first period, and these additions and extensions are generally built over and upon the original erections.

Zimbabwe is probably the oldest that, up to the present, has been found, and, as this style of architecture is iconographic, it is undoubtedly the work of the Stellar Mythos people, who used pure granite cement for flooring, etc. In the after periods this was not pure, but deteriorated during each successive period. This also is a characteristic of the temples and buildings of Central and South America.

The planes of the walls, like all the Stellar Cult people's work elsewhere, are even and smooth, the courses are regular, each stone closely fitting into the next, one stone always overlapping the joint between the two stones below it, and the courses are bonded from front to back. There are no false courses. The marks of edged tools can still be seen on some of the granite and diorite blocks at Zimbabwe. The sacred birds and other Zootypes could not have been sculptured with flint tools, and this applies to those found in other parts of the world. There is indubitable evidence that the ancients worked most extensively for iron in the district of Maka Mountain Pass, Lundi Valley, Bochwa Range and Muesa Mountains, where there is a chain of ancient iron-workings extending in an unbroken line for at least twenty miles, and which district is known to have carried an immense population of ancients. Iron chisels, wedges, hammers, and trowel-shaped instruments have been found at great depths in ancient workings.

The ancients also used stone hammers, especially in the reduction of quartz. These vary from ½ lb. to 50 lb. in weight, and have been found plentifully in the old workings; also stone axes, stone wedges, and flint tools have been discovered at Khami.

The orientation of all the old class of buildings was North.

I might mention here that the orientation of the Nauraghes in Sardinia was South. I mention this here because some writers have drawn attention to the similarity of the old structures found. The explanation is that these are both the Stellar Cult people's building, but those of the Sards were those of the first 52,000 years, when Set was primary God, whilst the Zimbabwe and other buildings of South Africa were North, and therefore the Stellar Mythos people at the time of Horus I. Probably an exodus to the Mediterranean Basin took place first, before the old Egyptians travelled back to the south of their own continent.

The construction of entrances varies. The earliest are always open to the full height of the walls; at a later date we find "lintels," but lintels were common to the later Stellar and early Solar people. In the Stellar we find rounded buttresses. In the late Stellar and Solar we find "square end" walls.

In the ruins of Stellar people we find carved birds and animals (Zootypes) representing the attributes of the Deity. These are iconographic. At Zimbabwe the figure of a Crocodile 16 inches long was found carved on a stone beam, and on the beams of the Rhodesian ruins soapstone birds are carved. The Eagle is found here along with the Crocodile. In the Solar these are absent.

Thus the decorations on outside faces and on the main and other walls, with mural pictures and patterns, are different in different ruins.

Amongst the Stellar people the pattern was chevron;

amongst the Solar we find that the check or chess-board and herring-bone patterns are commonest.

At Khami I have found no trace of the Stellar people as yet; these were Solar people—"Ma-Karanga" means the Solar people.

Zimbabwe therefore is some thousands of years older than Khami.

Also there is a tradition among the natives that a pioneer tribe of the Leghoya came down from the North and destroyed the people who built these ruins. The survivors of these are the Bechunas (which are an amalgamation of the tribes known as Ba-Sutu).

Messrs. Hall and Neal state that the people who built the first Zimbabwe were marvellously well versed in geometry, and, as is exemplified in the system of curving of walls and the elliptical form of the buildings and their orientation, must have possessed a magnificent knowledge of astronomy, especially that of the northern hemisphere, and also of the zodiacal science. In fact, the style of architecture of the oldest ruins proves that the people who erected these came down from Egypt with the Stellar Cult during the time that Horus I was primary God. These are analogous to those we find in Tunis and other parts of the world.

The second distinct period of architecture is evidently that of the Solar Cult people, and many thousands of years had passed between the building of the first and that of the last.

Of the Khami ruins we find traces of three different ages:—

1. Flint or stone age (found underneath the main wall, p. 18).

2. Builders who work in granite, and

3. Builders of clay dwellings—recent natives probably (Plate IV, vol. v).

On the Khami ruins, flint and stone instruments were found underneath the main wall.

Mr. Leo Frobenius, in "The Voice of Africa," has shown by his discoveries of ancient remains which he unearthed some twenty feet deep, that the Stellar Cult was fully established at Illife, Ille (the house of Ifa), and other places in West Africa, that for some reason the old Stellar Cult people have been overcome, destroyed, and that the natives have at the present time the old Totemic Sociology with Hero Cult.

We have undoubted evidence of the primary God Set, here named Olokum, who was succeeded by Ifie, or Horus, as we find him in some parts of North Africa (Tunis). The bronze head of Olokum, full face and profile facing, p. 308, is the God Set, the same form of him as we find in the ancient terra-cotta masks in Cagliari, Sardinia. On p. 319 we have Set in the form of the Great Serpent, bound and with head fastened in a basket—to represent Set bound and chained in his hole.

The monolith in this form found here, viz. ▽ (see p. 299) is the ideographic symbol for the name of God Horus, and the traditions of the natives state that this was surrounded by a circle, thus showing him as Horus in the Stellar Cult, God of the Celestial Paradise of the North.

The two monoliths at the entrance of the temples, both in Ifie and Modeke, were the representative columns of Set and Horus. The crocodiles (two) found in the so-called Crocodile Temple also prove iconographically that it was the temple of Horus of the Stellar Cult. Again, the Mighty God of Thunder, clothed in a transparent drapery with his double-headed axe and a symbol of a Ram, is Horus of a Double Power or Double Equinox (see other part of this work).

There is also portrayed on p. 30 a mask of Tat-Aan.

Mr. Leo Frobenius has materially assisted to unravel the mysteries of old Africa and the evolution that took place. He finds help to prove the high state of civilization these old Stellar Cult people had attained; the discovery here of glass, terra-cotta vases, and work in iron and bronze, similar to those found all over the world where these old Stellar people travelled, only proves my contention set forth in this work. He does not know where to place the origin of the peculiar Bow and of the style of Horn found here, but if he will cross old Africa to the land of the Nilotic Negroes and Masai he will find the originals. The present Yoruba have the same mythical ideas and traditions as are found amongst the Hero Cult Masai, and some remains of the Stellar Cult left by the people who once reigned here but are now no more. The people of the lower Niger speak a language resembling the Masai.

The "Chief Way" of the Yorubans being the same as the "Via Decumana of the Etruscans" is explained by the Stellar Cult only, and the interpretation is the 8 Great Powers, $7 + 1$, with their attributes or consorts added $= 16$.

Mr. Leo Frobenius is not acquainted with the key to his great discoveries, not knowing the gnoses of the evolution of man and Egyptian wisdom, and attributes all he finds to the Mythical Atlantis, whereas it is all African and came from Old Egypt. I trust that what I have written in this work will answer all his questions in "The Voice of Africa"; more especially would I draw attention to his erroneous interpretation of the "belief of individual descent." He states that "all members of the same family are the posterity of the same God. In so far as they return to the same Godhead when they die, and in so far as every new-born babe represents the rebirth of a predeceased member of its family, they are all parts of that Godhead. "In brief, every god is the founder of a family, it being no moment whether he be the god of the tempest, or the forge, or a river, of the earth, or the sky, etc." Our friend has evidently not studied Totemism. These so-called gods are the original "Totems" of the family—and all the children of the Mother Totem belong to that family; if her totem is "Lord the Lion," then all are young lions, and they do not mate with each other, for that is taboo; they must mate with "Lord the Crocodile." Neither do they believe in reincarnation as here stated. I have fully set forth all these points in another part of this work, showing how they propitiate the elementary powers, etc., and the difference between this latter and the propitiating of their Ancestral Spirits.

CHAPTER XXV

EVIDENCE OF ANCIENT EGYPT IN NORTHERN EUROPE

Dr. Fridtjof Nansen, in his elegant work "Northern Mists" (a most appropriate title), quotes the Greeks as his principal authorities for the "legends and myths" that he has discovered in unearthing many old writings whilst trying to solve "the geographical way" to the north of North America, the first discovery thereof, and the people of the North. He has not studied the religions, beliefs, and cults of these people, and to dismiss this point by stating that they are "heathen" is to dispel the best proof of their origin and identity. What most of the Greeks or "Christian Fathers" may have written on the subject is not worth reading, and might be consigned to Hades with many other untruths about the human and his development. Only by living amongst these people where they are still in their primitive state, gaining their confidence and proving to them that you understand their sacred signs, symbols, and so-called magic, and by taking into consideration their anatomical features, can one find the truth.

The oldest remaining people in the North are undoubtedly the Eskimo, who call themselves "Inuit" (human beings), but the Norsemen called them Skrælings; Claudius Clavers calls them Greenland Pigmies, and in Olaus Magnus' works we find the same name.

They vary in stature.

The nose is small and flat, narrow between the eyes, broad and expanded alæ below.

Eyes, small and slanting.

They are broad in zygomatic parts. They tattoo their

women's faces. In appearance, colour, and language, which is monosyllabic, they are related to the North American Indian tribes of the North. The next tribes in antiquity are the "Gann Finns." These are probably the same class and of the same exodus as the Ainu, mostly small people, very hairy on the body, and with the Bear as their great Mother Totem. The present Finns and Lapps are the descendants of the above.

It is more than probable that the above Finns are the offspring of those who were driven back South by a Glacial Epoch, and again driven North by the Stellar Mythos people who came after. Another interesting point to solve is: Did the Eskimo people exist here before the last two Glacial Epochs? that is, were they the sole inhabitants of the North, driven South by the Glacial Epoch and forced North again by the Finns, and then at the next epoch obliged to come South again, returning North after the cold period had passed away, always keeping North of the Gann Finns? They are both undoubtedly descendants of the Nilotic Negro, but of two different exodes, the Eskimo being the elder; their totemic ceremonies, signs and symbols, and religious conceptions prove this. The Finns and Lapps of the present day, however, have probably mixed and intermarried with some of the earlier people, as Mongols, for instance. Their folk-lore tales and legends prove this. As a rule they only intermarry amongst themselves, and at the present day, if they marry "outside," they who do so and their children are looked upon and treated as outcasts by the rest of the tribe. Dr. Nansen may be quite sure that the first humans to discover and cross to America were the Pigmies. These probably crossed by two routes from North Europe to America and from North Asia to America; of course the geography of the world was not the same then as now. After the Pigmies the Nilotic Negro came North and no doubt crossed by the same routes, and the earlier Stellar Mythos people followed. These were the first people who discovered America and peopled it from North to South. This, I contend, I have proved in the foregoing chapters. These people left no written language, but have left certain signs and symbols and sacred marks common to all, still treasured by their descendants here, which can only be explained and translated through the Egyptian. There can be no doubt whatever that during the earlier and later Stellar Mythos exodes a vast period of time must have elapsed. Manetho states that the Great Pyramid was built during this period, "at the end of the reign of the Gods and the Heroes, by the followers of Horus" (see *supra*). In the Ritual there is a passage which gives an approximate time for the Stellar Mythos, over 250,000 years. There is every proof at least of the *earlier* Stellar Mythos people crossing by one of these two routes after the above peoples, i.e. Pigmies and Nilotic Negroes.

Dr. Nansen need not be in doubt about their being able to build long and big boats; one has only to see the natives of Borneo, Solomon, and New Hebrides Islands to prove what these early humans could accomplish in this way. These people will go very long distances out to sea in their boats and back again, and think nothing of it, and from still extant ancient pictures we can see that the Egyptians had large boats with sails.

There is a tradition amongst the North American Indians that all the tribes were formerly one and dwelt together, and that they came across a large water towards the east or sunrise. They crossed the water in canoes, but they knew not how long they were in crossing, or whether the water was salt or fresh.

The Dakotes possess legends of "huge skiffs," in which the Dakotes of old floated for weeks, finally gaining dry land—a tradition of ships and a long sea voyage. Their language was the same as the Welsh.

We have here two different traditions, one leading to the supposition that some came via Japan and North Asia, and the other via the British Isles and Greenland.

It was during the time of the early Stellar Mythos that a great exodus must have taken place to the North, East, and West. These people travelled, and gradually pushed their way in each direction, most probably living chiefly by hunting and fishing, settling and establishing primitive towns or villages in some places, as agriculturists, the remains of whom are still found. No doubt first they travelled North by the great rivers and along the coasts, and must have crossed to America via Greenland and north-east Asia. But I am of the opinion that the *later* Stellar Mythos people, who built the cities and temples in Central and South America, crossed to North America via Asia only, and then went South; and I form this opinion from the fact that although we find numerous remains of the *early* Stellar Mythos people all along both the northern routes, i.e. Europe, via Greenland and Asia, we do not find in Northern Europe or Central Europe the remains of those beautiful huge temples and cities, with glyphs and writings of the old Egyptian Ritual, but we do in Southern Asia, Central America, and some parts of South America; neither do we find them in America far north of Mexico. The reason may be that these latter people crossed from Asia during a Glacial Epoch, and so could not possibly go North to America via Europe and Greenland. How far North these people lived and travelled it is not quite possible to say; there is no evidence discovered up to the present time of their having gone *far* North. Of course it might be said that the Glacial Epoch would destroy all if they had crossed before the last period, but then why should we find the evidence of the existence of the *earlier* Stellar people and not the *later exodus?*—that is, if the later exodus had crossed this way. Then after these we have the evidence of the Solar Mythos people who crossed to America, but not via the northern route, or yet from Asia to America. There is also *distinct* evidence that the Egyptians traded with the Central American east coast people, viz. the Mayas. The Solar Cult is not found in Northern Asia, extreme Northern Europe, or North America much above Mexico, and not on the west coast of Central America, only amongst the Mayas, and then south as far as Peru. The Egyptians knew America and traded there, and they called the Mayas Haûi-Nibû (People beyond the Seas). So that from the time of the early part of the Stellar Mythos there is no record of the passage between *Europe and America direct.* During the later

Stellar Mythos there are records of a passage via *Asia*. During the Solar Mythos there was only a passage direct to the Mayas, except by the Chinese. For some reasons unknown the others had been cut off during the Solar Cult, just as—from the time of the downfall of the Egyptian Empire there is no record of the passage across to America until the rediscoveries by Cabot, Columbus, and others, and the evidence which Dr. Nansen has brought forward. As regards his quotation of "Homeric Conception of the Universe," I do not think he would have mentioned these, and probably would have written his first few chapters differently, if he had been conversant with the history of Ancient Egypt, even to the extent that many people know it at present. The Egyptians not only knew that this earth was round, but the records still extant, written in stone on the Great Pyramid itself, prove that they knew as much of the universe as do our present astronomers, if not more, and must have known as much geography as many people of the present day.

Dr. Nansen quotes Herodotus largely, but has left out that "the Egyptians told him that they had sent out colonies all over the world from the earliest times."

His "five divisions of the earth" does not refer to this earth any more than those other myths, but to Amenta, or the Earth of Eternity. Herodotus could not speak the Egyptian language; he had to depend upon a translator for what the old learned priests told him. They never told the Greeks much, and even this they perverted and did not understand.

None of the Greeks ever understood their Eschatology, hence the "absurd muddle" they made of their own religious conceptions, hence the dark and degenerate ages that have elapsed for so many hundreds of years. If you will learn the Egyptian and then read Herodotus, you will understand this—if you read Herodotus and not the Egyptian, you will not understand anything.

Dr. Nansen must go back further than Eratosthenes (275–194 B.C.) for scientific geography as well as for astronomy. Eratosthenes studied at the Alexandrian

Museum and Library, where many records of the Egyptians were kept, and these no doubt would be open for his perusal as far as he could read them. The records he has left prove how much greater was the knowledge of the Egyptians than most people of the present day can comprehend.

Language, or the difference in language, can be no proof on the subject Dr. Nansen writes about, against their religious ideas and customs, their beliefs and the remains found. That you find these people at the North now speaking different languages is accounted for by the fact that from the time they left their "old home," where they had separated, they had travelled by different routes, and probably it was thousands of years before the descendants of the original Egyptians met again in the North; yet there is sufficient evidence left in some of their words to prove a common origin, still more so in their myths, legends, and religious beliefs, the latter being the greatest proof, because, above all things, these people would still carry with them from generation to generation their religious rites and beliefs, as their forefathers had handed them down to them and they to their children. This is what we find. The Cults have not altered, and the sacred signs and symbols remain the same and are common to all. Craniological proofs must also be taken into account, as Professor Sergi has rightly demonstrated.

Dr. Nansen in his chapter ix, "Wineland the Good," has brought forward many old legends and myths, considerably mixed, but which, when one comes to analyse them, can easily be distinguished one from the other. The story of the Stranded Whale (p. 344) is very old, and dates back to the totemic Sociology of the human. Dr. A. C. Haddon has mentioned the same tale which exists amongst the Torres Straits natives (see *supra*). The story of the Sow or Pig is also of the same date. The Sow, or Pig, was a type of the old Earth Mother, and the totemic ceremonies connected with her are still carried on amongst the New Guinea natives (see *supra*). All their myths and legends can be read through the mythology of Egypt, whence they arose, and were carried over all lands. His Wineland is evidently "Amenta," the "Land of the West."

His opinion that much of the folk-lore and tales was brought from Ireland I think is correct, as far as those connected with the Lunar and Solar Mythos are concerned, but this was Egypt via Ireland. Probably all the Stellar was brought by way of Asia and Europe. Some of the figures from the rock drawings, as well as the depicture on the stone from Stenkyrka in Gotland (pp. 243 and 236), are certainly of Egyptian origin; both in the first and the third figure, p. 236, it is the Sektet boat of the Egyptians; this need only be compared with the Egyptian for proof. There is the house for the spirit or mummy in the upper one, and a rough drawing of "Kepera Shrine" in the third one.

That the men are different in anatomical form, with the exception of the cranium, is only what would be expected; these people would not draw or paint "Egyptian men," whom probably they had never seen and perhaps never heard of; therefore those who drew these figures would depict them in the form of the humans as they knew them. The tales and legends, as well as these drawings of the original, would be passed on from one generation to another, much would be lost of the original and other elements substituted, as we find in all parts of the world; at the same time there is quite sufficient of the original left to identify them, if you can read and understand the language. Dr. A. M. Hauson, in my opinion, is unquestionably right in attributing these to the Egyptian.

The bronze knife depicted on p. 238 is probably early Solar. There is portrayed on it the Egyptian Sektet boat which carries the spirit from Amenta to the Circumpolar Paradise. The seven Pole Stars are shown at the end towards which the boat is travelling—with the Shrine portrayed in the centre of the boat. The Pole Stars are here shown North and South. The North is the hieroglyphic representation of Horus, and that of the South as well (i.e. King of the North and South), whilst the hieroglyphic symbol for Set is depicted at the end, as sunk down in the Abyss. The boat is portrayed as passing through the starry heavens, going towards the seven Pole Stars and Northern Paradise, typically Egyptian.

The other picture portrayed in Nansen's book is also Egyptian, with the two boats, Maatet and Sektet—the former below, the latter above. The North and South are depicted by two serpents; the Sektet boat is sailing to the North, to the Land of Paradise, depicted in front of the prow. The boat below is going to the western lands (Amenta), or rather it is the meeting of the two boats. As regards "the representation of the ship's sternpost does not correspond to any known type of ancient boat or ship," etc., if Dr. Nansen will look at "Signs and Symbols of Primordial Man" he will find two ancient boats there

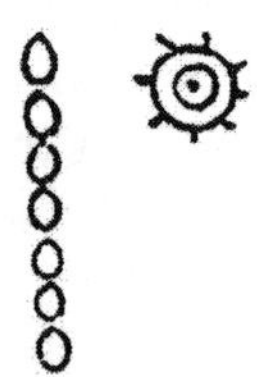

precisely as these are depicted; also a boat with sails and rigging, which were Egyptian long before anything in the North existed. Dr. Nansen has taken much trouble to study the Greeks, and all he could gain from them is not worth much. If he had studied the Egyptian he would have understood many things which he thinks incomprehensible, and his "northern mists" would have become "bright sunshine." He would have solved many riddles contained in his beautiful book. The Greeks distorted and perverted the whole of the Ritual, at least as much as they were told, and were ignorant entirely of the meaning of what they tried to copy. They are much too late to quote as authorities on some of the subjects about which he has written. This was the reason for that dark and degenerate age which has occupied about 5,000 years from the downfall of the old Egyptian Empire to the present time, when we find the human race emerging out of the night of the northern mists into the light of day and knowledge.

The earliest Stellar Mythos people, who had been developing in evolution in Egypt after the latest exodus of the "advanced Nilotic Negro," spread over all Europe and Asia and across to America. No doubt as they travelled North, East, and West they broke up into tribes, and some formed into nations known as the Thyssegloe, Turcæ, Aremphaans, in the northern and middle part of Europe and Asia, and the earlier Botyans. People of a later exodus were the Scythians, and to the north of these the Issedonians, and the Tibetans in Asia. They spoke the agglutinated language of the Egyptians when they left Egypt, and all the remnants found at the present day of their descendants still speak this language in a modified form. But in all the thousands of years that have passed from the original exodus to the present time we could not expect to find much of the original left, after the breaking up of one nation or tribe, the new nations formed out of the old, and the intermarriages that must have taken place. Still, we do find several Egyptian root words, and these have remained throughout all these ages "words of their home." But the Cults have not altered; their sacred signs and symbols and their religious beliefs remain the same and are common to all, which is the greatest proof you can have, because marriages and intermarriages, environment, and the conquering of the weak by the strong, would alter their appearance, etc., in many ways, but their religious beliefs would not alter much, and these "heathen" still believe in and practise what was taught them in their "old home." Names have changed because "language" has changed, probably many times during all these years, but when you translate this, and explain their myths, there is only one key to unlock the hidden past, and that is the Ritual of old Egypt. The rise of one nation and its decline, either by internal socialism, or by being exterminated or absorbed by wars—evolves a new nation, new people, new languages—this is the history which is being perpetually repeated in the development of the human race, and will continue to be so for many thousands of years more. But up to the time of the perfected Eschatology of the Egyptians, at the end of the empire, before this Cult had fallen into decay by the backslidings and dissensions of the priests, all the religious Cults had been strictly kept —by those who left the old Mother Country—as practised at the time they left, except when the further developed doctrine had displaced the former Cult; that is, when the Stellar people wiped out or drove farther away all the totemic people. The Solar, in the same way, in some places exterminated, destroyed, obliterated, or absorbed all they could of the Lunar and Stellar.

These points of evidence are so definite, and the proof of this still extant is so positive, that no one who can read them could have any doubt. With the fall of the Egyptian Empire and the commencement of Christianity, evolved from the Eschatology of the Egyptians, by the Copts, there commenced a dark and degenerate age. The Greeks only knew the little they had learnt from the Egyptians, and when the Romans who followed these spread partly over the world, it was an age of blood and conquest only, and after the Romans a still darker age for "the poor humans," characterized by the bigotry and ignorance of the priests, with infliction of torture, murder, and burning, in all the horrible forms that vicious, ignorant, and vindictive minds of men could conceive and put into execution, all in the name of Christianity. The Mohammedans also, in the name of the Prophet, carried fire and sword through many lands.

PART V

LUNAR CULT

CHAPTER XXVI

LUNAR MYTHOS CULT

WHEN our old forefathers discovered, after years of observation of the precession of the Pole Stars, that the time gauge from these was incorrect, they changed their reckoning of time from the Stellar to the Lunar. They had observed and marked down the revolution of the Pole Stars for 300,000 years, and had been patient for 52,000 years in order to observe if the Southern Pole Star rose again on the horizon as at old Apta. Finding that it did not, the change took place from the Cult of Set to that of Horus, whose symbol—the Pole Star North—remained on high, and they could see the precession of the Seven Glorious Ones (Ursa Minor), which always completed the circle without disappearing. Still the time gauge by reckoning from these would not come right, so they changed and commenced to reckon time by the moon—Lunar Time, and the Lunar Mythos resulted as a natural consequence, just as the original primary Cult of Set and the Southern Hemisphere was changed from South to North, and Horus became primary God, with all the preconceived attributes of Set given him, which originally belonged to Set, and to which they had added others.

Once in every 25,827 years the world's Great Age begins anew: another year has passed, and another year begins again in the great cycle of precession. The same cycle of time and the same phenomena occur simply as a matter of chronology; for how many hundreds of thousands or millions of years this has been repeated it is impossible to say. It was asserted by Martianus Capella that the Egyptians had secretly cultivated the science of Astronomy for

40,000 *years before it was made known to the rest of the world* (Lewis, "Astronomy of the Ancients," p. 264). The astronomers of Egypt had thought and wrote, observed and registered on the great scale of the Great Year of the Universe. The circuit of precession first outlined by the movement of the celestial pole was their circle of the eternal, or seven eternals, and was imaged by the Shennw ring and also by the Serpent of Eternity, when this was figured with tail in mouth and one eye always open in the centre of the coil. The Ritual of Ancient Egypt (ch. 114 and 123) not only proves the Ancient Egyptians to have been acquainted with the precessional movements: it also gives us an account of the actual changing of its pole star. The God Taht, the measurer of time, by means of the Moon and the Great Bear, is to be seen in the midst of the Mysteries, which are here described as those of *keeping the chronology for the guidance of posterity*. There is a change in the position of the Maat, or Judgment Hall, which in the Stellar Cult was at the station of the pole, and was shifted with the shifting pole. On account of this change, Taht comes as the messenger of Atum Ra in the soli-lunar mythos, to make fast that which was afloat upon the Urnas water; to readjust the reckoning and to "*restore the eye*" (Rit. ch. 114) by making it "firm and permanent" (Rit. ch. 116) once more for keeping time and period correctly on the scale of the Great Year. The backward movement of precession is described when Taht says to Atum Ra, "I have rescued the Atu from its backward course, I have done what thou hast prescribed for him." Renouf remarks, "I do not think any astronomer would hesitate to say that precession is meant" by this "backward course" (Rit. ch. 123, Notes). The Atu is a mythical fish with some relation to the course of the Solar bark—that is, to its backward course, the course of Argo Navis. Taht has "rescued the Atu from his *backward course.*" He has allowed for this retrograde motion in precession, and has made the eye firm and fixed once more by means of his reckonings as a guide of posterity.

So our forefathers merged the old Stellar Cult into Lunar, only changing names, adding fresh ones, and adjusting the correct time.

But there is evidence of so much overlapping of the Stellar and Solar Cults into the Lunar that it is very difficult to judge how long this Lunar mythology lasted, and how far it was distributed from its old central home —Egypt.

We find, however, Lunar temples in Egypt, the Mediterranean Basin, Asia, and South America, so that one may put it down at about from 50,000 to 100,000 years, seeing that these people had to traverse on foot all these vast regions and to establish this cult in these places. This they must have done, as witnessed by remains of Lunar temples found, as well as other evidence of this cult. In the Lunar Mythos it was Sati and Hathor who took the places of Sekhet-Bast and Sebek-Nit as the two Goddesses, "as one who conceived and the other who gave birth" to Horus. Horus here passes into the name of Khensu-Tehute, as the man of thirty-three years, and as the child of twelve years he is called Khensu-Hennu.

One name as Moon God was Xuns-Aah (Fig. 110, p. 504, Bunsen's Dictionary), or Ah (Bunsen's Dictionary, p. 341), and T. ti-Aah (Fig. 164, p. 508, Bunsen's Dictionary).

P-nb-Ta, Lord of the World, God, Son of Horus, represented by a child crowned with triple reed crown, and Goat's Horns (Fig. 101, Bunsen's Dictionary, p. 503), is the earliest representation in Lunar of the Child Horus, as was Horus eldest son of Ammon in Stellar, and Khensu-Pa-Khart, i.e. Khensu the Babe. As Khensu-Nefer-hetep he appears on the stele of Pai in the form of a mummied man seated on a throne; over his head is the uræus of royalty, and by the side of his head is the "Lock of Horus," or the Lock of Youth; in his hands the Flail, the Crook, a Tatt, and the Sceptre; he wears the Crown of the two feathers (the Amsu of the Stellar Cult).

These people wrote the old hieroglyphic language with the commencement of the linear writings, but probably only spoke the agglutinated form. Glyphs, signs, and symbols are found on various monuments in Africa, the Mediterranean Basin, Asia, and America.

As far as their anatomical features proved by the picture-graphs on monuments associated with this Cult, I should certainly class them as Turanians. Their skulls are all of the same formation. A late form of Horus in the Lunar Cult was Menthu, and in this form we also see the commencement of the change to the Solar as Menthu-Ra.

It was Hathor who descended into the underworld and there met the Solar God. The result of the meeting was Horus the Child. In the first part of this Mythos his Father was not known; hence the name of "Bastard"—this was attached to him until Thoth, the Moon God, discovered that he had a Father. Let us now examine some of the ancient remains found and still extant, and compare with Egypt and the Ritual for explanation of the same, because nowhere else can we account for their origin and meaning.

This figure represents the Mexican Isis as Hathor, the Moon Goddess, with the Maat feathers on her head, holding up or presenting the Child-Horus as her legitimate offspring to his Father—Atum. The great Evil Serpent stretched at the foot of Atum has accused her of having a Bastard. On the opposite side is Thoth, represented by the Jaguar, the recorder of Truth, who refutes the scandalous tale brought by Set and proves that the child was legitimate (see Egyptian text). Atum is represented as the central figure symbolically, with his foot crushing the head of the Great Serpent and treading down the lie that Horus was a Bastard. The ithyphallic symbol is here portrayed as a sign of resurrection.

I will quote only one passage more, to prove that it was Horus from the first period of the Stellar Mythos to the last of the Christus—Horus, and only Horus, under various names from the first to the last. In Ritual, Chapter 78, it states: "Thy Son Horus is seated upon thy throne, and all that liveth is subject to him. Endless generations are at his service, endless generations are in fear of him: the cycle of the gods is in fear of him, the cycle of the gods is at his service. *So saith Tmu*, the sole force of the Gods; not to be altered is that which he has spoken."

"Horus is the offering and the altar of offerings; twofold of aspect, it is Horus who hath reconstituted his father and restored him. Horus is the father, Horus is the mother, Horus is the brother, Horus is the kinsman, Horus proceedeth from the essence of his father and the corruption which befell him. He hath carried off endless generations, and given life to endless generations with his eye, the sole one of its Lord—the Inviolate One!"—the eye of Horus which gleams as an ornament upon the brow of Ra" (Chapter 8).

Now, Tem was represented by the Great Serpent in the Stellar Cult and Tmu was another name for Ammon, brought on in the earlier Solar, thousands of years before Osiris. Thus we see how the priests changed one name for the other in Cult after Cult, but from the beginning it was only the one and the same Horus. In Hymn III, Chapter 15 (Ritual), Tmu is called "the Unknowable, the Ancient One, the Mighty in thy mystery—Father of the Gods, who is united to his Mother in Manu."

In the city of Tchert "Menthu" was worshipped under the form of a man with the head of a bull, but instead of the Solar disk he wears on his head the Lunar crescent and disk. This is one of the names for Horus in their Lunar Mythos, and shows one of the connecting links. When the Lunar was merging into the Solar he was given the name of "Menthu-Ra."

The priests at the commencement of the Solar added "Ra" after Menthu.

That it is Horus I is proved by the pictures reproduced by Lanzoni, where he is represented standing upright, with the "head of a hawk" and holding in his right hand "an ear of corn."

The name Khensu-Tehute alludes to the man of thirty-three years, the twice great, the Lord of Khemennu, and as Khensu-Pa-Khart was represented as Khensu the Babe, and as Khensu-Hunen, i.e. Khensu the Child of twelve years. This in the Solar Cult was brought on as Horus the Child, son of Isis, as the Babe, and Iu-em-Hetep, as child of twelve years. The Ritual states: "Khensu-Pa-Khart caused to shine upon the earth the beautiful light of the Crescent Moon, and through her agency woman conceived, cattle became fertile, the germ of the egg grew, and all nostrils and throats were filled with fresh air." He was also called "the messenger" of the Great One and Lord of the Maat; also the Great God, Lord of Heaven, etc. The forms in which Khensu is depicted on the monuments vary, but whether standing or seated on a throne, he has usually the body of a man, the head of a hawk, and wears on his head a Lunar Disk in a Crescent, and at the latter end of the Lunar Mythos the Solar Disk, with a uræus or the Solar Disk with the plumes and uræus.

A study of Khensu is important, because it gives the key as to how these old Egyptian priests merged the Stellar into Lunar and the Lunar into Solar mythology. It is one of the great connecting links which must be studied if one wishes to know and understand how they brought on the first Trinity of the Stellar Mythos and added attributes to it in the Lunar and Solar Doctrines, and why those who have not studied this have made such a mixture. Volumes might be written on this subject, but I feel that what I have brought forward is enough for this work to prove the evolution of the Cults.

No Egyptologist up to the present, except the writer, has worked out these Cults and assigned to each its gods in their proper place and correct order. All other Egyptologists have mixed these all together, as they have done with the different Cults, and so have confounded "the numerous Gods of Egypt" and connected them with the Osirian Cult, which was the latest development before the Christian Cult, evolved out of these by the Copts at the downfall of the old Egyptian Empire, instead of being the oldest, as some have stated. Consequently we have been stumbling in the mire of darkness for the last 5,000 years.

Professor Maspero states that in the first Triad there were two Mothers or two Sisters and one Son (he leaves out the Father).

In the second Triad a Divine Father, Mother, and Son, and there it, or he, ends. But what is the truth which we find?

In the first, or Stellar, Cult we have Sebek-Nit (see suckling Horus) and Seket-Bast—the two mothers or sisters, "the one who conceives and the other who brings forth." Horus is the child, the son of the Father Ammon (see Ritual), and at a later time he is represented as Tem, the Son of Tmu, the Father, and the two Mothers are Nebthotpit and Iusaast.

In the second, or Lunar, we have Sati and Hathor as the one who conceives and the other who brings forth, the two Mothers or Sisters; as the child we have Khensu-Pa-Khart, i.e. Khensu the Babe; as the child of twelve years we have Khensu-Hunnu; as the man of thirty-three years old we have Khensu-Tehute. In another form Thoth or Aah was the Father and Ahi the Son, with Seshait-Safkhitabui and Nahmaûit (a form of Hathor) as the two mothers.

In another form we have Annit (Hathor of Denderah) and Satet, or two Sisters, as the two Mothers, the one who conceives and the other who brings forth, and Haroeris is the Child and Khnûmû the Father.

In the third, or early Solar, we have Bastet and Sokhet as the two Mothers or two Sisters, the one who conceives and the other who brings forth, with Ptah as the Father and Iu-em-Hetep as the Son.

In the early Solar a part of the Stellar was brought on in the form of Nebt-Hetep and Iusaaset, as the two Mothers, with Tmu as the Father and Tem the Son, but although this was one form of the early Solar, it was originally late Stellar.

In the fourth, or the Osirian part of the latest phase of the Solar Cult, Isis and Nephthys were the two Mothers or two Sisters, with Horus as the Child and Osiris as the Father.

In the fifth, or Christian, Cult, the Virgin Mary, and Mary the wife of Cephas were the two Mothers, with Jesus as the Child, and the Holy Spirit or Father God as the Father. These are the principal forms that we have traced out, but there were also others as attributes; for instance, a form of Hathor or Isis was Merit, the Nile Goddess—her son was Naprit-Sa-Mĕrit-su (a form of Horus). These were the representatives as Mother and Son, bringers or givers of corn, or grain, as food (representation may be seen on the tomb of Seti I).

We cannot find how many exodes there were of these Lunar Mythos people; they came immediately after the Stellar and the one overlapped the other. The same overlapping took place with the Solar Cult people, who came after the Lunar. This is amply proved by what we find in the Ritual and the signs and symbols found on monumental ruins of ancient cities in Africa, Europe, and South America.

Anatomically one cannot distinguish the Lunar from the Stellar people; we must class them osteologically as "present type." Their implements were similar to those of the latest Stellar Mythos people and early Solar. We can only distinguish them through the Ritual of Ancient Egypt and their signs and symbols found on the remains of their temples. In the Mediterranean Basin and some other places numerous remains have been found of the Lunar Cult people, to which I have drawn attention before, showing how they *typified* the "*Great Mother*" *partially developed in human form, as human.*

It was in the Lunar Cult that the first development took place from the zootype representation to the anthropomorphic, but it was a "crude form," as may be witnessed by the finds of the Great Mother in Egypt, the Mediterranean Basin, and other parts of Europe, which have been a puzzle to Professor Mosso, as stated and explained *supra.*

That the custom of dismemberment of the body before burial or mummification was still a practice of the Lunar Cult people is proved by the Ritual, wherein we read (Ritual, Chapters 16, 150). "Thou art Horus, the son of Hathor (Lunar Cult), the flame born of a flame, to whom his head has been restored after it had been cut off. Thy head will never be taken from thee henceforth. Thy head will never be carried away."

This proves that the custom of cutting the head off, which commenced at the time of their Totemic Sociology, was still practised during the Lunar Cult.

As I have already (*supra*) explained the reason for the formation of the Great Mother in a crude human form, and given the explanation of the types of Tree and Pillar Cult found in the Mediterranean Basin and other places, and that there was so much overlapping of the Stellar and Solar Cults with the Lunar, I am content to wait until further discoveries are made before forming a definite opinion as to the time this Cult lasted, and how wide was the distribution of these people.

In *Ancient Egypt*, Part I (1914), p. 9, there is an article by Lina Eckenstein. I think it is worth reading, for two reasons: first, it proves that this Lunar Cult existed, and second, it mentions Seneferu, the Great God, ravaging the lands before his ka Neb Maot (which do not represent earthly lands, as the writer supposes, but Amenta), but he wears *the head-dress of the Suk* and uses the Ht mace of the Masai.

The writer states that the "Horns worn by Seneferu are foreign to Egypt." It is a pity *Ancient Egypt* publishes such inaccuracies. *It was in Egypt that they originated*; the first old ideograph for the two horns was taken from the cattle of the Masai. Apt, meaning the brow of the God Iu, the special symbol of the Lunar Cult and attached to Hathor, the Cow. Hathor and Sati were the two Mothers in the primary form of the Lunar Cult. Again, when writing on *the Baboon*, Lina Eckenstein calls it the animal or the *incarnation* of Thoth (p. 11).

The Baboon was originally an Inner African type, which was continued in Egypt as an image of the Judge. In a Namaqualand fable the baboon sits in judgment on the other animals: the mouse has torn the tailor's clothes and blames it on the cat, the cat blames it on the dog, the dog on the wood, the wood on the fire, the fire on the water, the water on the elephant, and the elephant on the ant, whereupon the wise judge orders the ant to bite the elephant, the elephant to drink the water, the water to quench the fire, the fire to burn the wood, the wood to beat the dog, the dog to bite the cat, the cat to bite the mouse; thus the tailor gets satisfaction from the judgment of the wise baboon, whose name is "Yan" in Namaqua, whilst that of the Cynocephalus is Aan in Egyptian. We read in the Ritual of a golden dog-headed Ape which is "three palms in height, without legs or arms," and find the speaker in this character saying: "My course is the course of the golden Cynocephalus, three palms in height, without legs or arms, in the temple of Ptah" (Ritual, Chapter 42). The dog-headed ape was a symbol emblematic of the Moon as Ani, "the Saluter," in one form. In the Egyptian judgment scenes the Baboon, a Cynocephalus, sits upon the scales as the tongue of the balance and a primitive determination of even-handed justice; therefore, although the Baboon represented in an ideographic form a type of Thoth, the recorder of Truth, there was no such thing as incarnation. Lina Eckenstein has confounded the "horns" *on the Babylonian seal-cylinders, which were symbols of Solar descent, with the horns of Hathor, the first zootype form of the goddess of the Lunar Cult in Egypt* (see Horns, *supra*).

In the Tree and Pillar Cult of Evans (Lunar Cult) we have the Tree of Hathor, which was the Tree of Life, and it was the sycamore fig-tree, from the fruit of which a divine drink of the mysteries was made and drunk at a certain part or time of the ceremonies; therefore it was the typical tree to make one wise, and became a tree of abnormal knowledge. The divine drink is still made and used in the 18°. In Japan the divine drink is still taken and used from a palm, and travellers often make a hole in the base of a leaf and drink as it issues forth. It is of a very delicious flavour.

The Tree of Nut was the Tree of Heaven and Eternal Life; hence it was designated the Eternal Tree, as shown in the vignette of "The Book of the Dead"—the tree or eatable plant and the water supplied the elements of life to the manes in the lower Paradise, i.e. Amenta, Arru-Garden.

The Tree of Nut was Stellar, the Tree of Hathor was both Lunar and Solar.

A very important find, and proof of my contention that the Lunar Cult followed the Stellar, and that the former was gradually merged into the latter, is that of the discovery of a cist, or dolmen, of a new formation, found in Guernsey in October, 1912. I am much indebted to the Guernsey Society of Natural Science and Local Research, and particularly to Mr. A. Collenette, F.C.S., for supplying me with details and description of this dolmen, and permitting me to reproduce the photographs, etc., published in their *Proceedings.* Plate LXXV gives the general plan of dolmen found at L'Islet, Guernsey. This is the burial-place of a high priest and priestess, at the time of the change from Stellar to Lunar Cult. The reasons for this assumption are the following:

1. We have the two circles A and E situated North and South, representing the Stellar Division of the Heavens into two—North and South. We have the triangular stone for the cover of the tomb (see section), a symbol of Horus, God of the North and South. The orientation of the tomb, however, is not North, but either East or West (from the description given me this is not quite clear); *but it is not North,* which it would be if Stellar. Then the "Great Circle" which surrounds the principal cist DC is not a circle but indicates a Lunar Sign; this applies also to a smaller "circle" surrounding a second cist; there again we see that it is not a circle, but forms a Lunar Crescent merging into the circle E (South). Here we observe that the orientation of the cist is either East or West.

2. If this had been Solar Cult we should have found 3 *circles,* as North, Central, and South 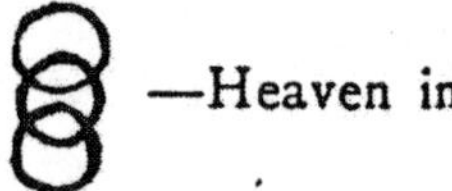 —Heaven in 3 divisions or 36 subdivisions.

3. If this were pure Stellar, only the two circles North and South ON. OS. would be depicted, but here we have the *half circle* of the Lunar Sign associated with the two circles of the North and South, clearly indicating and portraying that the Cult was changing from Stellar to Lunar. Now, as regards the orientation, the learned men of the Guernsey Society of Natural Science and Local Research state positively that this was *West.* I would not for one moment gainsay that this is not so, not having seen the original myself, but if it is West it further proves my contention, and the statements of the Old Egyptian High Priests, as quoted by Herodotus, who stated that "twice has the Sun risen where it now sets, and twice has it set where it now rises"; therefore, the burial took place at the time when the Sun rose in the West and set in the East (see *supra*), which, of course, means that they had observed and marked down the precession for at least two Great Years (52,000 years). The discovery of the above dolmen is a proof of what I have written and what has been recorded.

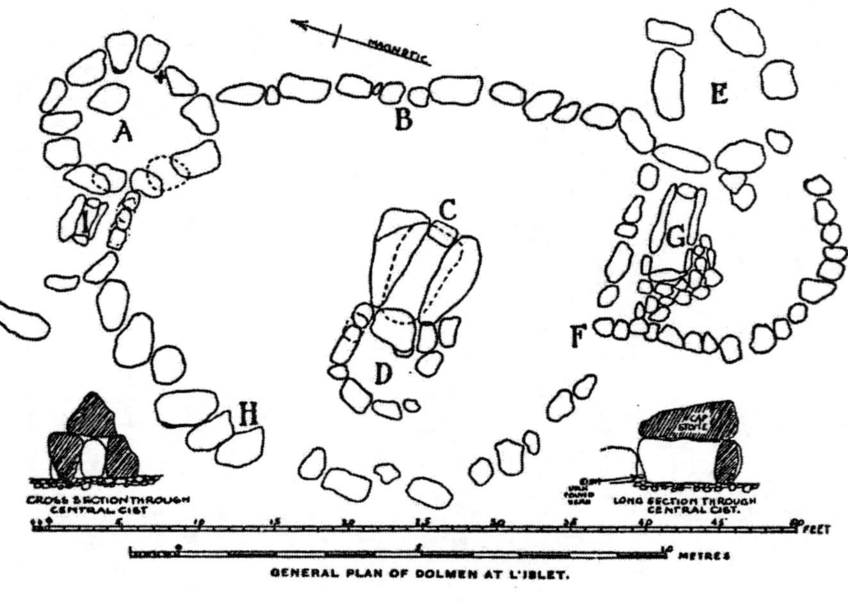

GENERAL PLAN OF DOLMEN AT L'ISLET.

PLATE LXXV.

Central cist from south-west, showing closing stone and antechamber.

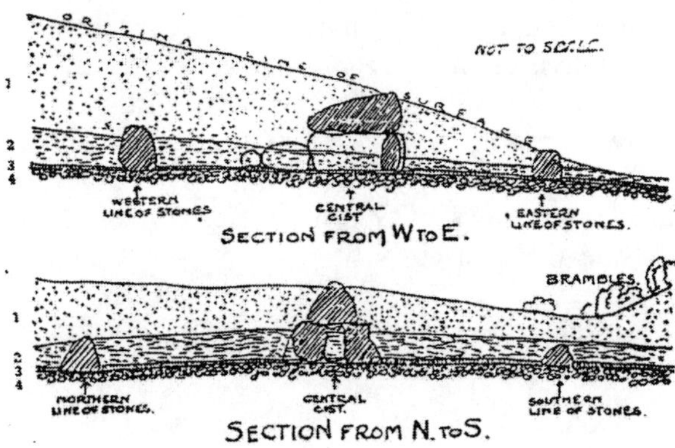

Stratification of mound through central cist.

1. Blown sand. 2. Peaty mould. 3. Clayey loam. 4. Old sea-beach.

PLATE LXXVI.

4. That these were the graves of the high priest and priestess is assumed because this was not a general burial-ground and is within the sacred precincts of the temples (circle N. and S.). Had this been pure Lunar we should have found the burial-place of the high priestess only, but we know that in the earlier changes Horus was still carried on from the Stellar Cult as the Primary God, and, therefore, would be represented in their Sacred Ceremonies by a high priest, who afterwards would take a subordinate position to the priestess.

This is one of the most important discoveries of late years, and proves that the Lunar Cult was brought to these Islands after the Stellar Cult and, therefore, before the Solar—many Stellar and Solar Cult burial-places have been found, but this is the first of the Lunar that has come under my observation, also the pottery and implements are quite characteristic. I reproduce (Plate LXXVI) geological section and central cist from S.W., showing closing stone and antechamber. The "urn" found here is similar to those in the *late Neolithic* tombs, which in itself is an absolute proof of what I have written here regarding these "cists."

For description of implements, pottery, and geological formation, I refer my readers to "An Account of the Discovery of a Cist or Dolmen of a type novel to Guernsey" in the Transactions of the Guernsey Society of Natural Science and Local Research.

CHAPTER XXVII

SOLAR MYTHOS PEOPLE

THE Solar Mythos people were those who came after the Lunar. They have been wrongly called "Sun Worshippers."

The old priests in Egypt (having carried their knowledge to perfection in astronomy and mathematics), finding that their Lunar time was not correct, began to reckon time by the sun, and simultaneously changed their mythology from Lunar to Solar, just as years before they had changed it from Stellar to Lunar, but there was much overlapping.

The first question that arises is, When did this occur? At Abydos, Professor Petrie has been excavating and has found well-preserved remains of ancient cities, which date at least from 20,000 B.C. The "finds" definitely prove that these old people practised the *Osirian Cult* at that time, and that they had this Cult then fully developed in practice.

The Solar doctrines commenced with Atum-Iu, the God of the Double Power, or Double Equinox, or Double Horizon, which was previous to Ptah, and Ptah was previous to Osiris by many thousands of years. Therefore the lowest estimate at which we can place the commencement of the Solar Cult must be 50,000 years ago at least. The Cult of "Horus of the Double Horizon" is not only one of the greatest importance in tracing the evolution of the human race, but it contains some of the profoundest secrets in the Egyptian Astronomical Mythology, because we have a form of Horus of the Double Equinox, or Double Horizon, in three Cults, namely Stellar, Lunar, and Solar. It was also carried on finally to the Eschatology. When Shu uplifted the sky and gave the eastern and western horizons to Horus, we have the most ancient form of Har-Makhu symbolized by the Disk of Aten. This so-called "Aten," or disk-worship, was the most ancient form of the God of the Double Horizon. This, however, was not Solar; it occurred in the Stellar. The disk was but an emblem of the circle made by Aten (one name for Horus as God of both Horizons, i.e. East and West). Here he is symbolized as a compound type of godhead, in which the mother was dual with the son, who was her child on one horizon, and her bull or fecundator on the other. If Dr. Breasted had understood the gnosis of the old wise men of Egypt, I think he would have written his "Development of Religion and Thought in Ancient Egypt" very differently (Plate LXXVII).

The word Aten, from At, was an ancient name for the child.

Horus-Behutet, God of "the Hut" or winged disk, the Chief of the Blacksmiths in totemic Sociology (see Nilotic Negroes), was the earliest form of Aten. This is the god who crossed from the western horizon to the eastern horizon upon the Vulture's Wings, which were an emblem of the motherhood, represented symbolically by Neith. "The Hut" was a dual symbol of the divine infant and the mother as the bearer of the child.

As the bird, she carried him over the intervening void of darkness, where the great Apap lay in wait.

Thus the Godhead of Aten consisted of the Mother, her child, and the adult male or bull of his Mother, in a Cult which preceded that of the Fatherhood.

The glory of Aten as the power that is doubled on the horizon of the Resurrection was the object of regard in this Cult, not the disk.

Another type is Neith, the suckler of crocodiles, which was an earlier form of the Virgin Mother, in Stellar.

When the autumn equinox occurred in Virgo, that was the place of conception for Sebek, the fish of the inundation. Six months later the sun rose in the sign of Pisces, and in the eastern equinox, when the fish, as the child and consort, or as the two Crocodiles, became the two fishes with Neith as the Mother, on one horizon, and Sebek on the other.

Thus, as we see and read the signs, the Virgin Neith conceived her child as Sebek-Horus, the fish of the inundation, which was depicted to express the adultship—these

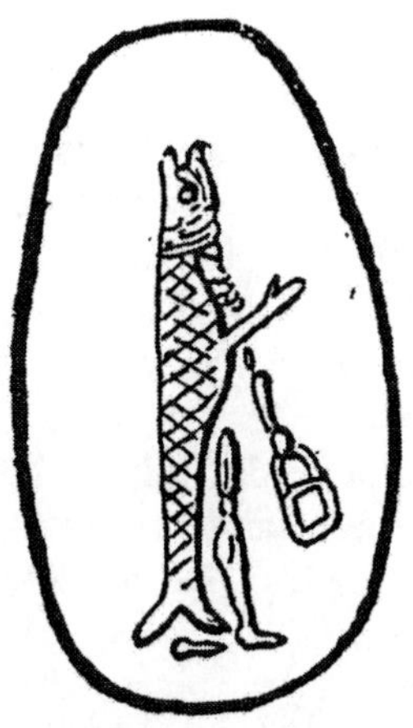

Chaldean representation of the Fish God. (Drawn by K. Watkins.)

were two typical fishes. This belief, or Cult, from Egypt spread throughout the whole of the world where the Stellar people went.

In the Stellar Mythos the birthplace was in Sothis, the star that showed the birthplace of the Child—the eight-rayed star. Both Mother and Child met in Sothis, both in Stellar and Lunar; in the latter as Hathor and the infant Horus. She was the House of Horus in the Lunar Cult, and the House was imaged as a tree or cone.

In the Lunar Mythos the cult was that of Hathor and Horus, the Mother and the Child, the Child who was the Calf on one horizon, and the Bull of the Cow on the other, which will explain why the Mycenæan figure accompanying the tree and pillar is at times a woman and at other times a child. They are the Goddess and the babe, identical with Hathor and the Child Horus in the place of birth, as Lunar Cult.

A well-known picture of the child Horus[1] depicts him standing on two Crocodiles, which express the double power or the double horizon.

In this representation Har-Ur is described as the old child who becomes young, that is, the elder who is transformed into the younger. This is represented and symbolized by the "Bes" head above Horus, as seen in the picture. Har-Ur, the elder first-born Horus, was the Child of the Mother when she had no husband. Horus on the Mount of Glory in the vernal equinox, standing on two crocodiles = Har-Ur, has now acquired the double power. The Bes head is depicted here in human form because this was now brought on from the Stellar Zootype to the Solar Cult.

In the Central States of America the portrayal is precisely similar, as may be seen from the photograph of a pottery fragment from Zaachilla and Cuilapa in "Mexican Antiquities," p. 305. The central figure there shows the young God Horus, and the head of the old Horus, as Bes, over it. This, if compared with Horus *supra*, cannot possibly be mistaken. This figure has the name of Horus attached to each side of his face, and the Lock of Horus represented on the right side of his head as the Central American portrayal.

Thus the Crocodile-headed Sebek, as the child attributed to Neith in Virgo, crosses the gulf of darkness, or the Abyss of Waters, from the West to the East, as Horus of the twofold Horizon, or as Horus of the Double Power in Stellar and Lunar Cult. That this was first worked out in the Stellar Cult and afterwards brought on in the Lunar, and then in Solar as Horus of the Double Equinox or Double Horizon, now called Har-Makhu or Harmachus or Atum-Iu, I shall prove. There can be no possibility of doubt, because the above is depicted first in Zootype forms and afterwards in the human, and as further proof we find the names Har-pi-Khart, or Horus the Child, and Har-Khuti-Khepra, the Stellar names in the first, transformed into the Solar in the later.

[1] See my "Origin and Evolution of Freemasonry Connected with the Origin and Evolution of the Human Race" (George Allen and Unwin Ltd.).

God of the Axe.

Another form of the evolution of this Cult after Aten was "Horus as God of the Axe, the Cleaver of the Way," from one horizon to the other. This must have lasted for a very considerable time, because we find it so widely distributed. Originating in Egypt, it spread throughout Europe, Asia, and America, as existing remains prove.

In his "Dawn of Mediterranean Civilization," Professor Mosso gives many illustrations of the Sacred Axe, which was the symbol and sign of this Cult. Primarily they were made of stone, then of copper, and later of bronze, all kinds having been found in Egypt and many other countries. I here refer my reader to some of them—Figs. 74, 75, 76, 77, 80, 82, 83, in Professor Mosso's work, D.M.C.

These Sacred Axes are also found depicted on many of the dolmens of Brittany, and are found in Germany, Greece, etc.

On frontispiece, Fig. 110, "Dawn of Mediterranean Civilization," is portrayed the Axe as the Sacred Symbol of Horus of the Double Horizon in the Lunar Cult in Crete. On the top of the Double Axes we see two birds, "the two Horus Hawks or two Doves," which are supported by two pillars (North and South). Horus is here represented as God of the North and South, and the two horizons East and West; the Sacred Axe appearing in the hands of a female divinity proves this to be Lunar Cult, the female divinity here representing Hathor.

The Double Axe, which has been called "the Mycenæan Tree and Pillar Cult," is an emblem of the Double Power, and the so-called God of the Double Axe is consequently a God of the Double Equinox, who was Har-Makhu, the Horus who passed into Atum-Iu as the son, and Atum-Ra as the man, in early Solar Cult.

The sun that made its way through the earth or abyss was known as the divider or the cleaver. This was the Solar power, which clove its way from West to East, and from horizon to horizon as Har-Makhu, God of the Double Horizon or Double Equinox in the annual round.

There is no doubt, in my opinion, that the title of God

of the Axe was first associated with Horus in the Stellar Cult, and afterwards brought on into the Lunar, and then Solar, as we find the sacred emblem in all three Cults. As Stellar, it would be first associated with Horus at the time Shu made him God of the East and West as well as North and South. The Cult of Aten was also another form.

A knowledge of the Cult of Horus of the Double Horizon is very important in tracing the evolution of religious conceptions and beliefs, as it gives the key to unlock the mysteries of the past. The present learned American Professor who is excavating the temple of the Sphinx has written that this "Sphinx is older than the Pyramids"! which is quite incorrect. The Sphinx was a symbol and sign constructed out of the solid rock, as an ideographic symbol to represent the passage of Horus from one horizon to the other, West to East. Therefore, figuratively, it is a symbol of Horus of the Double Horizon or Double Equinox, and an early Solar symbol before Amenta was constructed by Ptah, thousands of years later than the Great Pyramid, which was built during Stellar Mythos Cult.

Until the time of Har-Makhu the fatherhood of God had not been individualized in Ra.

Har-Makhu was the mother's child, when she was a virgin, represented by the white vulture of Neith or the sacred heifer of Hathor or as Tem = Horus as represented by the great Serpent. The child could be self-generated as the spirit of life, or vegetation, or of light, the phenomena being prehuman from the first. Child Horus in the Solar Mythos was typified as the little Atum son conceived upon the Western Mount. Adultship was attained upon the Horizon East with what was termed the double force, the sun being the symbol. In these two characters he was the double Horus, or the double Harmachis, the Solar God of both horizons and the fulfiller annually of the double equinox. The power of evolution was portrayed in Kheper, the transformer. Kheper showed the old beetle changing into young; the tadpole transforming into the frog; the human embryo developing *in utero*; the enduring spirit emanating from the mortal man.

Kheper was a form of Har-Makhu, as we learn from the inscription of the Sphinx. From Har-Makhu the father god Ra-Har-Machis was developed. Atum was Ra in his primordial sovereignty. The divine fatherhood was developed from Har-Makhu, who became the great God Ra in his primordial sovereignty as Har-Ur, the elder, first-born. Horus was the child of the mother when she had no husband, and he had no father, but in the Solar Mythos the "Father" takes the place of the mother, and as we see depicted, he is represented as both Father and Mother —a changing of the Cults was being enacted.

Thus we can prove that when Ptah with his seven Pigmies formed Amenta for the passage of the Sun, Moon, and Manes to pass through the earth, instead of going round as previously, the Solar Cult was fully established and the Fatherhood was established by Thoth. Horus was no longer a "bastard." Thoth proved that it was in the underworld of Amenta that Hathor (represented by the Moon) met his Father (represented by the Sun), and so the Fatherhood was established, and the anthropomorphic took the place of the zootypes as representations of God and his attributes.

In the Eschatology, the Child Horus always remained a child of twelve years until transformed into an adult of thirty years.

Imhotpu—he who comes in peace—was Ptah before he became incarnate as the third member of the Memphitic Triad, and Ptah at Memphis became Sokaris by dying (risen spirit name).

Anhûri (Tomb of Thinis) became Lord of the West as Khontamentit. Maspero states (D.C. p. 103) that Uapûaitû, the guide of the celestial ways, must not be confounded with Anubis, who guided human souls to the paradise of Osiris, and the sun upon its southern path by day and its northern path by night; but Maspero is wrong in this: it is the same God under different Cults with a double, as Shu and Anhûri. Uapûaitû of Sint became Anubis—it was he who guided the souls to Paradise *through the underworld* from West to East *and then to the celestial North.*

No. 1.

No. 2.

No. 3.

PLATE LXXVII.

I reproduce here the winged-dish of Horus I, under the name of Hor-Behutet, to which sometimes a crown and the two uræi are added, representing the Lord of the North and South.

Nos. 1 and 3 are Egyptian.

No. 2 was found by Mr. J. L. Stephens over the door of the temple at Ocosingo, near Palanque. It obviously corresponds to the Egyptian.

Drawn by Mrs. K. Watkins.

Solar Temple of Mitla, Oaxaca, Mexico.
Drawn by Miss Foreman.

Stellar Cult Iconographic Temple.
Drawn by K. Watkins.

PLATE LXXVIII.

The Solar people can also be distinguished from the Stellar by their myths and legends, which generally show two points of departure for the migrations of the human race.

One is from the summit of the celestial mount (Stellar), the other from the hollow underworld beneath the mount or inside the earth (Solar).

The races that descended from the summit of the mount, which was an image of the Pole, were people of the Pole, whose starting-point in reckoning time was from one or other station of the Pole Star, determinable by its type, whether as the tree, the rock, or other image of a first point of departure.

The tradition of the Pole Star people found in various countries is that they were born when no sun or moon as yet had come into existence. *They were pre-Solar and pre-Lunar in reckoning time.* In their legendary lore they try to tell us from which of the seven stations they *descended* as a time gauge in the prehistoric reckoning of their beginnings.

The oldest races that have kept their reckonings are descended from one or other of the seven stations in the mountain of the North.

Those who ascended from the nether world were of the Solar race, who came into existence with the sun, as it is represented in the legendary lore, that is, when the Solar Mythos was established, the sun being used as a type, and Solar time was reckoned.

In the Solar Mythos, they *ascended* from the underworld, which had been hollowed out through the mound of earth for the passage of the sun (represented as a symbolic ideograph—Amenta).

Thus there are two points of departure in the Astronomical Mythology, one from above and one from below.

In the "creation" of Atum, instead of being reckoned as the offspring of the old first Mother, or the group of the seven Pole Star Gods, men became the children of Ra, who in one account are said to have come into existence as tears from his eyes, or as germs of an elemental soul proceeding from the Solar God. This is the deity of the Eschatology in the Ritual, who says, "I am the self-originating force. Behold me, how I am no longer merely Solar or one of the seven elemental powers." He is the God in Spirit, the spirit that is divine, and a type of that which lives for ever. "And God spake unto Moses and said unto him, I am *Jahu*, and I appeared unto Abraham, unto Isaac, and unto Jacob as *El Shaddai, but by the name of Jahu I was not made known to them*" (Exod. vi. 3). This is the reason for the change we find in the Jewish traditions, i.e. the change from the Stellar to the Solar, and also that of name or title which follows the change in status.

Ra was known by other titles or names in the Mythos, but as Huhi, the eternal, he was previously unknown. In this character the god reveals his secret self as the Supreme one, whose name is then expressed as Huhi, the eternal, and Ra, the Holy Spirit.

The Hebrew deity, Ihuh, was not simply the one God in a single form of personality: *he is the Egyptian one God in his various attributes.*

He is the one God, both as the Father and the Son, who in the words of Isaiah (ix. 6) "is the everlasting Father and the Prince of Peace," who as Egyptian was Atum-Huhi, the eternal Father, and Iu, the ever-coming Son—Atum-Iu, or Atum-Horus, as opener of the horizon East, and Atum-Ra, as closer of the horizon West; the one God who was Solar in the Mythos and the Holy Spirit in the Eschatology.

That this was the same wherever we find the Solar doctrines I shall now prove; also that all can be traced back to Old Egypt. It was a change of Cult from the Stellar to the Lunar and the Lunar to the Solar, with change of names for the one great God and his attributes, one and the same from the beginning.

Menes, the supposed king of the First Dynasty of Upper and Lower Egypt, founded the city of Memphis, and built the temple there in honour of the God Ptah, but Abydos was many thousands of years older than Memphis, and Thebes was built before Abydos. Probably it was at Thebes that the Solar Cult originated, when the Lunar was changed into the Solar. (Proof of this will be found in the Ritual.)

This proves how many cycles of time we must retraverse before we come to the origin of the Solar Cult, and the discoveries of Professor Petrie will help to prove this historically.

Although Menes is commonly placed at the head, or as the first king of the First Dynasty, in my opinion this must be incorrect, because we have many remains of statues of kings dating long before Menes' time, with the Crown of the North and the Crown of the South portrayed on their heads, which I think is a sufficient proof of my contention, as these are known to be much older than the time of Menes, who was supposed to be the first king of united Upper and Lower Egypt.

The people by this time had greatly improved in knowledge. They had developed their old language and writings from the simple ideograph (monosyllabic) by adding prefixes and affixes (agglutinated), commenced and formed their alphabet, and then their Hieratic and Demotic (inflected). These Egyptians had been a progressive people since they commenced to form "the Nomes," at the time of totemic Sociology, up to the time they perfected their Eschatology, when dissension arose, first amongst the priests. Socialism and the destruction of the Empire followed.

In the Solar Cult we find that the nature powers, or elemental powers, which hitherto had been represented by Zootypes in the Stellar, and by half Zootype and half human in some of the Lunar, were now personified in human figure. Heretofore, in the Stellar and Lunar Cults, the primary representation of the elemental powers was of Zootype form. *There was no human figure personified* in the Stellar and only partly human in the Lunar.

The sun and moon were thus both represented as "had not come into being."

In their Astronomical Mythology, the nature powers which they propitiated were raised to the position of rulers on high, and this was the beginning of "the Gods" as a primary class of rulers, whose reign was divided into seven sections (seven stars of Ursa Minor), or heaven in seven divisions in the Stellar and eight divisions in the Lunar.

In the Solar Cult these Zootypes became divinities in the human form, nine in the Put-cycle of Ptah.

This occurred when Har-Ur, or the elder Horus or Horus I, was depicted as a child in human form in the place of the Lamb, Fish, or Young Shoot, and here we get a later Trinity in human form. The Father takes the place of the Mother now, as the primary, which hitherto had been given to the Mother.

But although various names are now substituted for the old ones previously used, their religious conceptions were not really much altered. It might be stated as a higher type of evolution of the older Cults.

It is only by tracing the evolution of man, and the evolution of his Cult, that we can distinguish the one from the other, but having learned this, we can interpret the signs and symbols found in various parts of the world, and state with absolute confidence to which of these people they belonged, and also to what parts of the world these people travelled. Man cannot be understood except through his own history.

There is a further proof which distinguishes these people the one from the other in a very definite and objective form, namely, *their buildings*. The Solar people can always be identified and differentiated from the Stellar, particularly in two points. They both built with polygon-shaped stones, beautifully finished and cemented together (Plate LXXVIII).

Monoliths were common to both.

The Stellar were iconographic, the Solar never, the reason being that all through the Stellar they represented their God and his attributes in zootype forms—prehuman. With the Solar people the anthropomorphic forms came into existence, i.e. the human form took the place of the pre-human. The Stellar people bonded their buildings, the Solar did not. The Stellar used finer cement than the Solar, so much so that many people, after examining their buildings, have stated that no cement was used, but that is not so (see *supra*).

The Stellar people never built arches—always lintels.

The Solar people used lintels at first, but at a later date built arches.

Amongst the Solar people we find certain signs and symbols which had not come into being in the Stellar, although they brought on and made use of many signs and symbols of the Stellar. Amongst the Stellar Cult people we find many signs and symbols which were not brought on and made use of in the later Cult.[1]

[1] For further proof, see "Signs and Symbols" and "Arcana of Freemasonry."

CHAPTER XXVIII

SOLAR CULT PEOPLE

THE type of the human race had now advanced in evolution to a higher standard, "face and features had become more refined," as depicted on the monuments; also if we compare the Mayas and the Incas with the old Stellar Cult people found in America, even at the present day, we can easily distinguish the differences.

In considering the surviving types of these peoples, however, at this period, all must be classed, so far as Osteo-anatomy would prove, as belonging to "the modern type of man."

We have to take into consideration, now more especially, the *intermarriages*, as well as environment, etc., all of which would affect both original types as they left Old Egypt and spread throughout the world. Their descendants were remote from their ancestors as regards time. The anatomical features of the old Stellar people depicted on the walls and sculptures in America, and elsewhere, portray a lower and earlier type of the human race than those of the Mayas and Incas, who were Solar people. We find the same two types in Babylonia and Assyria. An Egyptian type of feature, in each case, at the time of the two different Cults, it is not quite so marked, as the merging from one Cult to the other originated from here (Egypt), and the intermarriages would be one cause, and the demarcation would not be sudden; it took time in Egypt for the evolution to be accomplished from Stellar to Lunar, and then from Lunar to Solar. At the same time there is a very noticeable difference in Egyptian types of early Stellar and fully established Solar; this can be seen from the monuments and portrayals on the walls still extant, and from living types found there at the present day.

In Asia and America the two types are very distinct and decided, and the reason for this is obviously because from the time of the Stellar Cult exodus to the full evolution of the Solar, when other exodes left Egypt, an immense number of years must have elapsed, and there had been a progressive evolution of the human race here in its old mother home. If we sum up generally the two distinct types, we find that the first, or earliest, Stellar Cult people had either square, squat figures, coarse features, large flat noses, thick lips, or were short-statured individuals, with round head, sometimes deformed, oval face, high cheek-bones, large mouth, small oblong eyes—Turanian or Tartar types. All these and those that still remain had or have, the Stellar Mythos. In the late Stellar and the Solar people we find more regular, handsome features, with well-proportioned limbs, finely formed heads, high foreheads, well-formed noses, and smaller mouths with firm lips. The eyes were open, straight, and intelligent. Of course we find many modifications where intermarriages have taken place.

The Osteo-anatomy of these people differs in one form, namely, in the early Stellar Cult type we find that the *supra-orbital ridge is still rather thicker and more prominent than in the Solar*, but of much less extent than even in the early Nilotic Negro.

The next point to be considered is—who were the people in the world that believed and practised the Solar Cult, what proofs have they left by which we can trace them back to their connection with Egypt as "their home," and do any of these people exist and practise this Cult at the present time?

The answer is definite and conclusive, because we find that—

1. An exodus of these people left Egypt with all their religious doctrines and customs, spread over the greater part of Europe, and came to these islands and have left abundant evidence here. These we call the Druids.

2. An exodus went out to Asia, Sumarians, Chaldeans,

Babylonians, and Persians, passing and settling only in some parts of that continent. The Hindoos and Buddhists and later Japanese still practise part of the Solar Cult doctrines. The Japanese have much Stellar mixed with this, but probably the Solar Cult never reached the northern part of Japan, as there are no remains found there, only Stellar and Totemic (the latter amongst the Ainu). The early Sumarians and Chaldeans were probably Stellar Cult, and the Solar was introduced later.

3. *There was an exodus direct to Yucatan,* where these old Egyptians landed, bringing all the doctrines of the Solar Cult with them, and *this is the only way that it reached America.* It spread a little north of Yucatan, not quite to the west of Mexico, and then travelled down south and west to Peru.

This can be clearly proved by the remains of ancient buildings and cities, where they have obliterated as much of the old Stellar as they could; yet there are sufficient remains of the Stellar left to prove that Stellar people were there primarily. This can definitely be proved by the decipherment of what we find on the remains of ancient cities found there, by the difference in the anatomical features depicted on the walls, by the Zootype forms found on the Stellar, but not on the Solar temples, mural inscriptions and codices, by the differences in their calendars, language, and Cults. The reason why these people (Solar) did not go north was because they could not overcome the old Stellar people, as the evidence on the remains of these old cities proves. We can trace how far the old Solar people went by means of the temples and ruins found with only Stellar Cult portrayed on them, and not a trace of the Solar. The language differed as well; the old Stellar was an agglutinated form. The Solar had advanced to the earliest form of inflected. The Solar people travelled south and west as far as Peru, "the line of least resistance." The ruins left on their line of march and still extant bear ample evidence of this. But when they arrived as far as Chili the fierce old Stellar people stopped them.

Of the Solar Cult people found in America, as I stated above, these landed at Yucatan, and the old Maya population found there was subjugated to them. The Solar people destroyed much of the old Stellar here, both men and temples. They were stopped from going far north and so went gradually south and west via "the line of least resistance," as far as Peru in South America. Traces are found of them through Bolivia as far as the North-West Argentine, all along the high ranges, which appeared the easiest and safest road to travel from the old Stellar people. That they must have fought their way down against much opposition is proved by the fact that their line of communication was cut, as evidence still extant proves, namely, that although we find old Solar temples as well as old Stellar on this line of march, we also find that many of these Solar temples had been destroyed and replaced again by Stellar, and the evidence points to the fact that they founded in Peru a vast empire and were opening up their communication again to the north when they were overthrown by the Spaniards. These Solar people, so-called Incas, overthrew the Aymaras and Quechuas (Stellar Mythos people), but were stopped by the Stellar people of Chili and Bolivia. They absorbed the Aymaras and Quechuas in and about Peru into one great nation without imposing the Solar Cult at once. They replaced the old Stellar Egyptian hieroglyphic writings of the priests by the mnemonic system—the language of "agglutination" by "early inflected."

The Quechua language still remained the common language spoken throughout the empire, but the Incas, who were the rulers, spoke a language of their own (inflected) which had been developed in Old Egypt after these old Stellar people had left, thousands of years before.

The people in Colombia and Ecuador show much overlapping. The earliest Stellar people known then were the Quitus, who had established and extended a great empire until they were overthrown by the advent of the Solar people.

Amongst the Indians of the uplands of Peru, Bolivia, and Ecuador the queue is quite general and the anatomical features of the people resemble the Tibetans and Chinese-Japanese types. Many words used by them are pure old Egyptian.

The Incas pierced the lobes of the ears as the Masai do. Their religion was the old Solar Cult of the Egyptians, brought here via the Mayas and Yucatan.

Mr. Enock states: "We have here the remains of a civilization in Central and South America which until four or five hundred years ago was unknown to the world of Europe, a civilization so unique and beneficent in character that the historian of the conquest, Garcilasso de la Vega, exclaimed that 'laws so beneficent have never been enjoyed by any country under any Christian monarch.'" These things are not exaggeration; proofs and records of the Inca nation establish their truth.

There were the same laws as the old Egyptians had. The anatomical features of these Solar people found here from the Mayas to Peru show the higher type of the Solar people: they had well-formed, regular features, many were handsome and good-looking, with well-proportioned limbs, finely formed heads and foreheads, well-formed noses, small mouths and firm lips, eyes open, straight, and intelligent. No doubt there was much intermarriage between the Solar Cult men and the old Stellar Cult women; the result would of course, be a mixed type.

If we look at the map of Colombia, South America, we can see where the Incas came down and their line of march, because here commences the great Andean mountain system which runs north and south. To the north we find it divided into three ranges or Cordilleras, which are united at the extreme south. Also there is another great range of mountains to the east of them, running *due east and west*, which would form a barrier to prevent the Incas passing them, as the Stellar Cult people already inhabiting this district (see *supra*) would set up a very decided opposition to their passage south; whereas they had an easy passage down through the valleys and fertile plains between the ranges of the Western Cordillera, the Central Cordillera, and the Eastern Cordillera. After conquering the Stellar Cult people of one of these valleys, they would be protected on their flank until the Stellar Cult inhabitants of the other parts joined and came over the mountains in a sufficiently large body to cut off their retreat. This no doubt they did, because from the ruins of the temples found we can see that these countries were originally occupied by Stellar, afterwards by Solar Cult people, and that later they were again occupied by people of Stellar Cult. The Incas were opening up again north when they were destroyed by the Spanish.

The language question will help to prove this; my contention is, as will be seen if we inquire further into this point, that at the beginning the various tribes of South America may be divided into six or seven groups—the Uitoto, the Carab, the Tupi or Guarani, the Arawak, the Tukano, and those represented by the present Patagonians, the Fuegians, and the Pigmies. These were all anterior to the people who spoke Quechua, although amongst many tribes to-day Quechua is the only Indian tongue used. But this diffusion of Quechua is of relatively recent date amongst many tribes.

It was the language of the Incas or Solar Cult people, and where the Incas did not impose it upon the conquered the later missionaries did, because in territories like the Andaqui country, the upper Napo, the upper Amazon, which were never conquered by the Incas, we find the official language (of the priests) Quechua. The local language, in spite of the efforts of the priests, has not been completely supplanted.

For the regions like the Ecuadorian inter-Andean valley, which formed part of the Peruvian empire for nearly a century, and in which at the present time no other idiom has persisted, besides the Quechua, the fact, though less evident, is no less certain; indisputable documents published by the Ecuadorian historian Gonzales Suarez, reproduced in the *Journal de la Société des Américanistes de Paris*, 1907, prove that at the end of the sixteenth century Quechua had not yet become general throughout the upper plateau. At that epoch the local languages were still so widespread that the ecclesiastical authorities deemed it useful to have catechisms written in the divers dialects. The Inca empire reached as far south as the River Maule in Northern Chili and part of the Argentine provinces. All along the coast to a distance of about 250 feet above the sea-level are found mounds of shells containing bones of fish, birds, mammals, and pottery of various character, implements of bone, stone, and copper. The burials here reveal three different types of man. The earliest were the long-headed type, who made pottery of a rude description (Nilotic Negroes). These were followed by a round-headed type, who were buried in the thrice-bent position and whose culture was on a higher plane, their implements and arts proving them to have been Stellar Cult people; and later the Inca people—their remains, however, are not found in any considerable numbers south of Choapa. These long-headed people retired south before the Stellar Cult, and are now found as the Alaculuf of Tierra del Fuego.

"Urn burials" are found amongst the round-headed Stellar Cult people similar to those discovered in Egypt.

Mr. Joyce states, p. 511:—

"The Quechua name for the Creator was Pachacamac (the Soul of the Universe) [Horus of the Solar as Atum Ra], and this Pachacamac was identified with Uiracocha. As we have seen, an important centre of worship was the coast town of that name; but the fact that the name is Quechua seems to prove that the deity to whom the early temple was erected was not the Quechua Pachacamac, but merely identified with him by the conquerors of the coast."

Mr. Joyce is correct, because this was their name for him during Stellar Cult. The old Solar people knew that it was the same God, but they had changed the Cult in Egypt from Stellar to Solar, and when they found this temple of the Stellar God, here they changed it to the name of the Solar God, one and the same under different Cults, Uiracocha being the Stellar, Pachacamac the Solar name.

The prayer translated on p. 160 of Mr. Joyce's Book is part of the Ritual of Ancient Egypt, almost word for word.

In Asia the Solar Cult people did not travel farther than the south or middle of Japan, as is proved by the old Stellar Cult people still existing in the north of Japan and all the northern part of Asia. There is no trace of the Solar people found yet to prove that they went north of Japan, *therefore they did not get across to America by this way, i.e. from Asia to America*, as the Stellar Cult people did, and no trace of the Solar Cult can be found in the Pacific Isles. They (Solar) are not found in North America or the western part of Mexico, and not until we go farther south than this—and I have found no evidence that they ever reached the *extreme north of Europe*; as far as any evidence goes, I do not think that the old Stellar Cult people were interfered with here; they were pushed up here from the south, and there are still some living remnants of these who practise part of the old Stellar Cult.

M. Pierre Loti, in his book "Siam," mentions some wonderful old buildings situated in the depths of the forest of Siam, with enormous staircases and towers and colonnades with fine arches, at Angkor-Val, also at Bayou, which he considers much older.

It was from India that these people (Solar) emigrated and settled here after subjugating the Stellar people who were here before them, "men with turned eyes, worshippers of the Serpent," i.e. the Stellar god Tem. These Solar brought Brahmanism, and some centuries later Buddhism was introduced (both debased forms of the Solar Cult), and a swift decline took place, and the vast empire of the Khmers was overthrown more than 500 years ago.

CHAPTER XXIX

THE PEOPLE OF THE BRITISH ISLES

THE oldest human inhabitant of these Isles was undoubtedly the Pigmy, as his Implements have been found in large numbers. At present no part of his Osteo-anatomy has been found.[1]

It is not to be expected that any remains of these little primary men would be found after all these years, beyond their implements, considering that they buried their dead a few inches below the surface only, and in all the hundreds of thousands of years that have passed all would disintegrate, except where a possible accident happened and the body was buried in some preserving material such as sand, or in some cave or rock, etc. This would apply also to the Nilotic, who followed the Pigmy and whose Implements have also been found in large numbers. We must take these Implements as evidence of the existence of these people—they must have lived here or we should not find their Implements.

Of the primary and later Stellar Cult people we have sufficient evidence. (As regards the cave pictures in Wales, not having seen them, I am not in a position to say if they were made by the Stellar Cult people, or Nilotic Negro, or present man.)

The "artificial mounds" we find in these Islands, as well as in other parts of the world, viz. North and South and Central America and Asia, all bear a striking resemblance to each other, and in some instances are precisely similar, as, for example, the one at Salisbury Hill, at Avebury, 170 feet

[1] The Sussex or Piltdown *man* (or *woman*) I consider, as stated *supra*, to be Nilotic Negro remains.

high, which is connected with ramparts, avenues 1,480 yards long, circular ditches or dew-pans, and stone circles. These are the remains of the "towns" or dwellings of our forefathers, situated on hills or downs, with ramparts thrown up to keep off wolves and other wild animals from their cattle. The "avenues" were made for their cattle to pass through, morning and night, as they were driven from pasture, just as we find amongst the Madi Nilotic Negroes at the present day. Some stone circles mark the places of the "huts" for the "keeper and look out," and the "dew-pans," which were a peculiar construction, were made to contain their water supply. These were the earliest Stellar Cult people; they would thus have to protect their cattle from the ravages of wild animals. Having only stone axes, clubs, flint-headed arrows, and spears to use against ferocious beasts, many would therefore dwell together on the top of the down or hill, throwing up ramparts of earth, planting stakes, etc., on this, and digging ditches under them, so that wolves and other animals could not easily get up to the flocks, which, like the Nilotic Negroes, Madi, and allied tribes still extant in Af ica at the present day, they drove in at night to keep secure. These were the first formations of "towns" and "cities," and show the humans settling down to agricultural pursuits instead of leading the nomadic life of their fathers the hunters. Where we find the two circles of stones O N. / O S. we know that these people were Stellar Mythos and pre-Solar (see *supra*).

Twelve monolithic stones were erected to form each circle, the one circle to the North, the other to the South, representing the twenty-four Zodiacal stars, and as characters in the Egyptian wisdom, these earliest pre-Solar pillars were representations or symbolical of the pre-Solar powers and called "the old ones" or "the elders." They are of Egyptian origin wherever found in various parts of the world, and are traceable to two different groups of 12 = 24 mysteries of Stellar Mythos. These were the twelve who had their thrones as rulers (or æons) in the Zodiac and the twelve spirits with Horus-Khuti—Lord of the Spirits in the heaven of eternity. In the papyrus of Ani and of Hunefer we see the Judges in the Maat appear as twelve in number, sitting on twelve thrones. We find these two circles in Devon, Cornwall, and other parts of Britain, as distinguished from the three, which were Solar.

(I have thought it wiser to recapitulate somewhat in the case of Britain, because there has hitherto been so much mixture and confusion over the Druids that one cannot clearly set forth the case critically without doing so.)

Druids.

The name "*Druids*" of these Islands has hitherto included the *older Britons*, who had *the Stellar Cult*, and a *later exodus who came* here direct from *Egypt* with the *Solar Cult*. These have all been classed as *Druids* hitherto, and mixed together. The original Stellar Cult people travelled through the South of Europe from Egypt and went North by land, some of them three hundred thousand years, or more, before the Solar arrived. Some of the later Solar people came by sea, as records prove. I propose, therefore, to distinguish and differentiate the two classes bearing the *name* of *Druids* (although the Druids were only the priests) and bring forward evidence of the existence of the primary Stellar, the secondary phase of the Stellar Cult, and the Solar, it being more convenient for my readers to understand. After the "Nilotic Negro" (remains of which have been found in Britain), there was the exodus of the first Stellar Cult people from Egypt, who came here. This is proved by the position of burial in which the bodies have been found. The first religion of these people was the primary Stellar Mythos—as records prove, *their supreme God was Baal, i.e. Set of the Egyptians, God of the South Pole Star, El Shaddai of the Phœnicians.*

Another exodus from Egypt next arrived, with Horus as primary Stellar God; and a later exodus followed with the Solar Cult. This must have been so, as it would not be possible for the Nilotic Negroes here first to have developed by evolution all these different phases of religion, identical in form, names, signs, symbols and ceremonies with those as their old Mother Egypt, and with the original hieroglyphics of both Cults found carved on stones still extant. I shall enter into full details to prove this.

The people who were buried at Harlyn Bay, in Cornwall (see p. 40),[1] were Stellar Mythos people buried in the thrice-bent position with their face to the North, and a triangular stone over the cist, with apex pointing North. This burial position proves that they were Stellar Cult people at the time that Horus was primary God, and is identical with that found in the Tomb of Nagada. From the numbers found buried in Cornwall the population must have been fairly dense and the customs well established.

We find that their religion was similar in all particulars to the other Stellar Cult people throughout the world, if we compare the records of the past.

They all believed in one God, the Creator, Preserver, and Ruler of all things, the life and soul of the world, who endures for ever and exists throughout space.

These old people here worshipped God (first) under the name of Baal; they sacrificed in high places, adored in groves, planted oaks, intermarried with intimate relations—all of which was changed when the Solar doctrines were adopted here. They were all primary Stellar Cult people at first, having Baal or El Shaddai as their God = Set, the first Pole Star South God of the Egyptians. Later they were all Stellar Cult people with Horus (North Pole Star God of the Egyptians) as their primary God, he having superseded Set. The burial position of these people at Harlyn Bay, with face to the North and the apex of the triangle pointing North, and also their Osteo-anatomy, prove that they were "present types of man." They built temples on the same plan; they had all the same two circles with twelve pillars.

That the old Britons followed the doctrines of the Egyptians is plainly shown, as they changed from the primary phase to the secondary in the Stellar and then to the Solar.

We know that intercommunication took place between these Isles and Egypt at a very early time, therefore some of the old priests of Egypt with the Horus Cult must have

[1] See also pages 258 and 259 in "Signs and Symbols of Primordial Man."

arrived here, and later those of the Solar Cult—which were each adopted or forced on the former in turn.

There is no doubt that wherever the first Solar people went they tried to obliterate the Stellar Cult, but did not always succeed. They also married and intermarried with each other. They must all be classed as the present type of the human race as far as their Osteo-anatomy is taken into account. Their features would naturally alter in time owing to environment and intermarriage.

Their language was "inflected," it had passed the stage of "agglutinated," as we find amongst the Stellar Cult peoples.

CHAPTER XXX

COMPARATIVE WISDOM, ANCIENT AND MODERN

FINALLY, in tracing the origin and evolution of the human race, it might be of some interest to my readers to help to elucidate the immense ages of time that have passed, and to review "The New Knowledge" of recent years in combination with "The Old Knowledge" of the Wise Men of Egypt.

"In the beginning God created." We know that this earth emanated from our parent Sun, as also did other planets attached to our Solar system. Our Sun must have emanated from its parent Sun, and it takes "one great year" to perform its cycle, 25,827 years. Possibly the sun's sun, or centre of revolution, is only one of another, and how many times this is repeated it is impossible to say.

The next question is, are any "other worlds" inhabited by *human beings like ourselves*, and what reasons can be found to support this view? As far as our present knowledge goes the answer is negative. We can divide all "stars" into three main groups, Gaseous, Metallic, and Carbon stars, or those of highest temperature, medium temperature, and lowest temperature.

In the very hottest stars we find almost exclusively the gases hydrogen, helium and asterium; the latter is so far unknown on earth.

In the stars of medium temperature the gases become replaced by metals in the dissociated state, as thev exist in an electric spark of extremely high potential.

In the stars of lowest temperature the gases disappear almost entirely, and the metals exist in the state produced by the electric arc. Thus we see that a star has but few elements, when it is hot, but many when cold, i.e. we must conclude that many elements have been evolved from the few, and we find that the metallic elements appear first in the dissociated condition and afterwards in their normal form.

This proves that with a decrease of temperature the elements are evolved. Also we find that, as a rule, the elements of lightest atomic weight appear first. Now these atoms of elements are built up of nothing but corpuscles, and these corpuscles would form larger and larger aggregations as the temperature sinks. The stability of the atom depends on the arrangement of the corpuscles as well as on their number.

Corpuscles are little bodies laden with negative electricity thrown off from glowing metals, incandescent carbon, or gas flames.

They are, or have a mass equal to, one-thousandth of a hydrogen atom. Now, since a star has but few elements when it is hot and many when it is cold, the natural and reasonable explanation is that the many have been evolved from the few. If with decrease of temperature the elements are evolved by a combination of simpler substances, of course the dissociated forms would appear first. We do not find an element appearing first in the normal form and afterwards in the dissociated form.

The geologist, from an examination of the earth's strata from lowest to highest, finds an ever-increasing complexity in the organic remains which rocks contain. The astronomer, from an examination of the stars from hottest to coldest, finds an ever-increasing complexity in the so-called elements which they contain. They both deduce an evolution from simpler forms to more complex, and their deductions are equally valid. We accept the organic evolution, and we must accept the inorganic evolution. Organic evolution is measured by millions of years: inorganic by billions, from the hottest to the coldest stars.

But where does life come in?

The great law of continuity forbids us to assume that life suddenly made its appearance out of nothing, and tells us that we must look for the element of life in the very elements of matter, for the potentiality of life should exist in every atom.

The biologists and geologists tell us that life originated in the sea. If that is the case, then the constituents of living bodies should be the constituents of the sea-water and the air above it.

The constituents of sea-water are :—

Chloride of sodium	77·75
" magnesium	10·87
Sulphate of magnesium	4·73
" calcium	3·60
" potassium	2·46
Bromide of magnesium	0·21
Carbonate of calcium	0·34

The main constituents of air are nitrogen, oxygen, and carbonic acid. The elements constituting sea-water and air are thus oxygen, nitrogen, carbon, hydrogen, sodium, magnesium, potassium, calcium, chlorine, sulphur, and bromine.

We find from the ultimate analysis of the living beings that they consist of oxygen, nitrogen, carbon, hydrogen, sodium, potassium, phosphorus, sulphur, calcium, magnesium, iron, and silica.

Thus we see as a matter of fact that the elements constituting living beings are almost identical with the elements contained in sea-water and air, and these elements constituting sea-water and air are the first elements created in the process of inorganic evolution. This is easily determined by the examination of Sir Norman Lockyer's stellar temperature tables. Organic evolution is the last stage in the inorganic evolution. These tables inform us that in the running down of temperature from 30,000 degrees Centigrade, which is estimated as the temperature of the hottest stars, to the average temperature of the earth, the position of organic evolution is a point in the scale somewhere between the temperature at which water boils and ice melts. So we are akin to the stars. Untold millions of years ago, the tiny particles of the *materia prima*, the corpuscles or electrons, began their configurations which evolved into the atoms which constitute ourselves.

The question arises, did God, however long ago, start the circulating particles with the full plenitude of His energy, and then leave them to waste their energy in ever-multiplying configurations down to what, however far removed it may be, must be a state of rest and death, or did He give them this energy and perpetuity?

Sir Norman Lockyer has proved that whilst some stars are getting cooler, others are getting hotter, and this therefore answers the last question (see Lockyer's "The Meteoritic Hypothesis"), and one may deduce from this that other stars have no inhabitants.

Now let us see what the old Egyptians' ideas and beliefs were. In the latter part of their eschatology, I have already given their symbolic ideas of creation, but there was still an earlier belief associated with the primary Stellar Cult, which is found in the Hymn to Nut, preserved in the Pyramid texts. (Br. 95, 148.)

It traces the birth of life from air and fluid, i.e. from Shu, who represented the air, or breathing force, and Tefnut, who represented the dew, or liquid, so that our present scientists have not advanced in knowledge on this point any further than what was known to the old Wise Men of Egypt 300,000 years ago. We cannot "*trace the origin of the original corpuscles*" any more than we "*can trace the origin of life*," or reconstruct life from inorganic matter, or the original corpuscles out of nothing. All we know or can say at present is "God created." No human knowledge or science can fathom or go beyond that, not even metaphysics.

The discoveries recently made and published in *Ancient Egypt* have added very materially to our knowledge, and many other proofs are found therein proving incontestably the correctness of the statements made in this work.

Professor Flinders Petrie, F.R.S., F.B.A., etc., editor of *Ancient Egypt*, in his address, Part I, states: "When we try to grasp the prehistoric ages of Europe, it is solely to Egypt that we can turn for any definite scale of history with which the various periods can be connected. The thousands of years before classical writings can only be gauged by the Egyptian Dynasties."

In Part II, *Ancient Egypt*, p. 92, the reviewer of Paganism and Christianity in Egypt makes this statement: "The connection of Osiris with corn is classed as being only of late date; but the distribution of the cities where the corn-festival took place points to its being prehistoric before the Nome System was completed." I should not have referred to this except that Maspero also gives Osiris the presidential place as the Corn Spirit, which was simply copied and brought on from the original Horus in the Stellar Cult, the bringer of food and water. At Philæ Horus is represented as "Corn Spirit" by stalks and ears of corn springing from his mummy near running water; "thus Horus is represented as bringer of food and water." The old priests, however, tried to obliterate the original, and Osiris was given the title in the latest Cult, but he was not the original (see Stellar Cult chapter). Thus we see that the reviewer is correct in stating that it is prehistoric—very old—*but this title of Horus came into being during the Stellar Cult*, therefore after the first seven Nomes were formed, which was during Totemic Sociology.

"The Drew Lecture," p. 16, on Egyptian beliefs of a future life by Professor W. M. Flinders Petrie proves how little is really known or understood by even some of our most eminent Egyptologists on this subject. He is quite right in setting the Great Mother in the first place; here he calls her Nut and her consort Geb (Seb), and states that the hymn to Nut, preserved in the Pyramid texts, is regarded as the oldest fragment of the Ritual. It traces the sky from Shu and Tefnut, which he translates as space and fluid. Air, or breath of life, and fluid, liquid, dew, or drops of water, are more correct, the first thing that human life requires. Seb was the God of Earth and Father of Horus in one form. I have already given sufficient explanation (*supra*) of this. But I must put Professor Petrie right also over another point, namely as regards the various religious Cults. The Primary Sun worship, as he calls it, was not as he states. It was *Atum-Iu*, as represented by the young sun rising in the East, and setting in the West as Atum-Ra, the boy of twelve and the man of thirty. *That of Osiris was the latest Cult of all*, although it has been, and still is, much mixed up with the earlier; the reason is that the later priests always tried to blot out the previous Cult and place the name of their supreme God first.[1]

Although the Osirian doctrine, or Cult, was the predominant one in Ancient Egypt in the later dynasties, yet the evidence found both in the Ritual of Ancient Egypt, on monuments and papyri proves that this was the latest phase in their religious ideas and beliefs. The priests had tried to blot out every vestige they could of the existence of former Cults, and to merge everything they could into the Osirian, as they ever do in a new Cult. But Osiris, Isis, and Nephthys were only other names for Horus, Sebek-Nit and Sekhet-Bast in the Stellar; and Horus, Hathor, and Sati in the Lunar. Many different names of this combination do we find, but I maintain that what I have previously written demonstrates most critically the positions of the various Cults and how these followed on one from the other. The proofs I have brought forward are irrefutable.

The Egyptian belief in a future life was as old as the Pigmies. It had always been believed in from the earliest Homo: he believed, that when man died he rose again in spiritual form. These were the spirits of their ancestors, which they propitiated but did not worship; as, however, they believed that some were good and some bad, i.e. they could do good or bad actions, or have such influence over the living, they propitiated them and prayed that they should do them good and not evil. The righteous dead had the power of intercession with the great God to favour others in the judgment. "I will intercede for their sake in the Nether World," is the origin of the confession and absolution in the Roman Church. They believed that when the body died the spirit separated from the body, and became divided into a Ba and a Ka (Soul and Spirit); the Spirit had to go before the Great Judge, and if justified it returned and united with its Ba and entered Paradise. What the Spirit had to pass through "in Amenta" is fully set forth in the Ritual of Egypt.

There can be no doubt left in the minds of those who

[1] See "Origin and Evolution of Freemasonry."

will learn to read the old Egyptian Sign Language as to their belief in the future life. After the death of the Corpus, the Soul and Spirit left it. The Spirit then had to appear before the Great Judge. It was weighed in the balance against "truth and justice" as typified by the feather Maat—as we see in the Judgment scene in the Papyrus of Ani, the Spirit is in the balance against Maat, and his Ba is seen anxiously hovering above, waiting to see the result. The weighing here is in favour of Ani; he is then led by Horus, the Son of God, and presented to Osiris, receives his Crown of Glory, and is united with his Ba. Now he can come in and go out of Paradise at will. If the weighing had been against him—in other words, if the Great Judge had found that he was "not justified"—he would be given to the great devourer of Souls, Apapi, or, as we say, to the Devil, as he is seen waiting the result of the weighing. But also note in reading through the Ritual that there was another stage, which I believe has never been pointed out before, and yet it is quite distinctly stated in the description of Amenta. *It is when the Spirit is not justified in entering Paradise, and yet not given over to the great Apapi, or for ever damned.* It is an intermediate stage, where they can see the Glory of the Lord once in twenty-four hours for a short time only, and then relapse into misery and darkness again. *It is the purgatory of the Roman Church* and, from the evidence contained in the Ritual, one is led to believe that the Spirits of those who have been justified can intercede for them. "I will intercede for their sake in the Nether World" (Ritual). The prayer of their dear ones left here on earth sent through the priest, who was a medium, interceded, and, accompanied by one of the perfect Spirits, went to the Great Judge praying for his forgiveness, and craving admittance to Paradise for the loved one in the Nether World. This has never been pointed out before in any works that I have read, but it is clearly and distinctly set forth in the Old Egyptian texts (as *supra*). But the present priests are not mediums, and therefore can only act the part of frauds. They have lost, or never learnt, the true gnosis of the old Egyptian priests, and therefore their Spirits cannot leave them to accompany the perfect Spirits for intercession before the Great Judge.

The curious point about all this is that the Spirit of the dead Corpus has no knowledge of his existence, or of what he did here on earth until he is reminded of it; then he remembers. This is also borne out by the Ritual.

Professor Henri Bergson, lecturing at Edinburgh University on "The Doctrine of the Soul," states "there was a traditional doctrine which, in spite of change of garments, has remained the same throughout the ages until the present time. *It has been elaborated throughout Greek antiquity and received its finished form from Plotinus, a philosopher who has been too little studied, and whose importance has been too little appreciated.*" (Italics are mine.)

But Plotinus' doctrine was a bad copy of the old Wise Men of Egypt, who lived thousands of years before him. Nevertheless he does appear to have received and imbibed many of their ideas, more especially the theory of personality, and to have regarded the problems of the human soul as occupying the central position. But Plotinus went off into "metaphysics," whereas the old Wise Men of Egypt did not. "To know how the same being can appear to itself as an indefinite multiplicity of states, and nevertheless be a single and identical person," is to know the gnosis of the doctrines of the Egyptian priests, which was never understood and much perverted by the ignorant Greeks. The problem as stated by Plotinus was: "How can one person on the one hand be one and single and on the other hand multiple?" The one solution given is not the correct one, namely, "One solution was inevitable. It was that each of us was multiple in our 'lower nature' and single in our 'higher nature.'"

The Egyptian belief was that in our lower nature we were certainly multiple, i.e. we have a material body, and within this material body, which perishes and disintegrates with death, there is a Soul (Ba), and within the Ba there is a (Spirit the Ka); each of these on the death of the body (Corpus) leaves it. The Spirit then has to go before the Great God to be justified, or not, and if justified, it is united again with the Soul or Ba. But that is not its end. A male Spirit must for final happiness be blended with a female Spirit. It is not the unity of the *mind*, it is the union first of the *Soul and Spirit*, Ba with the Ka, and second, finally, the union of these as male with those as female, thus forming "one perfect spirit," yet a multiple, a pure spirit. Here we see how Plotinus had been taught, or received, the old Egyptian doctrines, but could not comprehend what he had been told; hence his question, "How can one person on the one hand be one and single, and on the other hand multiple?" And Professor Henri Bergson's metaphysics, I am afraid, would not help him, if he were still alive, to answer that question.

Life is an arrangement or combination of a certain number of corpuscles possessing energy, which produces what we term life, the combination and energy being received direct from the will of the Divine Creator. As a combination of male and female corpuscles this life is transmissible, to form other "lives," the life corpuscles themselves being transmitted from one organic being to the other through and by means of so-called protoplasm. The organic part of the original parents—or "matter"—undergoes disintegration as we term it, and "dies"—but the original corpuscles of the life energy do not die, or disintegrate; these exist for eternity as "Spiritual Life."[1] The "matter," the organic body, becomes again inorganic and reverts to the original elements of which it has been built up, compounds of corpuscles, when the male or female, as Spirit form, reaches and is received into Paradise. They are not perfect until the two are blended into one; this may take place almost immediately or it may not be consummated for an æon of years. It is necessary in the formation of the perfect spirit that the "corpuscles of energy" should be affinitively attracted, the one to the other, or, in other words, only a certain combination will combine with another certain combination, after which there is no change for eternity. That pure spirits can return to this material world and manifest themselves under certain circumstances, and influence the humans here, I

[1] See "Origin and Evolution of Freemasonry."

have not the slightest doubt. What becomes of those spirits eventually who have not been found "justified" I am not permitted to know, but that these can for a time influence the human for evil there is also no doubt.

PURE SPIRIT.

The Egyptians portrayed this as a compound symbol—the Ka and Akkhu bird.

CHAPTER XXXI

CONCLUSION

In Old Egypt only can we trace back the origin of the human race to the remotest past, the primitive Pigmy, with his belief in, and propitiation of, the elemental powers, Spirits of his Ancestors, and a Supreme Being. From the Pigmy we can trace the future development of the human race, Totemism without and with Hero-Cult, Fetishism with its elemental powers, its Zootypology, its Magic, its doctrine of Transformation, its Amulets and Charms, all of which came to its culmination in the Typology, the Mythology, the Magic and other religious rites and customs which finally ended in the Eschatology of Egypt, and nowhere else in the world can this be traced.

From the ashes and decay of the Solar Cult Egyptians, Christianity and new nations arose—the Assyrian, the Persian, the Babylonian, the Greek and the Roman, followed by the rise and fall of other nations, and a dark and degenerate age for thousands of years ensued, from the downfall of the Egyptians, as far as a higher spiritual type of the human was concerned. But for the last hundred years, and more especially for the last fifty years, a great stride has been made in the development and evolution of the human race. We have now reached a stage when men can trade, fight, love, make laws for governing their fellows; burn witches and tie martyrs to the stake; crucify the noblest or hang them; set beast to fight beast, and beast to fight with man; smear with oil and tar and make torches of those who differed from them in opinion; make gas and instruments with which to kill others miles away, and build hospitals to nurse the maimed; erect libraries, weigh the stars and measure their orbits, make ships to cross the sea and dive beneath the surface, and machines to fly the air, put a girdle round the earth, with lightning speed flashing their thoughts to the remotest ends of it, even without a connective wire, by harnessing the ether by which the world is surrounded; sculpture forms in marble which mock life itself; catch the glories of sunset skies and transfer them to canvas; make the sun assist them to depict the features of their friends; form societies to protect children and animals; rape innocent young girls and women and treat them with more vile shame than any other animal would treat its females; pile up riches and create the abyss of poverty; weep over tragedies acted and be indifferent to tragedies which are real; build churches to worship in; measure eternity into seconds, minutes, and hours; throw bridges across rivers; bring water from the hills to the towns miles away; develop science; sing songs of love, war, and worship—all this and more.

In the war lately waged against us by the Germans we find mighty proofs of the worst points of the above. To what end did the Divine Creator order this war to be? I believe His will is for the continuance of the British race—the highest type of the human developed at present—and that the war was ordained for the destruction of Socialism, or so-called Democracy, which was beginning to eat the life out of this country, and which if continued would in a very few years have destroyed us as a nation, as history proves this has been the cause of the downfall of all nations. To cause us to suffer that we should open our eyes, that we might commence again and go on to a higher type of the human evolution, and stamp out this Socialistic Vampire; to attain a higher spiritual feeling of brotherhood amongst us; to eliminate that bitter feeling of enmity of the various religious sects, each one of which professes to be the only true minister of the Divine will. For a permanent continued peace peoples will first of all have to eliminate the numerous false dogmas contained in the rituals and beliefs which dominate the various sects. It must be one united brotherhood of Christianity with the faith and belief in the One Great Divine Creator without dogmas. It is these false dogmas that cause so much difference and bitter feelings amongst humanity, and, as I have shown in this and other works, as "Christianity" is the highest point reached in the evolution of the Old Stellar Cult religion, it is the highest type of Christian faith to which we must look and by which we must live. Those who are not "Christians" have in the majority of cases remained in the faith of the types of the human evolution to which they have attained — types of all still exist at the present day. In each type of these humans the point to which religious evolution has attained can be associated with them, and as both the evolution of the human race and the evolution of religious doctrines travelled *pari passu* together, it has been a natural law, and in whatever part of the world these are now found each type is similar in every particular, and as the Divine Natural Laws or laws of nature are progressive by evolution we cannot retrograde and remain a great nation. Therefore the Divine Will has been to teach this nation the falsity of the course we were plunging headlong into before the war, and although we have sadly suffered, we shall emerge in time as a united brotherhood with higher and truer spiritual ideas. If not, and the nation returns to its "Socialistic reforms," then nothing will save us from a still greater calamity, and the British Nation will be one more added to the list of those "who have risen and fallen." The present Socialism is a retrograde movement in the evolution of the human race, originating in, and being carried on by, those who possess perverted Brain Corpuscles, just loose Brain Cells. Men—if we may call them men—who by their so-called oratory—which is in truth nothing but clap-trap terminology—have so much perverted the ideas of ignorant voters, so that they might be returned to power, a government which if it had continued would within a very few years have brought about the downfall of the nation by internal revolution, and the nation, as a nation, would have followed in the footsteps of those who had once been great and are now buried in oblivion. Therefore, as the white race has passed the stage of Socialism thousands of years ago (see *supra*), the Divine Creator will not allow any nation to exist if it becomes Socialistic. The Divine Laws are for a higher type and progressive evolution, not retrograde. Therefore for men (and Ministers to preserve to themselves their appointments and offices, with stipend attached thereto) to preach Socialism under the name of "Democratic principles, Democratic Nations and Democratic everything" is certain in the end to bring that nation to destruction. The fallacy of Democracy has been demonstrated over and over again in the histories of the destruction of past great nations. These Democratic Governments are simply a huge Bureaucracy, whose members think that its duty is to augment official power, official business, and official numbers rather than to leave free the energies of mankind. One of the later examples was well demonstrated in the case of the Incas, the downfall of whom was so easily and rapidly accomplished by the Spaniards—and what a striking object-lesson are their descendants to-day, sunk back into ignorance, downtrodden and degenerated, their glory passed into oblivion. For any nation to presume for one moment that it is going to be the exception and upset the Divine Laws of Nature will only prove once more, to future generations, how self-opinionated, incapable, and ignorant men who are elected to govern the destiny of any nation must fail. All who have the welfare and good of this nation at heart, and wish for the nation to continue still to a higher type of manhood, must see that those men and women who are trying to educate the ignorant masses of their fellow-creatures to vote for Socialism and Democracy are thus using their utmost endeavours to destroy the nation, as a nation, and if allowed to continue must by the Laws of Nature end in disaster and destruction. We have not come up through Palæozoic, Mesozoic, and Cenozoic times for nothing. If all the anguish of all the lives of all the past were to find one common vocal expression, what a cry to God there would be. Are we to believe that the great masses of men were nothing? *There must be a result in the world that is worth all the world, and for a future we cannot yet see or understand.*

In the beginning God created, and in His Creation, by evolution, man was evolved. He gave him hands to do and seven senses to apprehend, a will to drive and strive to know things, and a will to progress to some great unknown end in this world—for man to attain a much higher type—how much, or what the end?

Man has now won his way so far out of the dark conditions of pre-history that to know is to work and to do, and new knowledge gained is another step to the unknown summit to which man climbs. In evolution faith alone is not all we have for the truth of this, *we may not have a real knowledge of all the truth, but we have a recognition of it* on psychological considerations, and a great body of phenomena called "facts." These facts, independently of any theory connecting them, are useful, such as the comparative Anatomy and Physiological conditions, and causes of the rise and fall of nations; at present Physiological evidence is an unknown quantity, but the above will critically prove my contentions in this work, and if, like the proverbial ostrich, "we bury our head in the sands," shut our eyes to these facts, then there can be but one termination for the greatest nation that has so far ever existed.

The spiritual life hereafter at present is an unknown quantity to the majority; very few now living have been initiated into the Greater Mysteries as the old Wise Men of Egypt were, and the connection or reason of God's will being carried out by the means or agency of the Spirits of those who had once been human and now are not, is a Psychological fact that the majority cannot comprehend; but the human will continue to progress to a higher type, when the great mystery will no longer be a mystery and the truth will be revealed to them, and the Psychological laws will be found to be as accurate and true as the Periodic Laws of the Elements and Corpuscles, and in these laws will be found the solution of the Mystery of Creation.

Must Nature Perish?

What is meant by *nature* in the above question? If it refers to the organic part of the Earth, as we understand it, the answer must be in the affirmative in one phase, in another negative. In the form of organisms—organic life, or life incorporated in matter, which is called organic—this will perish as life in matter, but although this "phase" perishes, it is only as a particular form, or phase, that it dies or ceases to exist; a transformation then takes place into other forms. The chemical elements and Corpuscles of life, which make up the whole component parts of "Organic Matter," do not perish, or cease to exist; these assume other forms, caused by a regrouping of the chemical elements and corpuscles which compose the whole. It has been assumed from a physical point of view that the universe must come to an end, unless it receives *a new impulse of creation*, such as it must have had in the beginning. Every mathematical thinker during the last century has held that cosmic energy must decay by dissipation of heat, that is, the whole universe must come to the stillness of an everlasting death. This is the greatest and most mysterious of all the themes of pessimism. In some articles published in the *Observer* (1916), Sir Oliver Lodge seeks escape from the physical aspect of the problem and suggests that it may be found in the nature of life. In these articles Sir Oliver Lodge tries to define the distinction between high grade and low grade energy, and states that the higher energy is readily translatable and does work while it is being translated; *the lowest energy is untransferable.* But whenever the higher energy is used to do work it is converted into low energy, *because there is always waste, and that the fate of the Earth and of the Solar System is a very different thing from the fate of the Cosmos.* He rightly states that "expenditure of energy is only transference or transformation, and not destruction. That the life period of a Solar System, therefore, from its pristine nebula to its ultimate cold fate, may have been utilized in psychical and spiritual advantages of the utmost magnitude: and that the gain to the universe as a whole, though not to the material universe, by reason of the possibilities afforded by the temporary existence of that material collocation, may have been quite incalculable, and that material decay may conduce to spiritual uplifting," which to Sir Oliver Lodge appears to be the real object even of the existence of matter and energy.

The next question must be, Did the Divine Creator set in motion the circling particles with the full plenitude of His energy and then leave them to waste their energy in ever multiplying configuration, down to what, however far removed it may be, must be a state of rest and death, or did He give them His energy in perpetuity? The definite answer to this great question has been discovered by Stellar Photography and the Spectroscope, which show and prove that the devolution of the universe is balanced by its evolution. The next question is, What is this Energy? The answer is that this Energy consists in "Corpuscles," little bodies laden with negative particles of electricity. They have a mass equal to one thousand times less than an atom of Hydrogen and a velocity of 100,000 miles per second. They discharge electrified bodies. They are deflected by a magnet in the opposite direction from the Alpha-rays and positive Ions; they are also deflected by an electrostatic force. They cause phosphorescence and give rise to heat in the bodies which they strike, and communicate mechanical motion to these and in the striking give rise to X-rays and are absorbed by all bodies in direct proportion to their density. They act as nuclei about which atoms and molecules collect. They can exist alone; they cause clouds in moist air. They render the air a conductor of electricity and have a physiological and a chemical action, and can penetrate through a solid block of steel. These Corpuscles are "the atoms" from which all existing matter is made up; the Beta rays from Radium are Corpuscles, and for proof—

We will consider the stars, as to the composition of their elements, their birth, death, and rebirth.

The Stars[1] can be divided into three main groups:—

(1) Gaseous = Highest temperature. Example 1: Two in Argo (Zeta Puppis and Gamma); 2: Alnitam (Epsilon Orionis).

(2) Metallic = Medium temperatures, as examples: Achernar, Arcturus, Algol, Markal, Sirius, Procyon.

(3) Carbon = Lowest temperature; example, Pisces.

The first group are constituted, or made up, of a Strong Gas of the Helium family, or in other words corpuscles and Ions—Ions being positive atoms of electricity surrounding each corpuscle, which cannot exist by themselves—but the

[1] For further information I refer my readers to the works of Sir Norman Lockyer on this subject.

corpuscles or the negative particles can. These gases are called by some Proto-Hydrogen.

As the stars cool down from their highest temperature, 30,000° Centigrade, we find that these corpuscles unite in various combinations under their "Periodic Law," and form the chemical elements proto-magnesium, proto-calcium, proto-iron, oxygen, nitrogen, etc., and as the cooling process takes place Arc lines of other elements appear, proving that with decrease of temperature from 30,000° Centigrade to 0° Centigrade all the chemical elements that are known have been formed, and the chemical combination of these elements makes up or constitutes the material world and universe. At the temperature of from 80° to 100° Centigrade organic life begins to form on the earth by the union of corpuscles in an unknown combination and grouping, incorporating with the chemical elements. In the above we have what Sir Oliver Lodge terms "the higher grade energy" converted into his "lower grade energy," in one of its various phases. Sir Oliver Lodge states *that there is always waste and that the lower grade energy is untransferable.*

I contend *that there is no waste and that this low-grade energy is constantly being reconverted into high-grade energy.* Nor can there ever be any waste, as I will proceed to prove. My contention is that it is only a constant transformation, or transference, of the Corpuscles into different and various groupings which form the atoms of the chemical elements —or different groupings of the Corpuscles alone, which is brought about by the lowering of the temperature of the corpuscles, caused by a less intense bombardment of each other, and the formation of "groups," and that all the chemical elements and energy are formed from these Corpuscles by transformation, or transference, of one group into others. There can be no waste, because if the number of Corpuscles which form one chemical element, or one form of energy, does not absorb the whole, the other Corpuscles of a said group, by their Periodic Law, are then transferred into another chemical element or a different form of energy.

If the chemical elements, as we know them, are heated up to 30,000° Centigrade, all would resume the original Corpuscles again; this is proved by observing the formation of new stars and also when we heat metals or Carbon to a high degree of temperature. In the latter experiment a stream of Corpuscles is given off from the metal or Carbon, which has now lost part of its weight, but there is no waste of Corpuscles; these are transferred into another form of energy. Many given off in this way no doubt form with others (the moist air) clouds; and the so-called electrical disturbance in the Clouds, resulting in what is termed Lightning, is caused by the uniting of Corpuscles into a different combination under their Periodic Law, or a transference of one combination into another.

We have now followed the Star of the highest temperature down to what is termed a dead Star, No. 3, and in this cooling process we have seen that from its composition of pure Corpuscles and Ions it is now "a chemical combination of all the chemical elements" known. The next question is, What becomes of this dead Star which no longer exists to give light, and how was this Star formed, or how did it come into being?—this now untransformed energy of Sir Oliver Lodge. We know that Stars are formed, or begin with a Nebula, a swarm of cold lumps of matter originating from dead Stars, which I have shown (*supra*) contain the chemical elements as we know them on this earth, and we know that new Stars are constantly forming to replace the old ones. Thus, then, energy, although now in its lowest state, continues and, as we shall see, is converted into another form of energy. The Collisions resulting from the gravitation and attraction, each seeking the centre of the swarms, cause heat to be evolved and the temperature rises. We find as the meteorite, and lumps of cold matter, made up of disintegrated dead Stars, or the dead Stars themselves, bombard and clash with each other, the low-temperature Arc lines—which we saw in the cooling Star—begin to appear in the spectrum of the masses, and a new Star is now formed—we have such a Star in Antares.

But the bombardment and condensation continue and the temperature increases until it reaches its highest point (30,000° Centigrade), when all the chemical elements are reconverted into incandescent gas or our friends the Corpuscles.

We find by the spectrum that as the temperature increases so the chemical elements that had been formed by the cooling process gradually disappear in the inverse order of their formation or appearance in the cooling period, until the temperature has risen to 30,000° Centigrade, when all are converted or return to the original atoms—the Corpuscles and Ions. A chemical decomposition, figuratively, has taken place in the elements, and nothing has been lost and no waste has occurred in this inorganic universe, and the lower energy has become again converted into the highest energy. *This proves that we have a devolution and evolution which balance each other, and this is always in operation.* The same holds good on this Earth in the organic life. Trees, Flowers, Fish, Animals and Insects of all kinds are born, or come into being, live and die—but there is no waste; the life of these has departed, they decay and their corpuscles regroup again into some other forms of low energy, first into inorganic matter, this again into organic; there is no waste or destruction, but a transformation of the Corpuscles again from low energy to the higher.

Some of the organic corpuscles incorporated in matter, however, are transmitted, into a similar form of the original, in various ways from the different forms of organic life I have mentioned above—trees, flowers, etc., by seeds or cuttings from the parent—these are organic and contain organisms, as a higher form of energy. The same is true with animals, insects, etc.; by the fertilization of the female by the male the organic life is carried on, and inorganic matter which had become inorganic by the death of the parent is now absorbed again into organic. There is no such thing as destruction or waste of energy, but it only becomes latent or "low form." The ultimate basis of all things is the Corpuscles, and these never cease their energy; either in a low form or higher form, it persists; they are always active in one combination or another. I have no doubt that a great part of the heat and temperature of this Earth is kept up by the continual discharge and bombardment of Corpuscles constantly being emitted by Radium, Uranium, and other elements which are Radio-active, the highest form of energy. It is the same with the Human; part of his corpus changes by decay into inorganic elements when he dies, and part of his body, during life, is converted or transferred into other humans by the fertilizing of the female by the male, and the organic corpuscles of life are carried on from one generation to another, in a high form of energy. In the Human we have another form added which only exists on Earth in the Human—"Spirit form," but the Spirit is composed of Corpuscles only, in Corpuscular form; these are united or grouped together, under the Periodic Law, in the likeness of his Corporeal body, and in this Spirit form the Humans have to return to the Divine Creator to be judged. So we see throughout the universe there is no waste or destruction but only transformation, the higher energy being converted into the lower and again the lower into the higher. Here we see the fallacy of Sir Oliver Lodge's argument when he states that the fate of the Earth and of the Solar System is a very different thing from the fate of the Cosmos. I contend that it is not so. At some future time, how many millions of years hence no one can say, the Earth and our Solar System will become as the "dead Stars," after which a new Sun, or Star, will be reformed again out of the debris of our Solar System. The debris will, by attraction and gravitation, come together, and the bombardment and clashing together of the present elements will by heat thus generated be again converted into the original Corpuscles, the original atoms, from which we have been formed. There will be no destruction or waste either of Energy or the Chemical element. As far as the Human on this earth is concerned he, similarly to the inorganic Cosmos, follows the Divine laws of devolution and evolution, and this may be said to apply both in the evolution of the Human race in the greater sense, and indirectly, as an element, in the lesser sense. Only it is organic in one form and inorganic in another, that is to say, the Human race commences life as a Pigmy and in a million or two millions of years we find that by evolution it has passed through various types to the present white man; how far higher it will continue to progress in evolution is impossible to say. The different types can be compared with the various chemical elements, these being formed at different temperatures (see *supra*), and the various types of man have been formed by measurement of evolution in years.

The Spirit of the Human in Corpuscular form will not change, it will exist for eternity, for all those that have been, and will be, accepted by the Divine Creator, i.e. all those who have followed the Divine Laws. Other Spirits of the Human who have not followed and obeyed these laws will be transformed into a different grouping and combination of these Corpuscles by the will of the Divine Creator.

"Never the Spirit was born, the Spirit will cease to be never;
Never a time it was not, end and beginning are dreams.
Birthless and deathless and changeless, remaineth the spirit for ever,
Death hath not touched it at all—dead though the home of it seems!"

CHAPTER XXXII

TABULATED SUMMARY

TABLE OF THE CHARACTERISTIC DISTINCTIONS OF THE DIFFERENT GROUPS

PRE-TOTEMIC OR NON-TOTEMIC AND NON-ANTHROPOPHAGOUS.

Pigmy Bushman Hottentot Masaba Negro —all little men	1. Primitive Implements, chipped on one side only. Spears and bows and arrows. 2. Large abdomens, some steatopygous. 3. Long fore-arm and wide between ulna and radius. 4. Hair discrete peppercorn. 5. Thin upper lip. 6. Cranial capacity from 850 to 1,100 c.c. 7. All believe in Great Spirit, and propitiate elemental powers. They believe in after life and propitiate departed spirits. 8. Have no magic. 9. Have a limited monosyllabic language which they cannot write, about 200 words. Have a sign and gesture language. Can draw signs and symbols. 10. Nomadic. Live by hunting and fishing. 11. Have no tribal or totemic markings. 12. Make no pottery.

The Totemic Group—Anthropophagous.

Divided into Non-Hero Cult, from whom were developed the Hero Cult people, i.e. those who first divinized the elementary powers.

First, the Great Earth Mother, Ta-Urt, was divinized into Apt, and their Heroes were Set, Horus, and Shu, as the three first-born who were called "Bulls of their Mother." Four others divinized later.

Non-Hero Cult.—Very prominent supra-orbital ridges, giving an appearance of great beetling brow.

Nose.—Very depressed at roots and wide extended alæ, large, coarse mouths, with simian upper lips when young, but developed to thick lips at puberty and later. Large abdomen in child, disappearing into normal conditions when older. Long fore-arm.

Cranial capacity, 1,000 to 1,250 c.c. Dolichocephalic.

No Mythology. Do not tattoo, but produce cicatrices on body, circumcise, and subincise. Make rude pottery; nomadic life; live by hunting and fishing.

Hero Cult.—Some very tall and fine in appearance. Long legs with calves undeveloped. Nose in some cases finely developed, but never Pigmoid; mouth better formed, not such thick lips. Beetling brow disappears. Cranial capacity, 1,100 to 1,500 c.c. Heads range from Dolichocephalic to brachycephalic. Tattoo in place of cicatrization generally. Have Mythology and Hero Cult. Work in metals, wood, leather, weave fibre into cloth and make fine pottery; some became agriculturists and cultivated the soil; tamed wild animals.

All had totems, totemic ceremonies, and magic, and were at first anthropophagous. Implements chipped on each side roughly, not polished; had learnt art of hafting.

Language in both types monosyllabic. Non-Hero Cult people. Have about 900 words.

Stellar Mythos People.

Much higher type. Divinized all elemental powers. Believed in Supreme Spirit. Formed or worked out all the Astronomical Mythology. Built temples, sacrificed to Supreme Spirit, and founded the Eschatology.

Implements chipped on both sides and polished. Worked in all metals; made pottery with ideographic pictures. Formed hieroglyphic writings first, and introduced the commencement of linear writings, forming thus the first alphabet.

All their buildings iconographic.

Could not build arches—always lintels. Built with highly finished polygonal stones and monoliths.

Always depicted the gods and goddesses by Zootypes. Built huge temples (the Pyramids and others) in every part of the world to which they went.

Buried in the thrice-bent position, either in cists or lying on the side with face to the south first, afterwards to the north.

Were highly developed in astronomy and mathematics. Cranial capacity, 1,200 to 1,500 c.c. Osteo-anatomy, present type of man with supra-orbital ridges, a little larger in earlier type. Language agglutinated.

Solar Cult People.

Present type of man.

Cranial capacity, 1,250 to 1,600 c.c.

Language inflected and written alphabetically. Temples and buildings non-iconographic; orientated east. First depicted the human as the human, and all the gods and goddesses in human form. Buried with the face to the east and not in thrice-bent position. First built with lintels, but later could build arches. Non-anthropophagous.

FIRST HOMO

AFRICAN PIGMY (Hunters only)

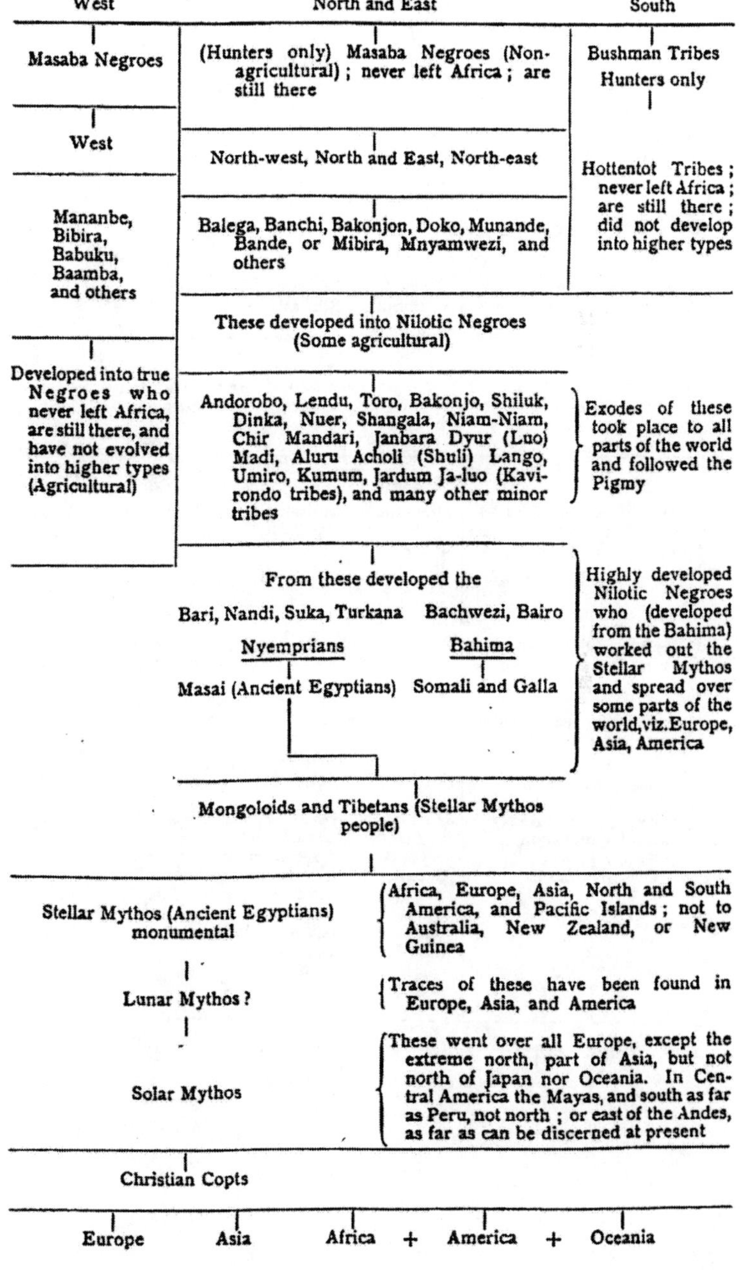

PIGMY (evolved in Africa)

African Pigmy	Exodes went out all over the world			
Developed into Masaba Negroes and Bushmen, only found in Africa	Europe Died out	America Existing in South America	Asia Existing there	Oceania Still existing there
Masaba Negro developed into the				
Nilotic Negroes				
An exodus of Nilotic Negroes all over the world took place after the Pigmy	No progressive evolution from these ; they will all die out			
Africa — still existing — and these developed into the	Europe Died out	America Still existing	Asia Still existing	Oceania Still existing
Monumental Egyptians, Masai and Turkana, and Suk, Bahima, Somali, and the Gala	No progressive evolution from these outside Africa ; some may have married Pigmy women, and the later exodus men married the women of the primary exodus			
Stellar Mythos people	Europe	America	Asia	Oceania
Earliest Tibetans and Mongoloids. Types are still extant in South America	Died out, giving place to the Solar Mythos people	Representatives still exist	Still exist	None; did not come here
Developed into Lunar and then Solar Mythos peoples	Married and intermarried, and some development in evolution took place as the white race came into existence during this period. Offshoots from the yellow, with intermarriage and climatic conditions, would account for the change			
Ancient Egypt at its zenith				
Solar Mythos people	Europe	America	Asia	Oceania
Ancient Egypt in decay and degeneration	Died out as Solar Mythos people, and Christians took the place of Solar Mythos	Types—Mayas in Mexico and South America in Peru. Never in North America (or south of Peru) or east of the Andes	Still exist	None
Christian Copts				
Christians. Africa	Europe	America	Asia	Oceania

GEOLOGICAL FORMATION TYPES FOUND IN THE DIFFERENT STRATA

	FORMATIONS	PERIODS	TYPES OF MAN
	Alluvial	Historic and Neolithic Flints	Modern and Barrow-men
	Buried Channel of Thames Lowest Terrace	Magdalenian Solutrean Aurignacian	Cromagnon Combe-Chapelle Grunaldi
	4th Glacial 30-foot Terrace	Mousterian	Neanderthal Spy, La Chapelle, La Quna, Jersey
	3rd Glacial 100-foot Terrace	Acheulian Chellean	Bury St. Edmunds, Moulin Quinon, Dennise, Grenville, Gallery Hill
	Boulder Clay 2nd Glacial	Worked Flints	Ipswich
Pleistocene	Mid Glacial Contorted Drift Cromer Beds Norwich Beds Plateau Drift	Worked Flints	Landudno Foxhall Heidelberg (Nilotic Negro)
	1st Glacial Lenham Beds Red Crag Series	Worked Flints	Pithecanthropus (Pigmy)
Pliocene	Coralline Crag Series	Worked Flints	Casteuldoloman (Calaveras)

The above is a supposed Geological Chart showing the probable sequence of strata formed in England during the recent Pleistocene and Pliocene Periods, which our present geologists have divided into four glacial epochs, and as the glacial epoch occurs every 25,827 years, all this would have been formed during 103,308 years, an utterly absurd and preposterous supposition, especially when we have the fact recorded still extant that the Stellar Mythos people (those who first kept time and reckoned it by the revolution or precession of the Pole Stars, Ursa Minor or Little Bear) had kept and recorded *ten such cycles during Stellar Mythos alone*—i.e. 258,270 years. This was after the Nilotic Negroes, Masaba Negroes, Bushmen, and Pigmies had come into existence, and before Lunar and Solar Mythos and Christianity.

APPENDIX TO CHAPTER II

SINCE I wrote the foregoing, Professor Arthur Keith, F.R.S., etc., has published his monumental work "The Antiquity of Man," certainly the most erudite work that has yet seen the light of day on this particular subject, and I would advise all my readers who are interested in the Antiquity of Man to obtain a copy and study this great work.

There are, however, one or two points on which we differ, but there are a great many facts he has brought to light which help to prove my contention very conclusively. In speaking of the Grimaldi and Cromagnon race, he seeks a comparison with "the tall races of the Punjaub of India," more especially with regard to the length of the lower limbs. These I contend were Hero Cult Nilotic Negroes, people who came up from the Nile Valley, where their descendants are still to be found—as examples, the Turkana and Bor-Jieng, many over 7 feet high (see *supra*). The tall Punjaubi and tall Sikhs are descendants of these Hero Cult Nilotic Negroes, and that accounts for the similarity in length of lower limb to the Cromagnon race; these died out in Europe: some may have married some of the women of the Non-Hero Cult Nilotic Negroes which they found here when they arrived, but they would not be allowed to marry the Stellar Cult women, so we thus find modifications in remains of the true type.

Professor Keith justly states that there were many types of the Neanderthal man in Europe; these were all Non-Hero Cult Nilotic Negroes, who were not a tall race, and their bones and beetling brows are characteristic. They have, as a rule, longer forearms than the Cromagnon, and, as I have shown, and it is open still for all to confirm, there are various types of these still existing in Africa, some with great development of supra-orbital ridges and bones generally, and others with less development of bones, yet all have "larger and coarser bones" than the Hero Cult race who developed from these, and the reason I have attributed to the changes that took place in the Pituitary Gland in the Brain.

Both Non-Hero Cult and Hero Cult peoples probably existed for some considerable period in Europe, as elsewhere, at the same time, but they would not mix. This is proved by what we find at the present day both in Africa and in Australia, where the Hero Cult natives have confined the Non-Hero Cult people to certain districts, which they must not leave on pain of death. In giving some of his comparisons Professor Keith has used the Skull and Brain of a native Australian (with capacity 1,450 c.c.); this is a *Hero Cult* skull and brain, and if he had used a low type of Non-Hero Cult native he would have found a much more primitive type, which would have corresponded with the Piltdown Skull in much closer relationship—one must take into consideration that although these natives have lived here in isolation for probably many thousands of years, yet they must have advanced somewhat in the evolution of development since their Ancestors left Africa. If Professor Keith wishes to find primitive man he must go to Africa, where he will still find the Masaba Negro of much more primitive structure than the Non-Hero Cult Nilotic Negro (these Masaba never left Africa), and yet still further back the Pigmy, both highly developed, and also a very low type; these very low type he can only find in Africa, where they still exist, and in these he will recognize "Original Man."

Comparison with the true negro of West Africa is of no use in trying to trace the evolution of the origin of man direct from the original—the West African developed from the Masaba, and are a type which never left Africa and did not advance in evolution. This type dies out.

Professor Arthur Keith states (p. 397) that Professor Elliot Smith is of opinion that a brain must reach a weight of 950 grammes (or about 1,000 c.c. in volume) before it can serve the ordinary needs of a human existence—before it can become the seat of even a low form of human intelligence —and further that "most of us are content to accept the Professor of Anatomy in the University of Manchester as our leading authority on this matter. As Sir Ray Lankester, F.R.S.'s, "writer of no credentials," I state absolutely that this opinion is erroneous, and the proof of such a fallacy can be shown at the present day by the examination of the head of a low class Pigmy, also by the low class Mai Darats, where the cubic capacity certainly does not in many cases exceed 900 cubic centimetres. I have examined some Pigmy skulls myself where the cubic capacity certainly does not reach 900 cubic centimetres, and one has only to read what I have written above to say what class of humans these are. Therefore, I emphatically say that Professor Elliot Smith, the great Professor of Anatomy of the University of Manchester, is obviously incorrect in his opinion, and the proofs I have given are critical evidence of my contention. Personally, I have not the slightest doubt that the Piltdown Skull belonged to a Non-Hero Cult Nilotic Negro.

INDEX